"Anything written by Dennis Johnson is well worth reading. He has been thinking about, preaching and teaching from Philippians for many years, so this work is all the more worth reading. It is filled with valuable insights into the biblical text and its practical consequences. This book is a true gem just like the man who wrote it."

—**S. M. Baugh,** Chairman, Department of Biblical Studies, Professor of New Testament, Westminster Seminary California

"Dennis Johnson's *Philippians: To Live is Christ* is a treasured contribution to P&R's Reformed Expository Commentary series. Johnson's experience as a preacher and teacher of preachers not only makes these messages wonderful models of exposition, but also his expertise in biblical theology permeates every page. The Christ-centered focus is beautifully and powerfully interwoven throughout."

—**Bryan Chapell,** Chancellor, Covenant Theological Seminary

"Pastors aspiring to preach expositorily (as all pastors should preach), take note! Philippians by Dennis E. Johnson is as good a source for such preaching as you'll ever get—and as good an example of it as you'll ever get, for it represents his own preaching. The commentary has everything you'd want: not only uncluttered exegesis, cross-references to the rest of Scripture, and theological integration, but also historical background and parallels, literary references, illustrations from and applications to modern life, allusions to pop culture, local color, and warmth of personal testimony. The chapter titles alone will whet your appetite. Use this commentary both for Philippians and also as a template for pastoral exposition of other biblical books."

—**Robert H. Gundry,** Scholar-in-Residence, Westmont College

"As a student of Dr. Johnson's, I came to admire his insight into Scripture and his pastoral heart. Exemplifying the model of proclaiming Christ from all of the scriptures, this outstanding commentary on Philippians unpacks the treasures of the apostle's teaching with profound skill that captivates the heart."

—**Michael S. Horton,** J. Gresham Machen Professor of Systematic Theology and Apologetics, Westminster Seminary California

"I am always glad to see commentaries that will help preachers. This one is sure to do that. It is beautifully and clearly written; that will help a preacher's style. It is theologically, historically, and canonically accurate; that will help a preacher's faithfulness. It pays great attention to details in the Greek text without being pedantic but it also shows how these little parts fit together to make a whole; that is what preachers are called to do. It is rich with individual insights that preachers will enjoy sharing. Its titles, captions, illustrations, and applications model what preachers should aspire to. Its courageous tackling of contemporary aberrations such as prosperity teaching will stiffen pastoral backbones to do the same. It condenses some of the best insights from other commentators and that will save a preacher's precious time! I highly recommend it for preachers and those who listen to them."

—**Greg R. Scharf,** Professor of Pastoral Theology, Trinity Evangelical Divinity School

"It is truly a delight for me to recommend Dennis Johnson's *Philippians: To Live Is Christ*. Only rarely does one come across a work that so successfully blends scholarship with clarity of exposition and pastoral application. One quickly becomes aware that the author has done his homework and done it well, carefully examining the exegetical options and reaching thoughtful conclusions. No fanciful flights of imagination here. In an admirable way, however, the commentary itself is free from distracting technicalities, so that the reader comes away with a clear understanding of each section as a whole. Moreover, Johnson refuses to treat Philippians as an antiquarian document to be interpreted in isolation from contemporary culture. Instead, he approaches the epistle from the point of view of the modern reader and, as a result, manages to communicate its contents as a living message. In contrast to most commentaries, this work can easily and with pleasure be read through in a few sittings. Anyone who does so will not be disappointed."

—**Moises Silva, T**ranslator or Advisor, The New American Standard Bible, The New Living Translation, The English Standard Version, The Message, and the Nueva Versión Internacional

"Wonderful piece of work: sure-footed, technically competent, instructive, warm, and practical."

—**David F. Wells,** Distinguished Research Professor, Gordon-Conwell Theological Seminary

Philippians

REFORMED EXPOSITORY COMMENTARY

A Series

Series Editors

Richard D. Phillips
Philip Graham Ryken

Testament Editors

Iain M. Duguid, Old Testament
Daniel M. Doriani, New Testament

Philippians

DENNIS E. JOHNSON

John,

May you rejoice in the Lord always!

Dennis Johnson

P&R
PUBLISHING
P.O. BOX 817 • PHILLIPSBURG • NEW JERSEY 08865-0817

Italics within Scripture quotations indicate emphasis added.

Page design by Lakeside Design Plus

ISBN: 978-1-59638-200-8 (cloth)
ISBN: 978-1-59638-657-0 (ePub)
ISBN: 978-1-59638-658-7 (Mobi)

Printed in the United States of America

Library of Congress Cataloging-in-Publication Data

Johnson, Dennis E. (Dennis Edward)
Philippians / Dennis E. Johnson. -- 1st ed.
pages cm. -- (Reformed expository commentary)
Includes bibliographical references and index.
ISBN 978-1-59638-200-8 (cloth)
1. Bible. N.T. Philippians--Commentaries. I. Title.
BS2705.53.J64 2013
227'.607--dc23
2013007377

For Robert B. Strimple Jr.,
whose exposition of Philippians 2:5–11
in the classrooms of Westminster Theological Seminary
and Westminster Seminary California
has enriched our awe of Christ,
whose preaching of Philippians
has deepened our joy in Christ,
and whose service to Christ's church
shows us the way to seek not our own interests,
but those of Jesus Christ.
Thank you, Bob,
for your friendship over four decades.

Contents

Acknowledgments

It is my joy—an apt word for this epistle!—to thank the team of pastor-scholars who oversee the development of the Reformed Expository Commentary (Philip Ryken, Richard Phillips, Daniel Doriani, and Iain Duguid) for their invitation to contribute this exposition of Philippians. I am grateful for Dan's and Rick's frank and helpful editorial feedback, which improved the work in many ways. I thank the board of trustees of Westminster Seminary California for granting me a study leave in the spring of 2010, during which much of the commentary was converted from spoken sermons into a print form more conducive to reading. I also want to thank John J. Hughes, my friend from seminary days together in the 1970s, for his efficient management of this project on behalf of P&R Publishing; and Karen Magnuson, whose insight and skill as copyeditor has strengthened the commentary in countless ways.

I have been privileged to preach portions of Philippians not only to the congregation that I serve as associate pastor (New Life Presbyterian Church, Escondido, California) but also to others. I thank the Lord and the elders who shepherd these flocks for the opportunity to teach and preach the "word of truth" in their worship. In 2009 it was my joy to teach on Philippians to the overseas and home-office staff of the Rafiki Foundation. The Rafiki "family," which serves orphans and widows in Africa in Jesus' name, exhibits many of the qualities that caused the apostle Paul to love and long for his sisters and brothers in Philippi. Seminary students and spouses enrolled in my fall 2009 elective on Philippians helped me to sharpen my exegetical attention to the text. More than that, our evenings together over those three months led us deeper into the joy in Christ, of which Paul speaks so often in this little letter. Brothers and sisters, you have waited long to see this fruit of our labors together: here it is at last!

My wife Jane's encouragement has sustained me through yet another prolonged writing project, and her editorial astuteness caught much that needed repair before it met the eyes of other editors. I am thankful to her, and to the Lord for her, every day.

Preeminently, I praise the triune God for answering the prayers of many for the completion of this exposition and for its usefulness to Christ's church. The divine Son who became the Suffering Servant is indeed alive and exalted above all. As this small human work reaches completion by the aid of his Spirit, I thank him that he has begun a more amazing "good work" in all who trust him and that he will most certainly "bring it to completion at the day of Jesus Christ" (Phil. 1:6). This attempt to open a petite but powerful portion of his holy Word is offered to the church with a wish echoing Paul's, that "Christ will be honored" through it (1:20).

Series Introduction

In every generation there is a fresh need for the faithful exposition of God's Word in the church. At the same time, the church must constantly do the work of theology: reflecting on the teaching of Scripture, confessing its doctrines of the Christian faith, and applying them to contemporary culture. We believe that these two tasks—the expositional and the theological—are interdependent. Our doctrine must derive from the biblical text, and our understanding of any particular passage of Scripture must arise from the doctrine taught in Scripture as a whole.

We further believe that these interdependent tasks of biblical exposition and theological reflection are best undertaken in the church, and most specifically in the pulpits of the church. This is all the more true since the study of Scripture properly results in doxology and praxis—that is, in praise to God and practical application in the lives of believers. In pursuit of these ends, we are pleased to present the Reformed Expository Commentary as a fresh exposition of Scripture for our generation in the church. We hope and pray that pastors, teachers, Bible study leaders, and many others will find this series to be a faithful, inspiring, and useful resource for the study of God's infallible, inerrant Word.

The Reformed Expository Commentary has four fundamental commitments. First, these commentaries aim to be *biblical*, presenting a comprehensive exposition characterized by careful attention to the details of the text. They are not exegetical commentaries—commenting word by word or even verse by verse—but integrated expositions of whole passages of Scripture. Each commentary will thus present a sequential, systematic treatment of an entire book of the Bible, passage by passage. Second, these commentaries are unashamedly *doctrinal*. We are committed to the Westminster Confession of Faith and Catechisms as containing the system of doctrine taught in the Scriptures of the Old and New Testaments. Each volume will teach, promote, and defend the doctrines of the Reformed faith as they are found in the Bible. Third, these commentaries are *redemptive-historical* in their orientation. We believe in the unity of the Bible and its central message of salvation in Christ. We are thus committed to a Christ-centered view of the Old Testament, in which its characters, events, regulations, and institutions are properly understood as pointing us to Christ and his gospel, as well as giving us examples to follow in living by faith. Fourth, these commentaries are *practical*, applying the text of Scripture to contemporary challenges of life—both public and private—with appropriate illustrations.

The contributors to the Reformed Expository Commentary are all pastor-scholars. As pastor, each author will first present his expositions in the pulpit ministry of his church. This means that these commentaries are rooted in the teaching of Scripture to real people in the church. While aiming to be scholarly, these expositions are not academic. Our intent is to be faithful, clear, and helpful to Christians who possess various levels of biblical and theological training—as should be true in any effective pulpit ministry. Inevitably this means that some issues of academic interest will not be covered. Nevertheless, we aim to achieve a responsible level of scholarship, seeking to promote and model this for pastors and other teachers in the church. Significant exegetical and theological difficulties, along with such historical and cultural background as is relevant to the text, will be treated with care.

We strive for a high standard of enduring excellence. This begins with the selection of the authors, all of whom have proved to be outstanding communicators of God's Word. But this pursuit of excellence is also reflected in a disciplined editorial process. Each volume is edited by both a series editor and a testament editor. The testament editors, Iain Duguid for the Old Testament and Daniel Doriani for the New Testament, are accomplished pastors and respected scholars who have taught at the seminary level. Their job is to ensure that each volume is sufficiently conversant with up-to-date scholarship and is faithful and accurate in its exposition of the text. As series editors, we oversee each volume to ensure its overall quality—including excellence of writing, soundness of teaching, and usefulness in application. Working together as an editorial team, along with the publisher, we are devoted to ensuring that these are the best commentaries that our gifted authors can provide, so that the church will be served with trustworthy and exemplary expositions of God's Word.

It is our goal and prayer that the Reformed Expository Commentary will serve the church by renewing confidence in the clarity and power of Scripture and by upholding the great doctrinal heritage of the Reformed faith. We hope that pastors who read these commentaries will be encouraged in their own expository preaching ministry, which we believe to be the best and most biblical pattern for teaching God's Word in the church. We hope that lay teachers will find these commentaries among the most useful resources they rely on for understanding and presenting the text of the Bible. And we hope that the devotional quality of these studies of Scripture will instruct and inspire each Christian who reads them in joyful, obedient discipleship to Jesus Christ.

May the Lord bless all who read the Reformed Expository Commentary. We commit these volumes to the Lord Jesus Christ, praying that the Holy Spirit will use them for the instruction and edification of the church, with thanksgiving to God the Father for his unceasing faithfulness in building his church through the ministry of his Word.

Richard D. Phillips
Philip Graham Ryken
Series Editors

Preface

Biblical commentaries characteristically begin with an introduction that addresses important questions preliminary to the interpretation of specific passages in the book under consideration, particularly matters on which there has been significant controversy among scholars or challenges to the authority or reliability of Scripture since the rise of historical criticism. Such issues include the question of authorship, the general life situation (including sociocultural context) of the author and the original recipients and the specific occasion that prompted the author to write, and the book's purpose(s), possible literary or oral sources, genre(s) and structure, and theological motifs.[1]

This introduction will be brief. For the epistle to the Philippians, questions of authorship and the location of the readers are moot, since a consensus exists across the theological spectrum that Paul the apostle authored the words of this epistle and that its destination was Philippi in Macedonia. There is disagreement, however, regarding whether the epistle was originally a compositional unity in its current order or existed previously as several brief missives that were subsequently combined—out of chronological order—resulting in the text as we now have it. Diversity of opinion also exists regarding the location (provenance) from which Paul wrote—whether it was his imprisonment in Rome, as he awaited Caesar's hearing of his appeal (Acts 28:17–31); the two-year custody that preceded it in Caesarea Maritima, capital of the province of Judea (Acts 23:23–26:32); or an earlier period of custody, not explicitly mentioned in

1. See, for example, Gordon D. Fee, *Paul's Letter to the Philippians*, NICNT (Grand Rapids: Eerdmans, 1995), 1–55; John Reumann, *Philippians: A New Translation with Introduction and Commentary*, AYB (New Haven, CT: Yale University Press, 2008), 3–20.

the New Testament (but perhaps alluded to in 1 Corinthians 15:32) in Ephesus. Epistles in the Greco-Roman world served various purposes and tended to follow different conventions, depending on the author's aims and relationship to the recipients. Scholars do not agree regarding what type of letter (or letters!) Paul has written to the church at Philippi, and diversity of opinion on this genre question yields differing analyses of the structure and flow of the epistle's argument.

For the purposes of this commentary series, most such introductory questions can be treated helpfully—though by no means exhaustively—in the course of expositing the individual pericopae (text units) of the epistle.[2] For example, Paul's references to his chains, the "imperial guard" (Praetorium), the uncertain outcome of his current captivity (Phil. 1:12–26), and "those of Caesar's household" (4:22) provide the opportunity to clarify the venue and situation from which the apostle wrote. Along with most interpreters over the centuries, I believe that Paul wrote from Rome as he awaited Caesar's decision on the appeal that brought him there from Judea (Acts 26–28). His mention of the Philippians' sufferings (Phil. 1:29–30), his gentle admonitions to unity (2:1–4; 4:2), and his expression of thanks for their generosity (4:10–20) provide glimpses into the Philippians' situation, and therefore suggest parallel circumstances in our hearers' experience that direct us to the text's most appropriate application today.[3] Passages in which Paul invokes "citizenship" terminology distinctive to this letter (1:27; 3:20) are the appropriate places to introduce Philippi's honored status as a Roman colony. In preaching the Word, pastors need to be discerning and strategic in deciding when, how, and how much of the scholarly

2. Since there is no extant manuscript evidence to suggest that the epistle ever existed as anything other than a single, unified document in the form in which we have it, I touch only briefly on the critical speculations that the abrupt transition at Philippians 3:1 or the late discussion of the Philippians' contribution in 4:10–20 might imply that the epistle as we have it was compiled from several pieces of earlier correspondence.

3. Bryan Chapell, author of the Reformed Expository Commentary volume on Ephesians, correctly emphasizes that it is necessary to identify any biblical text's Fallen Condition Focus as we move from exegeting the text's meaning and purpose in its original historical context to expositing and applying it in our contemporary setting. Chapell defines a passage's Fallen Condition Focus as "the mutual human condition that contemporary believers share with those to or about whom the text was written that requires the grace of the passage for God's people to glorify and enjoy him." Bryan Chapell, *Christ-Centered Preaching: Redeeming the Expository Sermon*, 2nd ed. (Grand Rapids: Baker, 2005), 50.

discussion should be introduced into sermons. My purpose is to model how to address introductory and background questions in preaching and teaching the church in such a way that the meaning and power of the text are illumined, or objections that our hearers may have encountered are answered, in order to facilitate humble listening to God's Word.

Dennis E. Johnson
Escondido, California
October 2012

Abbreviations

ACCS NT	Ancient Christian Commentary on Scripture: New Testament
AYB	The Anchor Yale Bible
BAGD	Walter Bauer, William F. Arndt, F. Wilbur Gingrich, and Frederick W. Danker, *A Greek-English Lexicon of the New Testament and Other Early Christian Literature*, 2nd ed. (Chicago: University of Chicago Press, 1979)
BECNT	Baker Exegetical Commentary on the New Testament
ESV	English Standard Version
KJV	King James Version
LXX	Septuagint
NASB	New American Standard Bible
NCBC	New Century Bible Commentary
NICNT	New International Commentary on the New Testament
NIGTC	New International Greek Testament Commentary
NIV	New International Version
PNTC	Pillar New Testament Commentary
WBC	Word Biblical Commentary
WCF	Westminster Confession of Faith
WSC	Westminster Shorter Catechism

Philippians

To Live Is Christ

1

Captivated by Christ Jesus

Philippians 1:1–2

Paul and Timothy, servants of Christ Jesus,
To all the saints in Christ Jesus who are at Philippi, with the overseers and deacons:
Grace to you and peace from God our Father and the Lord Jesus Christ. (Phil. 1:1–2)

What do you hear in the opening lines of the apostle Paul's letter to the church at Philippi? Are these words just stock boilerplate "preliminaries," to be skimmed over quickly to get to the meat of the matter? Should we process them the way we do a form letter's impersonal "To Whom It May Concern," or the fake familiarity of "Dear Valued Customer" in computer-generated mass mailings, sent by marketers who consider us "dear" and "valued" only because they want our dollars?

The openings of Paul's letters do sound alike. Their basic components can be found in almost any piece of first-century Greek correspondence: author, recipients, and a greeting (good wishes or a blessing). It would be a mistake, however, to dismiss Paul's handling of this standard template as though it were the thoughtless product of a mechanical "mail-merge" function. As similar as they seem, each of Paul's letter openings actually introduces key themes to be developed in the rest of the epistle, just as the opening lines of John Milton's *Paradise Lost* foreshadow the tragic story that follows:

> Of man's first disobedience, and the fruit
> Of that forbidden tree, whose mortal taste
> Brought death into the world, and all our woe,
> With loss of Eden, till one greater Man
> Restore us, and regain the blissful seat,
> Sing Heav'nly Muse.[1]

As these words give a premonition of Adam's fall and its dire effects, while promising rescue through a second Adam, so Paul begins his "conversation" in correspondence with the Philippian congregation with a preview of his agenda for writing. The apostle "tweaks" the Hellenistic epistle template to lay the groundwork on which he will build his pastoral counsel to his friends in Philippi.

The Backstory of the Church at Philippi

Chains and armed guards prevented Paul from carrying on a face-to-face conversation with the Christians of Philippi, so his epistle had to serve as his side of a dialogue between himself, this congregation's founding father, and his beloved children in the faith. Paul and the Philippians shared a history that had forged a strong bond between them. These believers would have heard every word from Paul's pen against the backdrop of that relationship. To pick up the subtle previews embedded in Paul's opening greeting, we need to do some detective work to place ourselves, as much as possible, into the context that the Philippian believers inhabited day by day. We need to comb through the epistle, the book of Acts, and other ancient records reflecting life in Philippi, picking up clues to the situation that prompted Paul to send this missive of warm love and surprising joy.

By the time that Paul, Silas, and their team reached Philippi, this city in eastern Macedonia already had a colorful history. Four centuries earlier, the city had been taken over by King Philip II of Macedonia, father of Alexander

1. John Milton, *Paradise Lost*, 1.1–6.

the Great—hence the name *Philippi*. In the century before Paul arrived, Julius Caesar's nephew Octavian and the general Marc Antony defeated Caesar's assassins in a decisive battle fought just outside Philippi, and the victors celebrated their triumph by constituting Philippi a Roman colony. That meant that citizens of Philippi had the same legal rights and privileges as citizens of Rome, the capital of the empire. Many retired army veterans settled in Philippi, adding to the city's "Roman flavor," which was reflected in its architecture and its language. Although surrounded by Greek-speaking communities in the eastern Mediterranean, Philippi had Latin as its official language. Not surprisingly, Philippi prided itself on its religious devotion to the Roman emperors, in addition to worshiping indigenous pagan deities. Yet one choice was missing from the smorgasbord of religious options offered in Philippi: there was no synagogue, apparently because the Jewish community was so small that it lacked the minimum quorum of ten males required by rabbinical tradition.[2]

These influences molded the Philippian mind-set that Paul and Silas met as they traveled west along a major Roman road (Via Egnatia) to this significant Macedonian city, located north of the Aegean Sea on the eastern side of what is now Greece. Outside the city gate they found a riverbank where women whose hearts hungered to know the God of Israel had gathered for prayer. One of these was Lydia, a textile importer from Thyatira in Asia, across the Aegean Sea. She believed the gospel as the Lord opened her heart, and offered her spacious home as the missionaries' ministry base (Acts 16:11–15). Later, Paul's exorcism of an evil spirit from a slave girl enraged her owners, who had profited from her "gift" for fortune-telling (16:16–18). The owners gathered a mob and played on Philippi's pride in its privileged link to Rome by accusing Paul and Silas of advocating "customs that are not lawful for us as Romans to accept or practice" (16:21). To quell the disturbance, Philippi's magistrates ordered beating and incarceration. By the next morning, however, an earthquake and an urgent midnight conversation had brought the jailer and his family from spiritual death into everlasting life (16:25–34).

2. Mishnah Megillah 4.3: "They do not recite the Shema . . . , they do not pass before the ark, [the priests] do not raise up their hands, they do not read the Torah, they do not conclude with a prophetic lection . . . and they do not invoke the name of God in the Grace, [when there are] less than ten." Jacob Neusner, *The Mishnah: A New Translation* (New Haven, CT: Yale University Press, 1991), 322.

When Paul wrote his letter a dozen years later, some who heard it read aloud had probably lived through those (literally) earth-shaking events. Was Lydia still hosting the church in her home, as she did at first? Was the jailer sitting in the congregation with his family, recalling Paul's bleeding back as the words "the same conflict that you saw I had" (Phil. 1:30) were spoken? Was he replaying in his mind the missionaries' surprising songs in the night as he heard Paul's new report of his current chains and contagious joy (1:18–26)? Was the slave girl there, too, in her right mind, set free by the name of Jesus, to whom every knee will bow, in heaven and on earth and under the earth, as every tongue confesses that Jesus Christ is Lord (2:10–11)? Were there Roman citizens who had once praised the emperor as lord and savior but who now rejoiced in a higher citizenship and awaited a greater Savior and Lord: for "our citizenship is in heaven, and from it we await a Savior, the Lord Jesus Christ" (3:20)?

Paul had a deep affection for this church. The letter is laced with terms of endearment and expressions of longing for reunion with his friends, to whom Paul says, "I hold you in my heart . . . . I yearn for you with all the affection of Christ Jesus" (Phil. 1:7–8), and whom he calls "my brothers, whom I love and long for, my joy and crown" (4:1).

On the other hand, the members of the Philippian church would also be aware that their congregation had problems. One flaw, which Paul will address later in the letter, was a subtle self-centeredness that showed itself in competing priorities and interpersonal frictions. He keeps returning to this concern:

> Do nothing from rivalry or conceit, but in humility count others more significant than yourselves. Let each of you look not only to his own interests, but also to the interests of others. (Phil. 2:3–4)

> Do all things without grumbling or questioning, that you may be blameless and innocent, children of God without blemish in the midst of a crooked and twisted generation . . . . (2:14–15)

> I entreat Euodia and I entreat Syntyche to agree in the Lord. Yes, I ask you also, true companion, help these women, who have labored side by side with me in the gospel . . . . (4:2–3)

Such rivalries and misunderstandings jeopardized the Philippians' unity at the very time when external pressure from persecution made it all the more imperative that they be "in full accord and of one mind" (Phil. 2:2). Although the physical threat of suffering (1:27–30) and the spiritual threats of Judaizing legalism (3:2–11) and lawless sensuality (3:18–19) lurked in the background, the frictions and fissures that divided these believers weighed most heavily on Paul's heart. Putting it bluntly, the members of this otherwise wonderful church were not jumping for joy at the prospect of being *slaves*, which is precisely the way that Paul unapologetically characterized himself and Timothy. Slaves, after all, had to do what other people wanted. Greeks spoke of them as "talking tools" or "thinking tools," like a plow or a hammer, only more versatile and able to perform a variety of tasks. Slaves had to submit their personal preferences, opinions, convenience, schedules—even their physical health and safety—to the agendas and whims of their masters. Who would volunteer for such a powerless position, unless compelled by armed force or economic necessity?

Later in this letter Paul will explicitly correct the Philippians' self-centeredness. In these opening sentences, he takes a very gentle approach to the sensitive subject of their resistance to the calling of slaves. He presents himself and Timothy as men who have found freedom in being slaves, captivated by Christ. Then he gives reasons to believe that becoming Christ's slave is the road to lasting joy.

Paul makes these points by mentioning one name three times in these two verses: *Christ Jesus . . . Christ Jesus . . . the Lord Jesus Christ*. This threefold repetition foreshadows how thoroughly Paul will extol Christ as the only theme worth preaching (Phil. 1:15, 17, 18), the only master worth honoring (1:20), the only cause to make life worth living and death worth dying (1:21). To each mention of Jesus' name Paul attaches a distinctive phrase:

Servants of Christ Jesus
Saints in Christ Jesus
Grace . . . and peace from . . . the Lord Jesus Christ

These three phrases are keys that unlock the mystery of how Paul and Timothy could find joy in being captivated as Christ's slaves, and how we can experience that same joy.

Servants of Christ Jesus

The epistle's opening verse expresses Paul's first point: The heart of joy is selflessly serving King Jesus and others for his sake.

Slave-Authors

Paul's emphasis on servanthood can be seen in two small but significant variations to the standard opening of a first-century letter. First, with respect to authorship, Paul groups Timothy's name with his own, and then shares with Timothy the title *servants* or, more precisely, *slaves*.[3] In other letters Paul included the names of his colleagues with himself as virtual coauthors (2 Corinthians, Colossians, Philemon, and 1 and 2 Thessalonians). But when he attached titles to names, he affixed one title to himself and another to his colleagues. We read, for example, of Paul the *apostle* and Timothy the *brother* (2 Corinthians; Colossians), or of Paul the *prisoner* and Timothy the *brother* (Philemon). Only in Philippians does Paul open an epistle by associating a colleague with himself *and then link their names* with a shared title, "slaves of Christ Jesus." Why would he do this here and not elsewhere—and, specifically, why choose the title *slaves* to describe himself and Timothy?

The Philippians need to see dramatized in Paul and in Timothy the counterintuitive truth that these men bear God's *authority* because Christ has captivated them as his *slaves*. Paul and Timothy are living proof that those whom Jesus *saves* he *enslaves*. In their self-centered preoccupations and competing agendas, Paul's Philippian friends need to see what joyful slavery looks like, up close and personal.

The claim that Jesus enslaves those he saves may sound harsh and uninviting: what kind of "salvation" is it that deprives us of our cherished autonomy and subjects us to the will of Another? But consider the link between being saved and being enslaved by Jesus from this perspective: everybody is somebody's slave. Despite the inflated claim of William Ernest Henley's Victorian poem "Invictus," none of us can honestly say, "I am the master of my fate:

3. The New Testament writers employ various terms for those whose occupation entails the obligation to serve others' desires and obey others' directives: *diakonos*, *pais*, *hypēretēs*, *therapōn*, and so on. Most of these refer to servants who have liberty, at least potentially, to seek employment in a different household. Paul's term here, *doulos*, refers specifically to "one who is a slave in the sense of becoming the property of an owner." J. P. Louw & E. A. Nida, eds., *Greek-English Lexicon of the New Testament Based on Semantic Domains* (New York: UBS, 1989), sec. 87.76.

I am the captain of my soul." No matter how much you would like to think otherwise, your every plan and action is driven by a desire to avoid pain or achieve gain by pleasing or placating some "lord" or other. The master you serve may be success or money, or what money can buy. Your lord may be affection or romance, or reputation and respect. You may be enslaved by other people's opinions, terrified at the prospect of rejection or ridicule, or perhaps you are haunted by the specter of life alone.

You also have to face the fact that every master other than Jesus will exploit and disappoint you in the end. Not all are as obvious as the evil spirit that had seized the Philippian slave girl and forced words out of her mouth. Not all are as blatant as the slave girl's owners, who treated her as a moneymaking piece of property. But every master other than Jesus will use you and then discard you. When we realize that we all serve one master or another and that other masters inevitably abuse and fail us, suddenly we find that there is nothing as liberating as being a slave of King Jesus. The church father Chrysostom commented: "One who is a slave of Christ is truly free from sin. If he is truly a slave of Christ, he is not a slave in any other realm. . . ."[4]

Being Jesus' slave not only frees us from every abusive master, but also confers delegated authority. Roman society had taught the Philippians to hear nothing but powerless subservience in the term *slave*. But Paul had introduced them to the Old Testament Scriptures, where the title "slave" or "servant of the LORD" was applied to leaders such as Moses, Joshua, and David.[5] Those ancient servants were previews of the ultimate Servant of the Lord foretold by Isaiah, who would accomplish God's will through obedience and suffering. In this letter Paul uses the title "servant [slave]" just one more time, to describe the Christ who was in the form of God and then took "the form of a servant" and offered the ultimate obedience in death on a cross (Phil. 2:6–8). The Lord delegates authority to his slaves, to accomplish his will and shepherd his people. More than that, the Lord honors the slave's role by assuming it himself in his incarnation.

4. Chrysostom, Homily on Philippians 2.1.2, in *Galatians, Ephesians, Philippians*, ed. Mark J. Edwards, ACCS NT 8 (Downers Grove, IL: InterVarsity Press, 1999), 217.

5. Neh. 10:29; Josh. 24:29; Ps. 89:20. See Peter T. O'Brien, *The Epistle to the Philippians*, NIGTC (Grand Rapids: Eerdmans, 1991), 45; Gordon D. Fee, *Paul's Letter to the Philippians*, NICNT (Grand Rapids: Eerdmans, 1995), 63–64; John Reumann, *Philippians: A New Translation with Introduction and Commentary*, AYB (New Haven, CT: Yale University Press, 2008), 82; G. Walter Hansen, *The Letter to the Philippians*, PNTC (Grand Rapids: Eerdmans, 2009), 39.

So Paul starts by inviting the Philippians to follow his and Timothy's lead, tasting the freedom of bowing to Christ's lordship. Paul is in custody, probably in Rome, awaiting the outcome of his appeal to Caesar himself. Paul is going to show them how being a slave of Jesus has set his heart free to accept any outcome to his legal case, as long as Christ gets glory through Paul's response to his circumstances. He says in Philippians 1:20: "It is my eager expectation and hope that I will not be at all ashamed, but that with full courage now as always Christ will be honored in my body, whether by life or by death." Paul is so captivated by Christ that all he cares about is seeing his Savior exalted.

Timothy shares the same single-minded focus on pleasing the Master. Paul names Timothy side by side with himself because he intends to send Timothy soon to Philippi. Timothy is so captivated by Christ that he cares more about his fellow Christians than about his own comfort or safety (Phil. 2:19–24). In Timothy's coming they will experience Paul's love, for Timothy is Paul's spiritual son, and sons resemble their fathers. More importantly, Timothy seeks the interests of Jesus Christ and therefore expresses the compassion of Jesus himself.

What would it do for our unity as the body of Christ, for our patience with others who see things differently, if we were to think like Paul and Timothy, to see ourselves as slaves of Christ Jesus? How would it impact our personal and family priorities in the way we spend our free hours and our dollars?

The best way to learn the joy of being Jesus' slave is by watching it worked out in practice. In the midst of our seminary poverty, my wife gave me a book entitled *How to Keep Your Volkswagen Alive: A Manual of Step-by-Step Procedures for the Compleat Idiot* (now in its nineteenth edition).[6] It avoided mechanics' jargon and had clear, cartoon drawings. Its humor was entertaining. Yet this manual could not compare with standing alongside a real mechanic and watching him work on an engine. The same is true of the process of getting your heart inside the freedom of joyful slavery: you need to watch how "the pros" do it. The Philippians could watch Paul and Timothy "show how it's done," as could other churches. The competitive Corinthians needed to learn humility by watching Paul's and Apollos's collaboration in ministry: "I have applied all these things to myself and Apollos for your

6. John Muir and Tosh Gregg, *How to Keep Your Volkswagen Alive: A Manual of Step-by-Step Procedures for the Compleat Idiot*, 19th ed. (Berkeley, CA: Avalon Travel Publishing, 2001).

benefit, brothers, that you may learn by us not to go beyond what is written" (1 Cor. 4:6). The author to the Hebrews urged that congregation's readers to recall the example of past shepherds "and imitate their faith" (Heb. 13:7).

Do you have in your circle of acquaintances some "skilled mechanics" in serving Jesus, models in servitude, so that you can watch them and see how it's done? Who are the fathers or mothers, older brothers or sisters in following Jesus whom you are watching as apprentices watch a craftsman—those about whom you say to yourself, "When I grow up, I want to be like him or her, quietly caring for others' needs first"?

Paul stated explicitly to Titus that he expected older Christian women to pass along to younger women the wisdom and spiritual maturity that God had granted them through years of learning and practicing the Word of God (Titus 2:3–5). No doubt he expected older men—and not only those who held the office of elder—to fulfill the same modeling and mentoring roles as those who had learned spiritual maturity: "sober-minded, dignified, self-controlled, sound in faith, in love, and in steadfastness" (2:2). Although years do not automatically confer wisdom (which is ultimately God's gift, not our achievement, Prov. 2:6; James 1:5), Scripture honors the aged: "Gray hair is a crown of glory" (Prov. 16:31). Many churches today, in a commendable desire to meet the distinctive needs of different groups—children, youth, and adults in various life phases—run the risk of segregating generations, making it hard for those who are younger to get to know those who are more mature. As individuals, too, we may gravitate toward people like us, who share our current interests and issues. When we do, we forgo a rich resource of wisdom that the Lord has prepared in the lives of those who are walking the path of faith ahead of us. Both for our congregations and for ourselves, the biblical model of spiritual nurture through godly examples calls us to honor the elderly and to pursue ways to glean the life lessons that they have to share with us.

Overseers and Deacons

The second adjustment that Paul makes to his customary opening is that he addresses this epistle not only to the church but also to its leaders. This is the only letter that Paul opens with a greeting to the church's officers, its "overseers and deacons." Paul writes nothing randomly. Why, then, this greeting to the church's elders—*overseers* is another term for *elders*

(Acts 20:17, 28; Titus 1:5–7; 1 Peter 5:1–2)—and deacons? We cannot be sure, but the profile of the congregation that we have seen suggests that Paul's purpose is to send hints to the congregation and to the leaders themselves.

First, to the *members of the congregation*, Paul presents a reminder: "When you are tempted to dig in and insist on getting your own way, remember that Jesus has embedded you in a network of authority and accountability, for your own good. You have overseers who are charged to watch out for your well-being and to correct you when you stray. And you have deacons, servants (*diakonoi*), who show you how to care for others with the compassion of Jesus, who came not to be served but to serve (*diakoneō*) (Mark 10:45). Learn the joy of servitude by watching your leaders." Elsewhere Paul instructs Christians "to respect those who labor among you and are over you in the Lord and admonish you, and to esteem them very highly in love because of their work" (1 Thess. 5:12–13). In contrast to some in our day, who consider church membership to be optional or even suspect, the apostles expected the followers of Jesus to be recognizable not only by their public profession of faith, but also by their commitment to the Christian community and their glad submission to the shepherds that Jesus appoints for his flock (Acts 2:41; 5:13–14; Heb. 13:7, 17).

Second, to the *overseers and deacons*, Paul drops the hint, "Brothers, as you exercise the authority that Jesus has delegated to you, remember that, like Timothy and me, you are 'slaves of Jesus Christ.' To be leaders in Jesus' kingdom is to be slaves of all, serving those whom you shepherd." In Philippians 4:3 Paul will lay on the shoulders of one leader, whom Paul considered his "genuine yokefellow," the heavy burden of helping estranged sisters reconcile with each other. Such intervention demands a spirit of selfless sacrifice. As one commentator observed, "Paul directs his opening greetings to leaders in the church (overseers and deacons) because they were the potential solution to the problem of disunity in the church."[7]

So Paul and Timothy, as "slaves of Christ Jesus," are living proof that the heart of joy is selflessly serving King Jesus and others for his sake. This servant's heart must be seen in the church's leaders and in its members. But what makes serving Jesus so strong a source of delight that even Roman

7. Hansen, *Philippians*, 42. Similarly, O'Brien, *Philippians*, 49–50; and, tentatively, Fee, *Philippians*, 69.

imprisonment could not dampen Paul's joy? The answer is found in Paul's second use of the name of Christ Jesus.

SAINTS IN CHRIST JESUS

Paul's second use of Jesus' name suggests why being Christ's slaves generates joy: The heart of joyful service is being set apart to stand awestruck before the beauty of King Jesus.

When Paul calls his Philippian friends *saints*, he evokes a picture of privileged access into the very temple of God. We hear the word *saints* repeatedly, but do we pause to ponder what it means? We may say about someone with extraordinary patience, "Oh, she's a saint." But what does the Bible mean by *saint*?

In our English Bibles, the noun *saint* and the adjective *holy* are two ways of talking about the same thing.[8] Although *saint* and *holy* do not look alike in English, they represent the same family of words in the biblical languages, Hebrew (*qadosh*) and Greek (*hagios*). These terms describe the *purity* that befits the *privilege* of standing in the *presence of God*. When the Lord appeared to Moses at the burning bush, God's presence made the ground under Moses' feet "holy," requiring that Moses shed his sandals (Ex. 3:5). On the yearly Day of Atonement, Israel's high priest—with elaborate sacrificial and cleansing rituals—passed through the Holy Place, the sanctuary's outer chamber, into the inner chamber, the Most Holy Place (Ex. 26:33–34; Lev. 16). The turban on his head bore a gold plate engraved, "Holy to the LORD" (Ex. 28:36). The Lord himself is supremely holy, as his awesome seraphim chant thrice over (Isa. 6:3–5). Isaiah trembled to realize that the Lord's holiness—his consuming purity—was lethal to defiled people. Even the high priest's sons were consumed by fire when they treated regulations pertaining to the sanctuary in a cavalier way, for the Lord said, "Among those who are near me, I will be sanctified [treated as holy]" (Lev. 10:1–3).

We might say that *holiness* is "dangerous privilege": dangerous because the all-Holy God is not to be treated casually, but also privilege because we were created to be near him, beholding his beauty and attending to his desires. Our popular usage of *saint* contains a grain of truth: a *saint* is a

8. The words *saint*, *sanctification*, and *sanctuary* came into English from Latin, whereas the words *holy* and *holiness* have homegrown Anglo-Saxon roots.

special person, set apart by God and granted access to God's holy presence. Yet, amazingly, the Bible calls people who are not pure or free of defiling sin *holy* and *saints*. God called Israel to be "a kingdom of priests and a *holy* nation" (Ex. 19:5–6). Though those Israelites were stiff-necked and prone to wander, still he pitched his tent in the middle of their camp. He had picked Israel out of all the nations and separated them as his own property, so they were *saints*, a people holy to the Lord (Deut. 7:6).

Surprising Saints

Paul's personal pedigree included his belonging to this special people, Israel (Phil. 3:5). But in opening this epistle, Paul surprises us by applying this precious title of privilege to a congregation that was composed of all sorts of people from diverse ethnic and religious backgrounds. Since the Jewish community in Philippi was so small, most if not all of these "saints" must have been Gentiles, raised in pagan religions. Lydia, the first Philippian believer in Jesus, was a God-fearer (ESV: "worshiper of God") (Acts 16:14). God-fearers were Gentiles who embraced the Jewish belief in one God and tried to follow the Ten Commandments, but did not fully convert to Judaism's dietary and other ceremonial obligations. Then there was the jailer, whom no one would have called a *saint* before the earthquake at midnight. Only after Christ shook his world did he wash his prisoners' wounds. Before that, he hadn't cared.

Now Paul applies the glorious title *saints* to Lydia and the jailer alike. How could the Creator, who is pure clear through, allow soiled, sinful people such as Lydia and that jailer, or Paul and Timothy, or you and me, to stand in his presence, admiring his glory and attending to his wishes? How could people like us even *survive* in the presence of such all-consuming purity? The answer lies in Paul's second use of Jesus' name: we are *in Christ Jesus*.

United to His Holiness

We may be so accustomed to Paul's formula "in Christ" or "in Christ Jesus" that we fail to notice the momentous reality that it conveys. Paul uses this phrase or something like it ("in the Lord" or "in him") over twenty times in this brief letter, the highest concentration in any of his correspondence except Ephesians and Philemon.[9] He uses it to describe the Christian's reason

9. Reumann, *Philippians*, 59.

for rejoicing (Phil. 3:1; 4:4, 10) and source of encouragement (2:1). Being "in Christ" is the protective environment in which God's peace guards our hearts from worry (4:7). "In the Lord Jesus" is the atmosphere in which Paul lays his plans for the future (2:19). But at its core, "in Christ" is Paul's shorthand for the truth that men and women and boys and girls who trust in Jesus are bound tight to him, so that his obedience and sacrifice and resurrection life become theirs. His death on the cross becomes their death under sin's condemnation and their death to sin's domination. His resurrection declares their right standing before God the Judge and ushers them into a new life of freedom to love God. No wonder Paul's desire was to be found "in" Christ, not claiming a righteousness of his own but resting instead in the righteousness that comes through faith in Christ (3:9).

That little word *in* traces the source of our hope to the fact that God has united believers to his Son, and thus given us a share in all that Jesus has accomplished, including Jesus' worthiness to stand in his flawless integrity before his Father. One commentator interpreted Paul's audience as "all in Philippi . . . who are holy through their union with Christ Jesus."[10] That captures Paul's point well. Only because Jesus was holy straight through, from start to finish, can we stand in the presence of the all-holy God and delight in his beauty rather than being incinerated by his white-hot purity.

For All the Saints

God's grace, which makes us fit to bask in his beauty, embraces "*all* the saints." Paul will include "you all" in his prayers, his confidence, his gospel partnership, and his longing (Phil. 1:4, 7, 8). Each "you all" is intentional: Paul embraces every believer in Philippi, and they need to do the same to one another. The rifts in the church in Philippi were not as deep as the party spirit at Corinth, where Christians sounded like children arguing at recess: "I'm on Paul's team," "I'm with Apollos," "I'm all for Peter," "I'm on Jesus' side" (1 Cor. 1:12). Even though the fissures in Philippi were not the chasms of Corinth, the Philippian church needed Paul's call to unity (Phil. 1:27; 2:1–4).

When our priorities compete and our preferences clash in the church, we tend to reduce Paul's *all* to *some*. We may say to ourselves, "I find it easy to

10. Moisés Silva, *Philippians*, 2nd ed., BECNT (Grand Rapids: Baker, 2005), 38.

serve with *some* of the saints, give thanks for *some* of the saints, and pray for *some* of the saints. But there are others . . . I'm not saying they are not saints, of course. But we rub each other the wrong way. We need to give each other plenty of space. You understand."

Paul says, "*No*, I do *not* understand. Since your status as saints is 'in Christ Jesus' and in his grace alone, I insist on embracing *you all* in my love, and I expect *you all* to do the same to each other." Later he will marshal reasons for our commitment to unity: "If there is any encouragement in Christ, any comfort from love, any participation in the Spirit, any affection and sympathy, complete my joy by being of the same mind, having the same love, being in full accord and of one mind" (Phil. 2:1–2). Our backgrounds and life experience may be so different that we do not *naturally* fit together. But if we are "saints in Christ Jesus," our lives have been *supernaturally* and inextricably interwoven.

Yet this single-minded, single-hearted unity is not easy to live out in practice in the daily frictions that try our patience with one another. When things don't go our way, it is easy to pull up stakes and move on to the next congregation, rather than to stay and work through hurt feelings or competing visions. What force is strong enough to hold us together, when our self-centeredness and our culture's individualism threaten to pull us apart? Paul's third use of Jesus' name answers that question.

Grace and Peace from Christ Jesus

Paul's opening blessing shows that Christ's grace and peace have the power to turn selfless service into lasting joy.

The typical first-century letter followed the identification of author and readers with the Greek word *chairein*.[11] *Chairein* meant "Rejoice"; but as often happens with commonly used expressions, in epistle openings *chairein* had faded into a colorless "Greetings." (How many people think "God be with you" when they say, "Goodbye"?) Yet Paul doesn't write meaningless Greek. He replaces *chairein* with a like-sounding Greek word, *charis*, which captures the heart of the gospel: "grace."

On the part of his dear friends at Philippi Paul is invoking nothing less than the favor of God, the embrace of the Father, lavished as a free gift

11. Acts 15:23; 23:26; James 1:1. See Gerald F. Hawthorne and Ralph P. Martin, *Philippians*, rev. ed., WBC 43 (Nashville: Thomas Nelson, 2004), 12.

on those who deserve condemnation. Paul is pleased that these folks are "partakers with me of grace" (Phil. 1:7). Both their faith in Christ and the privilege of suffering for his sake are gifts of God's grace (1:29).[12]

Grace from the Father and the Lord Jesus Christ is the source of that astonishing exchange that Paul had described in 2 Corinthians 5:21: "For our sake [God] made [Christ] to be sin who knew no sin, so that in him we might become the righteousness of God." In his epistle to the Philippians he portrays each side of this exchange of grace. In chapter 2 we hear that Christ, who was "in very nature God" (NIV), took a servant's nature and died on a cross (Phil. 2:6–8). In chapter 3 we hear the other side of the exchange: Paul is found in Christ, receiving right standing with God through faith in Christ (3:9). This is the amazing "trade": Jesus the innocent condemned and punished, and we the guilty declared right in God's sight.

The result is *peace*, the reconciling reality that secures our place in God's heart. Nothing but God's grace could give us peace with God. Our insults to his honor created a chasm of antagonism between us and our Creator, and this terrible divide will not disappear just by our pretending it isn't there. This is also true in our relationships with each other. When someone has hurt you, it doesn't "make the problem go away" for the offender to ignore the pain he has inflicted, and to say glibly, "Well, let's just move on now." The injury has to be acknowledged, and the pain has to be dealt with. Peacemaking always has its price, even among human beings. The aggressor must pay the price of humbling himself and admitting the wounds that his words or deeds have inflicted. When possible, he makes amends. The victim, too, pays a price: the price of releasing resentment rather than holding the offender's guilt as a weapon to be wielded against him in the future.

The wonder of the gospel is that, though we must admit with grief that we have offended our good Creator, the God whose honor we have violated has come to absorb the pain that should be ours. Paul reminded the Christians at Ephesus that Jesus is the ultimate Peacemaker and that the price of peace was his death: he "reconcile[d] us both to God in one body *through*

12. The Greek verb *charizomai* (ESV: "granted") is a cognate of the noun *charis* ("grace"). The verb does not always connote the bestowing of an *undeserved* gift (see its use in 2:9 [ESV: "bestowed on"]). But the conceptual parallel between Philippians 1:29 and Ephesians 2:8–9 ("by grace you have been saved through faith. And this is not your own doing; it is the gift of God, not a result of works") supports the understanding that in Philippians 1:29 the verb expresses a granting that is by God's undeserved grace.

the cross" (Eph. 2:14, 16). The peace that Christ secured for us frees us to pay the price of making peace and keeping peace with each other, to put others' needs and interests above our own.

Grace and peace belong to those who approach God as "Father" and who bow to Jesus Christ as "Lord." God's grace and peace impart not only forgiveness but also transformation of the direction and affections of our hearts. God is not so incompetent as to leave us forgiven but unchanged in the poisonous self-centeredness of our hearts. The people on whom God the Father and his Son, the Lord Jesus Christ, set their invincible love will never again be satisfied to be locked into our own interests, fixated on our own reputations, or enslaved to our own self-image. Christ's glory becomes our heart's chief delight, and his love for others ignites our compassion.

To receive grace and peace from the Father and the Lord Jesus is to discover the joy of belonging to the Master who made and redeemed us for himself. *The Book of Common Prayer* captures the paradox of our status as slaves of Christ when it speaks of God "whose service is perfect freedom."[13] The authors of the Heidelberg Catechism were right to affirm that the Christian's only comfort in life or death is "that I am not my own, but belong—body and soul, in life and in death—to my faithful Savior Jesus Christ."[14] To belong to Another—to be captivated by Christ Jesus—is true liberty.

Subverting Our Self-Centeredness

At first glance, these two brief verses seemed so matter-of-fact, didn't they? They looked like the preliminaries that we could skim over quickly. Now that we have listened more closely, however, we discover that from his

13. "O God, who art the author of peace and lover of concord, in knowledge of whom standeth our eternal life, whose service is perfect freedom." "The Order for Daily Morning Prayer," A Collect for Peace, in *The Book of Common Prayer . . . according to the use of the Episcopal Church in the United States of America* (New York: Seabury, 1953), 17.

14. Heidelberg Catechism, Lord's Day 1, Question and Answer 1: "Q. What is your only comfort in life and in death? A. That I am not my own, but belong—body and soul, in life and in death—to my faithful Savior Jesus Christ. He has fully paid for all my sins with his precious blood, and has set me free from the tyranny of the devil. He also watches over me in such a way that not a hair can fall from my head without the will of my Father in heaven: in fact, all things must work together for my salvation. Because I belong to him, Christ, by his Holy Spirit, assures me of eternal life and makes me wholeheartedly willing and ready from now on to live for him." Accessed at http://www.crcna.org/pages/heidelberg_intro.cfm.

opening syllables Paul has gently brought us into the heart of the matter, subtly subverting our instinctive self-centeredness.

We like to be lords. Even if we cannot make others do our bidding, at least we want to call the shots for our own lives. But Paul and Timothy, slaves of Christ, turn upside-down our assumption that freedom is found in getting our own way. Have you experienced the liberation of surrendering to the mastery of Jesus the Christ, the eternal Son of God who became a slave, to free you from yourself and from the masters that drive you?

God designed us for togetherness and created us for community. But indifference, isolation, and competition have seduced us into thinking that freedom is found in "looking out for Number One," keeping options open, and avoiding long-term commitments. Paul and Timothy challenge our self-defensive individualism, throwing their arms wide to embrace "all the saints in Christ Jesus." None of us stands alone. Each needs the support and accountability of the rest of the body of Christ. Are there any "saints in Christ Jesus" whom you have trouble loving as brothers or sisters? Do you honor and heed the shepherds and servants in whose care God has placed your spiritual well-being? Do you pray for them, encourage them, and respect them as they protect the church's unity and purity?

Do you realize how much you need the grace of God in order to have peace with God? To enjoy a reconciled relationship with the holy God, we need grace that we have not earned and could never deserve. Does the matchless condescension of the Lord Jesus Christ so grip your heart that you are humbled and hope-filled at the same time?

2

He Finishes What He Starts

Philippians 1:3–8

And I am sure of this, that he who began a good work in you will bring it to completion at the day of Jesus Christ. It is right for me to feel this way about you all, because I hold you in my heart, for you are all partakers with me of grace, both in my imprisonment and in the defense and confirmation of the gospel. (Phil. 1:6–7)

In 1990, Public Broadcasting Service television stations across the United States aired Ken Burns's historical documentary series about the American Civil War. One gripping aspect of the series was the oral reading of correspondence between soldiers in the Union and Confederate armies and their loved ones back home. Voices speaking those old words, illustrated with early photographs of that conflict, took viewers behind the historical data and into the daily experience and emotions of ordinary citizens. Whether a letter was going to a farm in Pennsylvania or a plantation in Georgia, we heard gracious eloquence expressing the heart cries of people who felt the same affections and apprehensions that we experience today.

Reading Paul's letters to churches is like eavesdropping on somebody else's love letters. In some (Galatians, for example), we hear anger and frustration, as a pastor who loves his spiritual children passionately expresses perplexity over their gullibility. In others (such as 2 Corinthians), we hear pastor and people making up after a spat, and we sense his relief as he reaffirms their reconciliation. Paul's little epistle to the church at Philippi is overflowing with terms of deep affection. These expressions of love begin in Paul's opening thanksgiving for God's transforming work in the Philippian believers: "I hold you in my heart" (Phil. 1:7); and then "I yearn for you all with the affection of Christ Jesus" (1:8). Later he will call them "my beloved" (2:12) and "my brothers, whom I love and long for, my joy and my crown" (4:1). Not only does Paul speak tenderly to this church, he also spares them the sharp scolding that he felt compelled to speak to others. Clearly, the church at Philippi has a very special place in Paul's heart. Of course, even the fireworks that explode in the letter to Galatia were expressions of deep love. Paul would not scold so vigorously if he didn't care so much. But the Philippian believers, though far from perfect, were—how should we say it?—less painful to love.

There is a major difference between how we should receive Paul's 2,000-year-old letter to Philippi and how we might "eavesdrop on" a 150-year-old letter from the Confederate camp outside Gettysburg. What sets Paul's correspondence apart from other letters of an earlier age is that they are not merely expressions of an ancient pastor's love for his flock. Paul is writing under the inspiration of the Spirit of the risen Christ, so his letter to Philippi is also Jesus' expression of love to us today. So we are not just eavesdropping on somebody else's correspondence from a bygone era: Jesus the living Lord is looking us straight in the eye and saying, "This is how I love you: enough to embrace you—and enough to correct you, too."

Jesus says, "Those whom I love, I reprove and discipline" (Rev. 3:19). Even Paul's dear friends at Philippi would need his gentle correction. Later in the letter we catch clues that these believers were reacting to external opposition not by strengthening their bonds of unity with each other but by letting their fellowship fragment into competing pockets of self-centeredness (Phil. 1:27; 2:1–4). At the very time when they should have been giving and receiving each other's support and encouragement, when hostility and pressure from the surrounding society should have driven them together, the congregation was marred by individuals' preoccupation with their personal problems while ignoring others' needs (2:4) and by their grumbling against each other and their questioning of the goodness of God (2:14).

As we hear Paul's thanksgiving and prayer for his friends at Philippi at the opening of his letter (Phil. 1:3–8), we might well wonder whether the Philippians' problems—their suffering and the self-centeredness that it had brought to the surface—had even crossed his mind. If we listen very carefully, though, we will hear not only Paul's deep affection but also, more quietly, hints of their current struggles. Just as every letter from a Union infantryman to his fiancée in Connecticut or from a Confederate officer to his wife in Tennessee implied the shared story of their relationship and the pain of their separation, so Paul's opening thanksgiving sends subtle signals of the challenges confronting his friends at Philippi.

Later, Paul will urge the Philippians to blend thanksgiving into their own petitions in prayer (Phil. 4:6). Here he sets the pace, interweaving his thanks to God for the Philippians with his requests to God on their behalf. Yet Paul's thanksgiving (1:3–8) and his petitions (1:9–11) are so rich that each warrants a separate exposition. In this study we focus on Paul's thanksgiving, which reveals the heart of one who has received Christ's antidote to self-centeredness. Paul's thankful heart shows that as we entrust ourselves to Jesus, he gives two gifts: a love that stretches our hearts to embrace others, and a joy that places our pain into perspective, enabling us to see our suffering in the context of God's comprehensive plan to make us like his Son. Paul is experiencing these gifts of heart-stretching love and a perspective-restoring joy, and he is confident that God has started the same transformative work in his friends at Philippi. So he reports his reasons for thanksgiving in words that highlight their partnership with himself—and with each other—in the love-evoking, joy-inducing grace that Christ's Spirit conveys through the gospel.

A Heart-Stretching Love

Paul's thanksgiving for his friends in Philippi shows that Jesus frees us from the tyranny of self-centeredness by his love, an expansive love that stretches our hearts to embrace others. How does Paul make this point? He reports that his prayers to God overflow with thanksgiving and joy because they are fed by his memories[1] of the believers at Philippi and of their partnership in the gospel. His heart is so focused on his friends that he virtually forgets his own difficulties. In fact, we could read the first half of this text (Phil. 1:3–8) without picking up any hint that there is anything amiss in Paul's current situation. Only in verse 7, when he speaks of the Philippians' partnership with him "both in my imprisonment and in the defense and confirmation of the gospel," does he refer to his own situation, which he and the Philippians know well.

Paul was "imprisoned," but not in a cell block with iron bars and locked doors. The Greek word that the ESV translates "imprisonment" is actually "chains" (so NIV), and we should picture Paul in the situation described in the last chapter of Acts, as he awaited the disposition of his appeal to the emperor. In Philippi Paul had been incarcerated behind locked gates (Acts 16:23–26), but as he composed this letter he occupied rented quarters in Rome, probably near the Castra Praetoria, the barracks of the Praetorian guards on the eastern edge of the city. One or another of those guards was chained to Paul at all times, depriving him of privacy and freedom of movement; but he was free to receive guests (28:16, 30–31). Paul was a man of action, with a restless drive to take Christ's gospel to folks who had never heard the message of salvation (Rom. 15:18–24). By this point he had previously spent over two years in custody in Judea, waiting for his "case" to be resolved by provincial officials there. Then he endured a long and

1. The Greek prepositional phrase that stands behind the ESV's "in all my remembrance of you" could also be translated "in all your remembrance of me," and interpreted as an allusion to the recent gift that Epaphroditus had brought to Paul, which showed that the Philippians were concerned for Paul but had previously lacked opportunity to express their concern in a tangible exhibit of their "remembrance" of Paul (see Phil. 4:10). One commentator, for example, argues that "your remembrance of me" and "your partnership in the gospel" together give the grounds in the Philippian believers—and specifically in their financial generosity—for Paul's thanksgiving. Peter T. O'Brien, *The Epistle to the Philippians*, NIGTC (Grand Rapids: Eerdmans, 1991), 59–61. So also John Reumann, *Philippians: A New Translation with Introduction and Commentary*, AYB (New Haven, CT: Yale University Press, 2008), 102–3. On the other hand, other commentators argue persuasively that *Paul's* memories of the Philippians are the occasion of his thanksgiving. See, e.g., Gordon D. Fee, *Paul's Letter to the Philippians*, NICNT (Grand Rapids: Eerdmans, 1995), 77–80; Gerald F. Hawthorne and Ralph P. Martin, *Philippians*, rev. ed., WBC 43 (Nashville: Thomas Nelson, 2004), 19–20; Moisés Silva, *Philippians*, 2nd ed., BECNT (Grand Rapids: Baker, 2005), 54; G. Walter Hansen, *The Letter to the Philippians*, PNTC (Grand Rapids: Eerdmans, 2009). Those memories include the Philippians' financial partnership with Paul, but only as evidence of the fact that God had begun the "good work" of salvation in them, so that they had become partakers in grace with Paul. The flow and content of verses 3–8 favor the ESV's rendering: Paul remembers, and the Philippians are remembered, as Paul prays with thanksgiving.

harrowing sea voyage across the Mediterranean. Now at last he was awaiting the emperor's decision on his appeal, chained to a succession of Roman soldiers and confined to a rented residence. It had been years since he had proclaimed the good news in synagogues and marketplaces. Moreover, some Christians in Rome were capitalizing on Paul's restricted mobility to try to surpass him in soul-winning success, hoping to make Paul's chains chafe (Phil. 1:15–17). They saw themselves as Paul's rivals, so they imagined that their success would compound his frustration.

If anyone had a right to "vent" dissatisfaction over his circumstances, especially to a sympathetic audience such as his dear Philippian friends, it would have been Paul. Yet instead of self-pity, we hear his thanks to God for the believers in Philippi. His focus is on the generosity of God, and particularly on how God's grace is transforming these Jesus-followers. In a few sentences he will offer an update on his captivity. But even when he mentions the negative aspects of current circumstances, his tone will be one of joy. How can this be?

We begin to grasp Paul's secret as we notice that he directs his thanks to "my God." His relation to the Creator is not merely that of an object to its maker, or even a servant to his master. By the grace of "God our Father and the Lord Jesus Christ" (Phil. 1:2), Paul has come into a personal, intimate, covenantal bond with the living God. This God had promised his ancient people, "I will walk among you and will be your God, and you shall be my people" (Lev. 26:12), so their psalmists had addressed him in this confidence: "You are my God, and I will give thanks to you; you are my God; I will extol you" (Ps. 118:28).[2] Paul's reference to God as "my God" conveys the wonder that Paul's relationship with "his" God was intensely personal. In closing this epistle, Paul will assure his friends that as they have met his needs, so also "my God will supply every need of yours in Christ Jesus" (Phil. 4:19).

The Breadth of Christlike Love: "you all" (1:3–4)

Paul's intimate bond with the Father, whom he calls "my God," was different from the self-absorbed individualism that characterizes so much of

2. See also Pss. 31:14; 63:1; 86:2; 140:6; 143:10; Isa. 25:1; Hos. 2:23.

American "spirituality." In verses 3–4 Paul repeats the word *all* or *every* (the same Greek word) no less than four times: "*all* my remembrance . . . *al*ways . . . *every* prayer . . . for you *all*."[3] Paul gives thanks over every memory, at every moment, in every mention, for every member. The emphasis falls on the last *every*: every one of you—"you all." This comes on the heels of Paul's address "to *all* the saints in Christ Jesus" (Phil. 1:2). Moreover, Paul will repeat these words as he describes his confidence about God's work in "you all" since "you all" share God's grace (1:7), and then again when he expresses his longing for "you all" with the affection of Christ (1:8). By these words Paul throws his arms wide to gather in each and every one of his children at Philippi, to pull together brothers and sisters who may have drifted apart through misunderstanding or mismatched priorities. In effect, Paul's "you all" throws one arm around Euodia and the other around Syntyche (and their respective partisans), drawing them face to face, to see eye to eye once again (4:2).

People who have been around churches for a while are accustomed to Paul's saying things like this, and we are inclined to nod and then dismiss the remarks as meaningless sentiment. After all, it is unnatural to embrace a diverse group of people, who have grown up in very different backgrounds and inherited different ways of looking at life. So when we hear Paul, raised as a Pharisee in strictly observant Judaism, claiming to care for everyone in a church that included a Gentile businesswoman from Asia, a hardened Roman jailer, and people from other pagan pasts, we take Paul's words with a grain of salt. It is hard enough to get along with people with whom we have much in common, much less those formed by different ways of making sense of everything from food to philosophy!

So was Paul simply spouting a pious platitude when he opened his thanksgiving with words that say, in effect, "I love *you all*—all the time, treasuring my memories of all of you and bringing all your needs before my God"? No, he was both intentional and deeply sincere. He was even prepared to affirm his joy in the ministry of those competitive preachers who thought their success would frustrate him (Phil. 1:18). As long as they were truly preaching Christ, Paul was ready to rejoice. He had received the grace to love even those who were acting like rivals.

3. The deliberate redundancy would be even more evident as the Philippians heard the epistle read aloud in Greek: "*pasē* . . . *pantote* . . . *pasē* . . . *pantōn*."

The Intensity of Christlike Love: "I hold you in my heart" (1:7–8)

Paul's love for all the Jesus-followers in Philippi (and Rome and elsewhere) was not a grudging, foot-dragging "well, I really ought to" pseudo-love, a false front of politeness that masked a heart smoldering in resentment or frozen in sullen indifference. Whatever the Philippians' quibbles with each other, Paul's love for them was not only extensive but also intensive. He insisted, "I hold you *in my heart*" (Phil. 1:7) and invoked God as witness to his sincerity in affirming "how I *yearn* for you all [that fourth *you all* in our text] with the affection of Christ Jesus" (1:8). Since the days when the KJV spoke of "the bowels of Jesus Christ," our English versions have tidied up the Greek term that the ESV now renders "affection." But those earlier translators captured an aspect of the original that "affection," pleasant as it is, fails to capture. Paul is really speaking of the kind of deep emotion that you "feel in your gut." It's that emptiness in the pit of the stomach that parents feel when a son or daughter moves away from home and they know that things will never be quite the same. Matthew used the same terminology to convey the turmoil that Jesus felt when he saw the crowd's confusion: "When he saw the crowds, he had compassion[4] for them, because they were harassed and helpless, like sheep without a shepherd" (Matt. 9:36). The compassion that compelled Jesus to teach God's truth, like a shepherd who cares for harassed and helpless sheep, is the same restless longing that Paul now feels as chains and distance prevent him from caring in person for his Philippian friends. Toward the close of the epistle he will again express his yearning to be near his "brothers, whom I love and long for" (Phil. 4:1).[5] When the affection of Christ grips our hearts, absence not only "makes the heart grow fonder," but also makes the heart grow restless. Paul's love is not only expansive, embracing all his fellow Christians, but also intensive, filling the core of his being with a strong longing for their spiritual well-being.

As we observe the intensity of Paul's love, we also need to recognize that his feelings have solid foundations in reality. Paul's love, like Christ's,

4. This verb in Greek is *splangchnizomai*, a cognate of the noun *splangchna* (ESV: "affection"; KJV: "bowels"), used by Paul in Philippians 1:8.

5. The adjective *epipothētos* (ESV: "whom I . . . long for") in Philippians 4:1 is a cognate of the verb "yearn" (*epipotheō*) in 1:8. Note also Epaphroditus's longing for his fellow Philippians (2:26).

takes its stand on truth, as he implies in commenting on his confidence that God had begun a good work in these believers: "It is right for me to feel this way about you all" (Phil. 1:7). In our ears the word *feel* (ESV; NIV; NASB) connotes emotion, but Paul's Greek word has a strong mental component.[6] He uses this word frequently in this little letter. He will insist that the unified "mind" with which Christians regard each other (2:2; 4:2) must reflect the "mind" of Christ (2:5). The mature must "think" as Paul does, pressing on toward the goal of perfection in Christ (3:15), in contrast to those whose "minds" are set on earthly objectives (3:19). Paul feels great confidence as he thinks about the Philippians, but his feelings have a foundation in the fruit that he has seen the gospel produce in their lives. This fruit is the way in which these believers have stood with Paul as partners in grace "both in my imprisonment and in the defense and confirmation of the gospel" (1:7). Paul cherishes these Christians warmly, and he has good reason to do so.

In view of Paul's restrictive and tenuous circumstances, we may wonder how he could invest such emotional energy in others. The Philippians have been generous toward Paul monetarily and personally, and their delegate Epaphroditus has risked his life to help Paul. Yet it is still extraordinary that Paul's emotional equilibrium is linked more to these brothers and sisters than it is to his own living conditions or legal complications. Where does such love come from?

The Source of Christlike Love: "Partnership in the Gospel . . . Partakers . . . of Grace" (1:5, 7–8)

Twice in this section Paul uses the Greek word that we often see translated as "fellowship," or even transliterated into English as *koinonia*. It describes the bond that ties together people who have something important in common, who share a special privilege or treasure.

6. In contrast to recent versions, in Philippians 1:7 the KJV translated the Greek verb *phroneō* as "think." Commentators note that this verb, especially in the ten instances in which it appears in Philippians, "points to more than Paul's emotional feelings; it also connotes thinking, being concerned, having an opinion or an attitude about something or someone." Hansen, *Philippians*, 51. Hawthorne and Martin, *Philippians*, 26, conclude, "It is a word that embraces both feeling and thought, emotions and mind." See also O'Brien, *Philippians*, 65–67; Fee, *Philippians*, 89–90; Silva, *Philippians*, 47; Reumann, *Philippians*, 116.

Sometimes in the New Testament "fellowship" or "partnership" words refer to financial partnership, a sharing of material resources. In Acts 2 and 4, for example, Luke shows how the Holy Spirit created a spirit of open-hearted, open-handed generosity among the early followers of Jesus in Jerusalem, so that the affluent readily liquidated their assets to help the poor, devoting themselves to "fellowship" (*koinōnia*) by having "all things in common" (*koinos*) (Acts 2:42, 44; see 4:32). The financial overtones of fellowship appear at the end of this epistle (4:10–20), where Paul thanks the Philippians for their faithful partnership (*koinōnia*) in giving and receiving, exemplified in their recent contribution.

As he begins the letter, however, Paul focuses on two treasures that he shares with the Philippian Christians and that go deeper than their monetary support. These gifts from God explain why his friends have been ready to partner with Paul to support his preaching and sustain him in his chains. These same shared gifts would also cause the Philippians to grow in the expansive, intensive love that flowed from Paul's heart to embrace them all, despite their current frictions with one another. The source of heart-stretching love is *partnership in the gospel*—which is to say, *partnership in God's grace*, extended to us in Christ. Paul's brothers and sisters in Philippi are his partners in the gospel, not merely or primarily as contributors donating to a cause, but rather as recipients of the good news about the price that Jesus paid in love to rescue them from sin and death. They had been Paul's partners in the gospel "from the first day" (Phil. 1:5), when Paul's team met with Lydia and the other women by the riverside. Through that gospel, as the Lord opened hearts to respond to Paul's message (Acts 16:14), they had become "partakers with me of grace" (Phil. 1:7).

Paul will unpack the content of this good news later in this letter. He will tell the story of Christ, who was "in very nature God" (NIV) and equal with the Father, assuming human nature and the slave's low status, even stooping to endure death on a cross (Phil. 2:6–8). Paul will narrate the exaltation of Christ in resurrection glory, to receive the adoration of all creation (2:9–11). The apostle will show how Christ's achievement in obedience, death, and resurrection became personal in Paul's own experience, leading Paul to embrace the gift of a righteousness that was not Paul's accomplishment but God's gracious gift, received by trusting in Christ (3:7–10).

It is because Paul and his friends have a shared trust in this good news—relying not on themselves but on Jesus—that Paul loves these folks so extensively and intensively. God's grace has dealt the death blow to Paul's self-confidence and pride, exposing the misguided arrogance that lurked beneath the mask of his pious law-keeping. That same grace has answered the accusations of Paul's insecure conscience, assuring him that the gift of Christ's righteousness provides a firm foundation for confidence of God's approval. Such pride-shattering, shame-silencing grace has freed Paul to love all who stand with him on that same solid rock, Christ himself. So he thanks God for them and embraces them in the love that he and they have received through Jesus.

Moreover, he is confident that as the Philippians reflected on the grace that they have received, their hearts will be freed to love one another without reservation or competition. They might need help remembering how lavish God's gifts of grace have been, so Paul will gladly remind them that they have received encouragement in Christ, comfort from God's love, participation (*koinōnia* again) in the Spirit, affection, and sympathy, before he urges them point-blank to cultivate unity with each other (Phil. 2:1–3).

If we are partners in the gospel, it is because we are partners in grace—because the invincible Spirit of Christ has pulled us, in spite of ourselves, out of the pit of our self-centered self-reliance, made us face the ugly reality of our guilt and helplessness, and drawn us to trust in Jesus. God's grace evokes our gratitude, and Christ's love ignites our love, not only for the Lord who rescued us but also for people who need and have received his unmerited love right along with us. If you are honest, you must admit that it is sometimes hard to say, with Paul, to *everyone* in the church—to those sitting beside us, in front of us, behind us, or over on the other side of the sanctuary—"I *always* remember you *all* with thanks and joy, *always* pray for you, always yearn for *you all* when we are apart." Did she slight or ignore you? Does he sing too loud and off-key? Are their children too restless? When fellow Christians' offenses and defects loom large in our minds, it is because we've lost sight of the marvel that all of us who belong to Jesus are partners in the gospel and fellow recipients of his abundant grace. To love our neighbors as ourselves, and especially our Christian brothers and sisters, extensively and intensively, we need a source of love deeper than our own puny hearts. We need "the affection of Christ Jesus," imparted by his Holy

Spirit residing in us, constantly turning our gaze upward to the Lord who showed us compassion, and then outward to those who need to experience his compassion through us.

A Perspective-Restoring Joy

Paul's thanksgiving mentions a second gift that God has graciously granted: joy. In verse 4, Paul says that his every prayer for his Philippian friends is offered "with joy." Here he introduces a theme that will run through this letter like a golden thread in a tapestry. The theme of joy or rejoicing arises sixteen times in the 104 verses of this brief letter. That is once in every seven verses, far more frequently than in any other epistle.[7] Against the dark fabric of Paul's confinement and possible execution and the Philippians' similar sufferings, the thread of joy gleams like gold. Though some in Rome who should be friends want to increase Paul's pain, Paul will rejoice (Phil. 1:18). Though his death might be near, Paul will rejoice (2:17). He wants his Philippian friends to join in his joy (2:18), and he instructs them to rejoice in the Lord, again and again (3:1; 4:4). By God's life-changing grace, these believers themselves are "my joy and crown" (4:1).

Obviously, Paul's joy is not based on his circumstances. He is not writing from the comfort of a study on a country estate. Just a few words after his first mention of joy in Philippians 1:4 he will mention his chains, bringing into view the captivity and threat to his life itself that he is suffering (1:7). Nor is Paul merely putting a "happy face" on a bleak situation, in order to keep up morale among the troops. Paul's basis for joy is that he and his partners in grace are loved by an almighty, all-wise, all-faithful God. Paul's joy is utterly realistic, because it is grounded in God's track record in past and present. God will let nothing stand in the way of achieving his goal: he finishes what he starts!

7. Phil. 1:4; 1:18 (twice); 1:25; 2:2; 2:17 (twice); 2:18 (twice); 2:28; 2:29; 3:1; 4:1; 4:4 (twice); 4:10. By contrast, "joy" (*chara*) and "rejoice" (*chairō*) appear fourteen times in the (much longer) 2 Corinthians (once per 18 verses), five times in 1 Thessalonians (once per 17.6 verses), three times in Colossians (once per 19 verses), and even less frequently in Romans, 1 Corinthians, Galatians, 2 Timothy, and Philemon. The compound verb "rejoice with" (*synchairō*) also appears twice in Philippians (2:17–18) and twice in 1 Corinthians. If these are included in our frequency survey, one or another of these "joy" words appears every 5.7 verses in Philippians.

Notice, in verses 5 and 6, the sweep of Paul's panorama across history, from past to present to future. He recalls his friends' entrance into the gospel partnership on the "first day," when Paul and Silas arrived with their team in Philippi. Then he ponders their consistency as colleagues in the gospel enterprise, both as glad beneficiaries of God's grace and as grateful investors in its advance, "until now." Yet the source of Paul's confidence and joy goes deeper than the Philippians' record of reliability. As Paul looks from the present to the future, his confidence rests in the truth that "he who began a good work in you" (on that first day a dozen years ago) "will bring it to completion at the day of Jesus Christ" (Phil. 1:6). Paul knows full well—and the Philippians must remember it, too—that the only reason that they have been his partners in the gospel from the first day to the present is that *God* began a good work in them, and that *God* has sustained that work and is sustaining that work today. Because the sovereign work of the invincible God is the source of the Philippians' faith and Paul's faith, Paul can be confident that God, the Master Builder, will carry on his construction project until he has finished the job. Augustus Toplady sang the wonder of this confidence, blending the hope of this text with Paul's declaration of God's inviolable love toward us in Romans 8:38–39:

> The work which his goodness began,
> The arm of his strength will complete;
> His promise is yea and amen,
> And never was forfeited yet.
> Things future, nor things that are now,
> Nor all things below or above,
> Can make him his purpose forgo,
> Or sever my soul from his love.[8]

That point of completion is "the day of Jesus Christ," when the Father sends his Son, the Messiah Jesus, back to earth to complete our rescue by his resurrection power, transforming "our lowly body to be like his glorious body" (Phil. 3:21). Old Testament prophets foretold the coming of "the day of the Lord," when God would intervene decisively in history, primarily

8. Augustus M. Toplady, "A Debtor to Mercy Alone" (1771), in *Trinity Hymnal* (Suwanee, GA: Great Commission Publications, 1990), no. 463.

to judge rebels in his righteousness (Joel 1:15; 2:1; Amos 5:18–20; Mal. 4:5). Picking up their terminology, Paul spoke of "the day of the Lord," the day of Jesus' second coming, which will bring not only wrath to the unrighteous but also relief and resurrection to believers (1 Thess. 5:2; see 4:13–18). Paul assures the Christians of Corinth that the Savior "will sustain you to the end, guiltless in the day of our Lord Jesus Christ" (1 Cor. 1:8); and the apostle anticipates that the Philippians' persistence in holding God's Word fast will show that his ministry among them has been fruitful "in the day of Christ" (Phil. 2:16).

What, specifically, is the "work" that will be completed when Jesus returns in glory? It is the Holy Spirit's project of turning stone-dead people into living, loving replicas of Jesus, the radiance of the Father's glory. The apostle John promises, "Beloved, we are God's children now, and what we will be has not yet appeared; but we know that when he appears we shall be like him, because we shall see him as he is" (1 John 3:2). The Spirit begins with people who are part of "a crooked and twisted generation" (Phil. 2:14), opposed to God's authority and holiness, their hearts deeply programmed to place themselves above their Creator. Like them, we are spiritually dead by nature, unwilling and unable to reverse the inward "twist" of our hearts, like the valley of parched skeletons in the prophet Ezekiel's famous vision (Ezek. 37:1–10). But through the gospel, the Spirit of Christ breathed life into the dead. He began a lifelong process of new creation, transforming stone-cold hearts into warm and tender flesh (36:26–27) and turning ingrown hearts inside out. Now Christian believers' deepest desire is to put God's glory and pleasure first, and others' needs before our own. The verb translated "bring to completion" (Greek *epiteleō*) appears only ten times in the New Testament. Seven of these are in Paul's epistles, and four refer to Paul's commitment to complete the collection and delivery of the Gentile churches' offering for believers in Judea (Rom. 15:28; 2 Cor. 8:6, 11 [twice in this verse]). Paul's other uses of the term refer to the completion of the Holy Spirit's sanctifying work of transforming believers to resemble Christ's purity and love. In addition to this text in Philippians, Paul insists that the Galatians recognize that because their life in Christ began "by the Spirit" through faith in the gospel, they should realize that they will not be "perfected" (*epiteleō*) by the flesh—that is, by their own efforts to keep the Law (Gal. 3:3).

Of course, God's persistent (and ultimately invincible!) work begun in us on the "first day" does not make it effortless for us to exhibit Jesus' extensive and intensive love. Without minimizing in any way the Spirit's sovereign work in bringing Christians to complete holiness, the apostle also calls us to fulfill our responsibility, in dependence on God, to participate in the process of our sanctification: "Since we have these promises, beloved, let us cleanse ourselves from every defilement of body and spirit, bringing holiness to completion [*epiteleō*] in the fear of God" (2 Cor. 7:1). If love became "automatic" from the day we came to believe, Paul would not need to instruct, persuade, and motivate believers to "count others more significant than yourselves" (Phil. 2:3) or to "walk in love, as Christ loved us" (Eph. 5:2). Although God's Spirit *will* successfully complete his renovation of our hearts, we must still resist the ingrained habits of our hearts that contradict his holiness and love. In fact, certainty of the Spirit's ultimate success is what gives us reason to expect that our erratic efforts to redirect our hearts' focus away from ourselves and toward others can actually make progress. Paul spells out this dynamic in Philippians 2:12–13: "Therefore, my beloved, as you have always obeyed, so now, not only as in my presence but much more in my absence, work out your own salvation with fear and trembling, for it is God who works in you, both to will and to work for his good pleasure." This is God's good work: he makes us want what pleases him, and he enables us to do it. The fact that *God is working* does not make us passive puppets, waiting for the Spirit to move us. Instead, it ignites our hope and drive to "work out your own salvation"—God's comprehensive rescue from sin's tyrannical control, as well as from its guilt and penalty.

The God who finishes what he starts lets nothing limit the extent of the salvation he achieves for and in believers. His work of regeneration rescued us from spiritual death. His declaration in justification rescued us from guilt and condemnation. Through reconciliation and adoption, he has rescued us from alienation. But he will not stop pursuing his comprehensive rescue plan until he has liberated us thoroughly from the loitering presence and influence of sin. This aspect of salvation—sanctification—is the work of God's Spirit, every bit as dependent on his life-giving power as his first touch, which turned our hearts of stone into hearts of flesh. The Westminster Shorter Catechism rightly describes sanctification as "the work of God's free grace, whereby we are renewed in the whole man after the image of God, and are enabled more and more to die unto sin, and live unto righteousness."[9] Our ongoing transformation is unquestionably his work, and for that reason we can confidently expect its successful completion. But sanctification is also that aspect of God's comprehensive plan for our redemption and re-creation in which he engages us as collaborators. As Paul wrote to the Romans, because believers have been united to Christ in his death and resurrection, we have "died to sin" and its enslaving tyranny (Rom. 6:1). The Spirit's sovereign grace has broken sin's stranglehold on our hearts, so we believers can and must "consider [our]selves dead to sin and alive to God in Christ Jesus" (6:11). We must not let sin "reign in [our] mortal bod[ies]," but rather "present [ourselves] to God as those who have been brought from death to life" (6:12–13).

We still stumble, just as the Philippian believers did. Despite their constant and costly commitment to Paul, they were intimidated by the gospel's opponents (Phil. 1:28) and at odds with each other (2:3–4; 4:2). Perhaps Paul paused so early in the epistle to highlight his confidence that God would finish the transformation that he had begun in them because their fears and frictions tempted them to doubt whether they could survive in this long pilgrimage of faith until "the day of Jesus Christ." You, too, know the doubts and misgivings that haunt your heart when you have again fallen short of the Christlike love that you long to express. The last time you gossiped or complained about that fellow believer, you regretted your words almost as soon as they left your lips. Yet now you have blurted out your frustration to another person! The last time you blew up rather than calmly and gently dealing with the fact that your wife or son or daughter isn't yet perfect, your heart was pierced by the look of pain on your loved one's face. Yet here you are again, inflicting the same wounds on the tender hearts of those closest to you. You know the scenario, and you know the taunts of Satan the Accuser that echo in your mind after one more failure: "You call yourself a Christian? You hypocrite! You fraud! You blight on Jesus' good Name!"

But God's sure Word teaches you to reply to your prosecutor, "True; guilty as charged. But my place in the heart of God is secured not by my poor efforts to do or be good, but by Jesus' fulfillment of perfect love for the Father and of sacrificial love for all of us for whom he died. Your accusations have been

9. WSC, answer 35.

answered and overcome by the blood of the Lamb, shed for me (Rev. 12:10). Moreover, the almighty God who began a good work in me will take that project all the way to perfection at the day of Christ Jesus. Neither you, Satan, nor I can thwart his irresistible grace that *will*, someday, make me loving as my Savior is loving. True, my sins have pained his heart, and so they pain my heart, too. But my sins and selfishness and rebellion are not strong enough to stand against his almighty Spirit. So, having stumbled again, I will pick myself up, knowing that he is picking me up. I resume running the race in the strength his Spirit gives me, in the love his Son has shown me, in the joy he holds out before me and lets me taste even now." The wise pastors who formulated the Westminster Confession of Faith realistically described the ongoing spiritual warfare involved in our growth in godliness, and the ups and downs of our progress in sanctification.[10] They also affirmed, however, that the sanctification and "perseverance of the saints depends not upon their own free will, but upon . . . the free and unchangeable love of God the Father; upon the efficacy of the merit and intercession of Jesus Christ, [and] the abiding of the Spirit."[11] God's assurance that he will complete his work in you should sustain your striving after holiness in the preserving and purifying power of his Spirit.

Has God Begun His Good Work in You?

As we have eavesdropped on Paul's love letter to Philippi, have you heard it as Jesus' love letter to you? Are you trusting today in Jesus' blood and righteousness, rather than in your own achievements? That faith is the hallmark of those who have partnership in the gospel, who are partners with Paul in God's grace. Such trust is the sign of new life, showing that the living God has begun a good work in you, turning your ingrown heart "inside out," to adore him gratefully and to love others lavishly in the affection of Christ Jesus. Do you catch glimpses of the Holy Spirit at work, quietly rooting out your self-centeredness, stretching your heart to embrace others—even folks with whom you disagree—in the affection of Christ Jesus? Do you find joy

10. See WCF 13.2: "This sanctification is throughout, in the whole man; yet imperfect in this life, there abiding still some remnants of corruption in every part: whence ariseth a continual and irreconcilable war, the flesh lusting against the Spirit, and the Spirit against the flesh."

11. WCF 17.2.

in the assurance that the invincible God who began a good work in you *will finish* what he started, that he will amaze angels by making you a flawless replica of his Son's pure love?

If your answer to these questions is "Yes," our Master's liberating commission to you is: "Live out the salvation I have given you, am giving you, and will give you fully when we see each other face to face. Live it in a costly love that cares for others—extensively and intensively—before caring for yourself, in a joy that smiles at suffering, in confidence that your Father's power will transform pain into perseverance and purity."

If your answer to those questions is "No, but I wish I could say yes," I have good news for you. Jesus is still gathering people to be partners in the gospel and participants in his grace. The beginning of a new life is so simple that it is hard. It is to surrender your illusion of independence and recognize that God is the only person who can begin a good work in you. It entails asking God for the grace to shift the weight of your heart from yourself to Christ Jesus, and believing his promise that he welcomes all who turn to him with an awareness of their need. Ask Christ today to begin his lifelong work in you, flooding your life with heart-stretching love and perspective-restoring joy.

3

A Pastor's Passionate Prayer

Philippians 1:9–11

And it is my prayer that your love may abound more and more, with knowledge and all discernment, so that you may approve what is excellent, and so be pure and blameless for the day of Christ, filled with the fruit of righteousness that comes through Jesus Christ, to the glory and praise of God. (Phil. 1:9–11)

Prayer is a privilege and a problem. Most of the great world religions speak of personal deities, and their adherents try to communicate with those deities. Depending on the religion, God or the gods may be one or many, infinite or finite, remote and aloof or close and engaged. However the divine is conceived, worshipers speak to him (them) often and urgently, hoping their words will be heard. But prayer is not only a deeply ingrained response of the human heart. We sense that it is a privilege, and we know it is a problem.

The recognition that prayer is a privilege is expressed in the gifts and offerings, even bloody sacrifices, that often accompany words spoken in prayer. Just as peasants have no right to demand an audience with their king, so finite and flawed people have no leverage to *obligate* God or the gods to give ear to their words. Prayer is a privilege, not an inalienable right.

But that also poses the problem of prayer. How many offerings, how much fervor, how many words or turns of a prayer wheel will capture the attention and win the consent of the deity to whom we cry? Israel's God had open ears to hear the nation's cry for rescue (Ex. 3:7–8). Yet by Jesus' day, Israel's leaders had slipped into the pagans' solutions to the problem of prayer: "For they love to stand and pray in the synagogues and at the street corners, that they may be seen by others. . . . When you pray, do not heap up empty phrases as the Gentiles do, for they think that they will be heard for their many words" (Matt. 6:5, 7).

Christians, too, have problems in prayer. Often we neglect prayer, struggling to handle factors that we can see rather than casting ourselves on a God whom we cannot see. Sometimes we are distracted from conversation with God. Thomas Manton, a Puritan pastor, captured our distraction from prayer when he wrote, "Our thoughts stray in and out like a spaniel that runs up and down, and then returns to his master."[1] Sometimes our prayers don't seem to "work." Our heavenly Father does not give us what we want. For such cases James offers an uncomfortable but accurate diagnosis: "You do not have, because you do not ask. You ask and do not receive, because you ask wrongly, to spend it on your passions" (James 4:2–3). With disconcerting pastoral insight, James exposes the roots of our frustrations in prayer. Sometimes we imagine that we can go it alone, forgetting to ask our Father for our needs. At other times we ask, but our motives are askew, twisted in on our own cravings rather than aimed at his glory and our real needs.

How can we learn to pray? Instruction helps, but example is the key. Our prayer life must be informed by biblical truth about the God whom we address: his sovereign control, infinite wisdom, perfect holiness, and tender compassion. But we also need to *hear how prayer is done* by those who are on intimate speaking terms with the Father. We need to eavesdrop on their words and to absorb their hearts' priorities. As Jesus finished praying, a disciple asked him, "Lord, teach us to pray" (Luke 11:1). Overhearing one side of that conversation between two persons of the Trinity whetted the disciple's appetite for an intimacy akin to that which the incarnate Son enjoyed with his eternal Father. Jesus would let his friends eavesdrop on

1. Thomas Manton, *Puritan Sermons 1659–1689*, 1:401–10, quoted in *Voices from the Past: Puritan Devotional Readings*, ed. Richard Rushing (Edinburgh: Banner of Truth, 2009), 190.

other intratrinitarian conversations, climaxing in his High Priestly Prayer just before his death (John 17).

We can learn to pray not only by listening to Jesus' prayers in the Gospels, but also through the petitions that flowed from the Christ-formed heart of the apostle Paul. Paul opened most of his epistles with reports of his thanks and petitions for his churches. He did so not merely to fit the formula for Hellenistic letters, but to place his readers and their problems in the presence of God. He was showing them how to pray for themselves and for each other.

Paul's short, straightforward prayer in Philippians 1:9–11, like the thanksgiving before it, foreshadows themes soon to be addressed in the body of the letter. But Paul was genuinely praying the petitions that he reports here. Unlike prayers sometimes offered in church gatherings or at the family dinner table, these are not thinly veiled admonitions. Listening to grown-ups' prayers, children learn early how to use prayer to deliver the messages that, in the opinion of the person praying, others need to hear. They plead fervently (out loud, of course!) that the Lord would bring repentance to the sibling who refuses to share, or to classmates who play too rough at recess (by name, of course!). Pastors' "prayers" are subtler: "O Lord, give restraint to our sister who gossips, and give cheerful generosity to our affluent brother whose donation could balance the budget." We recognize a "to-do" list disguised as a prayer when we hear it!

Paul, however, is describing here the gifts he wants for Christians, not the performance he wants from Christians. He is not using the pretense of prayer to send directives to the Philippians. He wants their love to grow, and in due course he will urge them to prefer others' needs over their own (Phil. 2:1–4). He will even name names, calling on two treasured coworkers to be reconciled to each other (4:2–3). But he begins by stressing that selfless love can characterize their relationships only because God gives the gift of overflowing love. When Paul sees his friends mature, it will not be because they have been "guilt-tripped" or browbeaten into trying to stifle their selfishness. Love is a fruit produced by God's Spirit through Jesus Christ (1:11; see Gal. 5:22). So Paul wants his friends to overhear his heart's cry for their increase in love, which only God can produce. He is putting into practice the truth that their spiritual growth depends on the God who began a good work in them and will complete it at the day of Jesus Christ (Phil. 1:6).

What would you ask, if you could make any request at all to someone who had boundless—or apparently boundless—resources? Imagine a tycoon businessman calling you into his boardroom, and instead of firing or promoting you, he said, "All my fortune is at your disposal—ask for anything my money can buy." What would you ask for?

Better yet, if you were Solomon, newly established on the throne of Israel, and God appeared to you and offered, "Ask what I shall give you" (1 Kings 3:5), what would you request? Or suppose the Lord promised to do whatever you ask for the people you love most—your husband or wife, children or parents, a brother or sister, a boyfriend or girlfriend. What would you request for them? That is what is happening here: Paul has an audience with the *infinitely rich* Sovereign of the universe, and he can ask this King to give the most wonderful gift imaginable to his beloved spiritual children in Philippi. What will it be? Paul asks God to give the Philippian Christians *overflowing love* blended with *discerning wisdom*.

The order of Paul's prayer report is straightforward. First he indicates *what* he wants God to give his friends (Phil. 1:9). Then he explains *why* he wants this gift for them, mentioning three outcomes that will result from the gift (1:10–11). While profiling those outcomes, Paul indicates *how* God gives the gift of abounding love (1:11).

What Christ and His Apostle Want for You: Overflowing Love

Love Abounding in Selfless Giving

If your mental image of *love* has been painted by romantic movies and Valentine's Day cards, Paul's petition "that your love may abound more and more" (Phil. 1:9) may sound sweet, sentimental, and nonthreatening. It is popular to think of love as a mellow emotion, a warm glow associated with lovers' hand-in-hand walks by the sea. This is romance or "being in love," as we call it: a relationship of mutual attraction, of emotional give and take, your showing affection to another person as he or she returns the favor. It is a pleasant and significant experience, to be sure. God's Word itself—notably in the Song of Songs—celebrates physical attraction and romantic love. Scripture likewise commends the mutual devotion that gives friends delight in each other's companionship. Paul directs Christians: "Love one another with brotherly affection. Outdo one another in showing honor" (Rom. 12:10).[2]

Yet there is a higher love, a rarer love, that is less about getting than about giving. This type of love costs the lover dearly. It is less about fuzzy feelings that make us glow inside than it is about the inconvenience and heartache involved in placing others' needs before our own. The price that such love pays is seen most clearly in the God to whom Paul prays, who has demonstrated "his own love for us in that while we were still sinners, Christ died for us" (Rom. 5:8). This is the Christ who "loved us and gave himself up for us, a fragrant offering and sacrifice to God" (Eph. 5:2; see Gal. 2:20). So when Paul prays that God will give his friends the heart to *go on loving* and to *grow in loving*, "abound[ing] more and more," he is talking not about our getting, nor even about our sharing, but about our giving, and then giving even more. This capacity to love is God's gift, but it is a costly gift to receive. It involves investing your heart and hands in meeting others' true needs, whether or not they respond with emotional payback.

Paul is not the only apostle to link loving with giving, and then to link giving with dying. In his gospel, the apostle John recorded the well-known words: "God so *loved* the world, that he *gave* his only Son, that whoever believes in him should not perish but have eternal life" (John 3:16). In his first epistle, John explained the costliness of love: "By this we know love, that he laid down his life for us, and we ought to lay down our lives for the brothers" (1 John 3:16). And then, to keep us from evading the summons to self-sacrificing love by relegating it to unlikely life-or-death situations, John brings us back into the mundane: "But if anyone has the world's goods and sees his brother in need, yet closes his heart against him, how does God's love abide in him?" (3:17). Paul will also encourage the Philippians to have "the same love" toward one another by attending "to the interests of others" (Phil. 2:2–4). But before Paul commands them to love, he wants them to hear him ask God to do what only God can: to make them ready to pay the price of love that hurts.

God has already begun to give the Philippians this capacity to love, so Paul refers to "your love" as an already existent reality. God has begun his

2. The ESV's "Love one another" reflects the Greek adjective *philostorgoi*, and "brotherly affection" reflects *philadelphia*. Thus, twice in one brief sentence Paul commends the mutually beneficial love of friends (*philoi*) for each other.

good work in them (Phil. 1:6), and Paul has tasted its fruit in the Philippians' partnership with him (1:4, 7). But the Philippians' love is not all that it could be or would be. Like Paul himself, the Philippians have not yet reached perfection (3:12–13), so they could not be complacent about their current level of love for others. As Paul still strains toward the goal, so they should yearn to express an ever-growing, overflowing willingness to love each other to death—whether this would mean many little deaths (to their desires, agendas, or reputations) or laying down their lives for others in "the last full measure of devotion."[3]

Do you want to love like this? Is it at the top of your list as you pray for yourself, your family, and your brothers and sisters in the family of God? Or when you go to God in prayer, have you set your sights so low that you ask for things that mere money can buy? Why waste the precious privilege of addressing the Lord of the universe by asking only for tangible and temporary trinkets, when he stands ready to bestow a richer treasure: the transformation of your desires to align them with his beloved Son's perfect love?

Perhaps you feel that you already love enough. Your experience shows that love is costly, and your "costly love list" is as full as you can handle. You love enough people already, and caring about them causes enough inconvenience and heartache. Your schedule is full and your emotional reservoir is low, so you cannot take on more folks' problems. A Jewish legal scholar who once questioned Jesus apparently felt that his "plate was full," in terms of his obligation to love. He knew that God called him to love his neighbor as himself, but he hoped to confine that command within reasonable limits (Luke 10:25–29). So he asked Jesus, "Who is my neighbor?" How wide a circle of people must I love as I love myself? How tightly can I draw the boundaries of my neighborhood? Jesus told him a story of a despised Samaritan who showed compassion to a nameless victim of violence. Then Jesus turned his question around: "Which . . . *proved to be a neighbor* to the man who fell among the robbers?" (12:36). Jesus' implication that we are called to love without boundaries may make us squirm. Yet we should know that there is something wrong about carefully rationing love within a conveniently manageable circle. If God lavished love on us, his enemies, the love he instills in us must reflect that love without limits. For Christians in Thessalonica,

3. President Abraham Lincoln's "Gettysburg Address" (November 19, 1863).

not far from Philippi, Paul prayed: "May the Lord make you increase and abound in love for one another and for all, as we do for you, so that he may establish your hearts blameless in holiness before our God and Father, at the coming of our Lord Jesus with all his saints" (1 Thess. 3:12–13). Notice the scope of the abounding love that God gives: not only "for one another" but also "for all." As daunting as Paul's prayer for ever-abounding love is, its very grandeur should stretch your heart. God's love for you in Christ and his agenda for you, his child, should make you *long* to become a person who rejoices to put others first, without resentment or regret.

Love Guided by Discerning Wisdom

The love that Paul longs to see flourish in the Philippians differs from our common ideas of romance and friendship not only because it consists of giving rather than getting, but also because it is blended "with knowledge and all discernment" (Phil. 1:9). Christlike love is clear-sighted and sober-minded. We sometimes say, "Love is blind." Moonstruck couples are sometimes so infatuated with each other that they lose touch with reality. Whether teenagers or adults in midlife crisis, such blind lovers are oblivious to their own immaturity or incompatibility. Others see the warning signs, but the enamored couple dismisses friends' cautionary counsel, rationalizing that naysayers "do not understand how special our love is." Friends, too, sometimes cling to blind, misguided loyalty, rather than facing a friend's flaws honestly, daring to see and speak truth in love.

In religious circles, too, people sometimes pit love—construed as nonjudgmental affirmation—over against clear-eyed discernment. In the face of the church's sad history of conflict and division, ecumenical movements in the twentieth century exalted love, expressed in cooperative service, over truth articulated in creeds. In 1925 the Universal Christian Conference on Life and Work, meeting in Stockholm, Sweden, memorably expressed this perspective in its theme: "Doctrine Divides, Service Unites." No doubt many people feel that religion is useful when it makes people compassionate, but that when a religion claims to have the truth, trouble starts! They can even cite the apostle Paul himself, who seems to have pitted love against knowledge in his pithy contrast: "knowledge puffs up, but love builds up" (1 Cor. 8:1).

Admittedly, some churches have clung to truth and neglected love, or vice versa. Jesus praised the church at Ephesus for its doctrinal vigilance, but he rebuked its loss of love (Rev. 2:1–7). Speaking to the church at Thyatira, by contrast, Jesus commended its love and critiqued its compromise with falsehood (2:18–29). So also in our time, some churches fixate on doctrinal purity at the expense of love, showing little patience for those who are theologically imprecise and little compassion for those in need. Others follow the lead of that Stockholm conference, extolling a service-based love that refuses to make distinctions, lest anyone feel excluded.

To both extremes Paul says, "Nonsense." He did warn the Christians of Corinth against a loveless "knowledge" that breeds pride, but he also taught them: "[Love] . . . rejoices with the truth" (1 Cor. 13:6). True love loves truth: seeing it clearly, speaking it lovingly, hearing it humbly, and defending it firmly. So Paul tells the Philippians, in effect, "I am asking God to give you a love that *acts* for others' well-being—and a love that *knows* what others' well-being really is, because you see people as they really are and speak God's truth as it really is." Christ's love goes beyond good intentions, beyond well-meant affection. It is characterized by *accurate knowledge* and *insightful discernment*.

A moment's thought reveals why eyes-wide-open love is superior to the superficial affection that ignores reality, for fear of causing offense or distress. Suppose that you have a friend who is a world-renowned oncologist. When she reads your MRI and foresees a sobering scenario, does she put on a happy face and pretend that all is well, to keep you from feeling afraid? Not if she is a real friend! She will choose her words carefully, breaking the news gently and inserting seeds of hope wherever she honestly can. But her love for you and her medical insight will not let her leave you in a world of illusion, blissfully imagining that you can defer needed treatment. Genuine love lives in the real world and wisely leads the beloved to face reality, too.

Likewise, God's love for us moved him to give his priceless gift, his Son, and that gift was informed by his infinite wisdom. He knew what we needed, although we did not—so he took the necessary step that we could not imagine. Christ gave his life in abject weakness, bloodied by human cruelty and abandoned by his Father in a travesty of justice that seemed sheer folly (1 Cor. 1:18–25). Christ's cross diagnoses the severity of our spiritual condition in the direst terms, telling truth about ourselves that we would prefer not to hear. But God's wise love knew that no less radical remedy could rescue us, so his love toward us abounded in divine knowledge and all discernment.

Paul loves the Philippians so much and Jesus loves us so much that their longing is that believers will overflow with this costly, insightful reflection of divine love: love that sees and speaks truth for others' eternal well-being, love that grasps what must be done to bring healing, and love that is willing to do it. Do you want that kind of capacity for wisdom-laced love for yourself? You should, and Paul tells you why.

WHY CHRIST AND HIS APOSTLE WANT WISE LOVE FOR YOU: THREE OUTCOMES

Three outcomes will follow, when God grants Paul's petition by causing the Philippians to abound in discerning love. First, they will "approve what is excellent" (Phil. 1:10a). Second, they will become "pure and blameless for the day of Christ, filled with the fruit of righteousness" (1:10b–11a). Finally, their transformation will contribute to the greatest goal imaginable for any creature, "the glory and praise of God" (1:11b). Through God's grace, they will reach "man's chief end," that is, "to glorify God, and to enjoy Him for ever."[4]

Approving Excellence

The *first outcome* of God's gift of abounding, discerning love is that believers have the capacity to "approve what is excellent." Paul's choice of terms paints a picture of a discriminating buyer trying out competing products, putting their claims to the test, before purchasing the one that stands out from the rest. The term *approve*[5] includes a "proving" process that leads, in the end, to an informed preference for the superior option. Today we might think of test-driving an automobile before making so large an investment. In the ancient world, the prudent farmer would "test-drive" a yoke of oxen pulling a plow, to see whether they were of equal strength and stamina, before buying them.[6]

4. WSC, answer 1.

5. Greek *dokimazō*. *BAGD*, 202, offers the glosses "*put to the test, examine*," then "*prove by testing*," and finally "*accept as proved, approve*."

6. Such common sense shows the absurdity of the excuse offered by one invited dinner guest in Jesus' parable of the great banquet. That rude guest claimed that he had no time to attend the feast because he urgently needed to "examine" five yoke of oxen, but only *after* he had bought them (Luke 14:19)! The refining of precious metals also entailed the proving-to-approval process expressed in the Greek verb *dokimazō*. Intense heat burned away impurities and demonstrated the superior value

In our day of information overload and excess options, it becomes a survival skill to be able to sift the significant from the worthless, and what is superior from what is merely sufficient. Having too many choices is confusing at best and paralyzing at worst. How many cable-television channels can you access? At any particular hour, which has the most entertaining program? Of the dozens of breakfast cereals in the supermarket, which is the tastiest and most nutritious? Which mobile-phone service has the widest coverage and fastest access? How can one choose the best, most important, or most worthwhile option of even twenty alternatives, not to speak of a hundred or more?

When God gives you a heart that loves others wisely, setting you free from grasping selfishness and grieving self-pity, what is *really important* begins to come into focus. With practice you develop a taste for the things that count, things that last. You learn to make choices that align your priorities with God's wise purpose for you, and to exhibit Jesus' wise love toward others.

What does Paul have in mind when he speaks of "what is excellent"? Certainly Paul would include *moral superiority.* He wants his friends to make righteous choices that conform to God's Law, the standard that defines how love treats others (Matt. 22:37–39).[7] Later he will advise the Philippians: "Whatever is true, whatever is honorable, whatever is just, whatever is pure, whatever is lovely, whatever is commendable, if there is any excellence,[8] if there is anything worthy of praise, think about these things" (Phil. 4:8).

But Paul also wants them to have the sense to sort out the *more essential issues* from less significant ones. Jesus used the term that Paul employs here[9] to affirm that human beings are superior to sheep and birds (ESV: "of . . . more value") (Matt. 12:12; Luke 12:24). He was speaking of our *intrinsic importance,* not of our *moral rectitude.* Paul longs for the Philippians not only to choose the path of obedience, but also to *focus on more important issues* and keep less crucial points in proper perspective. Paul had to correct

---

of the gold (for instance) that remained—an analogy applied by Peter to the effect of suffering on genuine faith (1 Peter 1:7; see 1 Cor. 3:3).

7. In wording very much like Philippians 1:10, Paul observes in Romans 2:18 that the Jewish people, as those informed by God's Law given through Moses, were confident that they could discern and "approve what is excellent" (*dokimazeis ta diapheronta*).

8. Here the term is not *ta diapheronta,* "things that differ from and surpass others in value," but *aretē,* "*moral excellence, virtue*" (*BAGD,* 105).

9. Greek *diapherō,* in Philippians 1:10, in the form of a substantive present participle, *ta diapheronta.*

Christians in Rome and Corinth because they condemned fellow believers over ethically insignificant issues such as eating meat that had been offered to idols (Rom. 14:1–12; 1 Cor. 10:23–31). Such a diet broke no command of God, so condemning others over such food displayed an inability to distinguish issues on which God's Word was clear from personal preferences that could and should be left to the liberty of each believer's conscience before God. Perhaps the interpersonal friction in the Philippian congregation also entailed an inability to tell the difference between biblical principle, on which they must stand united, and lesser issues on which they could disagree for the present. Paul was mature enough to recognize that he was still "in process" and had not reached perfection, so he could live with others' disagreement on minor issues: "If in anything you think otherwise, God will reveal that also to you" (Phil. 3:15). Not every difference was a matter of life and death, and God's Holy Spirit would continue his work in them all (1:6) until they all "attain[ed] to the unity of the faith and of the knowledge of the Son of God, to mature manhood" (Eph. 4:13). Paul could wait for that complete unity to come. On crucial issues, he vigorously combatted error (Phil. 3:2). On less vital points, he could live with diversity of practice and conviction. He wanted his dear friends at Philippi to have the same perspective, valuing essentials and (to be colloquial) "not sweating the small stuff."

How often in the church are our estranged relationships and denominational divisions the result of a lack of *discerning* love that can test and treasure what is indispensable? Some, in the name of theological vigilance, "major in minors," progressively refining their communion by excluding any whose views vary in the slightest from a high benchmark of precision. Refusing to see any issue as of secondary importance, they find their unity under constant threat of further splintering. Others extol a love that devalues discernment. They cast a veneer of inclusive niceness over substantive disagreements on matters at the core of the Christian faith. Refusing to see any truth as of primary importance, they find their unity fragile also. Without a core of conviction regarding Christ and the gospel, their affiliation lacks the "gravity" to hold together people from diverse backgrounds, with diverse tastes, pursuing diverse agendas. Both extremes undermine the church's unity by subverting the foundation of that unity in the unique authority of God's inerrant Word. The wise love that God gives instills in us the good sense to commit ourselves to things that count.

Becoming Righteous

The *second outcome* that flows from God's gift of discerning love is oriented toward the future. We are becoming "pure and blameless for the day of Christ, filled with the fruit of righteousness that comes through Jesus Christ" (Phil. 1:10b–11a). God gives wise love not merely to impart effective interpersonal skills in this life. He is making us new people, who will finally be fit to meet him face to face at the climax of history.

In today's permissive Western societies, words such as *pure*, *blameless*, and *righteousness* are easily dismissed as boring, nerdy, stodgy, dour, prissy, or even self-righteous and hypocritical. Two thousand years ago, in the Greco-Roman cities in which Paul preached, things were not so different. Although Law-observant Jews and some Gentile ethicists were alarmed at that culture's slide into reckless self-indulgence, the lifestyles of the rich and famous, including the emperors themselves, set trends toward sensual experimentation and luxurious decadence for the society at large, just as many of today's rock stars and professional athletes send all the wrong signals to the fans who idolize them.

Paul was boldly countercultural. He was not preoccupied over the latest fad that set Greco-Roman high society abuzz. He loved his Philippian friends enough to lift their sights to a coming day, "the day of Christ," when Jesus, the Lord of the universe, will return in power and glory. On that day, it will be obvious that the treasures marketed by filmmakers, automakers, fashion designers, entertainers, the movers and shakers in business, or the in-crowd on campus are cheap trinkets that cannot last. Then everyone will stand before God's throne, compelled to look into his soul-piercing eyes and to realize that he reads every hidden thought and shameful secret.

Yet on that day there will be people who will not have to squint or cringe away from the King's searching gaze. They will be "pure and blameless" on the day of Christ. Of several Greek words that can be translated "pure," here Paul chooses one that focuses on sincerity, a lack of mixed motives and a transparent integrity that has nothing to hide.[10] Paul warned the Corinthian believers about tricky teachers who

10. Greek *eilikrinēs*. This adjective appears only here in Paul (also in 2 Peter 3:1), but he uses the cognate noun *eilikrineia* in 1 Corinthians 5:8; 2 Corinthians 1:12; 2:17.

concealed their real motives and watered down their message to win an audience. Unlike those religious hucksters, Paul and his colleagues preached and lived "as men of *sincerity*, as commissioned by God, in the sight of God . . . in Christ" (2 Cor. 2:17). Paul lived and served *coram Deo*, "before the face of God," so his motive was love toward others, untainted by concerns for himself. He asks God to give his friends that same sincere others-serving, self-forgetful love.

God's gift of love will also make us "blameless." Although we will not reach complete, sinless purity until we see our Savior at his coming and are made like him by that sight (1 John 3:2), Paul's term here (the Greek adjective *aproskopos*) presents a "blamelessness" that begins now, before we reach perfection. It describes an individual who does not "stumble" into sin, or one who does not cause others to stumble. It appears only three times in the New Testament. In Acts 24:16 (NASB: "blameless conscience") it may bear either sense, but in both uses in Paul's letters (1 Cor. 10:32; Phil. 1:10) the focus seems to be on avoiding offense (that is, stimulus to sin) to others. As God's Spirit makes our love abound with discernment, no defect in us will cause others to stumble in the pilgrimage of faith and faithfulness. Paul's use of this term in 1 Corinthians 10:32–33 shows what he has in mind here as well: "Give no offense to Jews or to Greeks or to the church of God, just as I try to please everyone in everything I do, not seeking my own advantage, but that of many." The command "give no offense" contains the adjective[11] that appears in our text as "blameless." The Corinthian Christians must not exercise their freedom so as to throw a stumbling block into anyone's path. Paul's prayer for the Philippians as well is that love will so permeate their relationships that they will stand united on the day of Christ, having furthered each other's trek toward holiness, rather than tripping each other up.

At Jesus' return, not only will believers be "pure" and "blameless," but they will also be "filled with the fruit of righteousness." Paul evokes the frequent biblical metaphor that compares godly people to fruitful trees and vines. Psalm 1 compares the person who soaks his heart in God's Word to "a tree planted by streams of water that yields its fruit in its season" (Ps. 1:3). Jeremiah describes trusting the Lord in the same

11. Greek *aproskopoi . . . ginesethe*.

imagery (Jer. 17:7–8). Jesus told his friends that they would bear fruit for the Father's glory as they remained united to him, like branches drawing their life from the vine (John 15:2–5). Paul wrote elsewhere of the fruit of God's Spirit—love, joy, peace, patience, kindness, goodness, faithfulness, gentleness, and self-control (Gal. 5:22–23). Both James (3:18) and Hebrews (12:11) identified the "fruit of righteousness" with peace, so it makes sense that Paul asks God to enable the Philippians to love each other selflessly, deepening their unity in the peace of God. The "fruit" metaphor teaches that just as a tree cannot bear fruit when no moisture nourishes its roots, so none of us can produce attitudes or actions that please God apart from the life-giving irrigation of his Spirit. John Chrysostom, a great preacher of the early church, said about this passage: "[Paul] is not speaking here of a kind of uprightness or virtue that tries despairingly to grow without Christ."[12]

Glorifying God

Because our purity, blamelessness, and fruitfulness in righteousness do not originate in our own willpower but in the work of God's Spirit, an even more wonderful *third outcome* will flow from his gift of overflowing love. On the day of Christ, all that he has done in us and through us will lead "to the glory and praise of God" (Phil. 1:11b). Our greatest joy will be that God himself gets all the credit for the beauty he has bestowed on us. God, not we, is the center of his universe and the focal point of his creatures' admiration. So we will join the whole realm of creation in one endless round of applause to God, and we will marvel at our privilege to be the venue in which "the manifold wisdom of God might now be made known to the rulers and authorities in the heavenly places" (Eph. 3:10; see vv. 20–21).

How Jesus Gives You Wise Love, for His Father's Glory

We have seen *what* Paul longs for God to give his spiritual children: overflowing love, informed by truth and discernment. We have seen *why* Paul begs for this gift from God: such love will enable them to value what

12. John Chrysostom, Hom. Phil. 3.1.8–11, in ACCS NT 8:222.

is valuable, to reflect their Savior's blameless purity, so to enhance the glory that goes to God. The remaining question is: "*How?*" How does God turn self-centered people into people who care more for others than for themselves?

The short answer is simply "Jesus Christ." Paul says in our text that the fruit of righteousness "comes *through Jesus Christ*" (Phil. 1:11a). In the first eleven verses of this letter, Paul has mentioned the name of Christ seven times: twice in verse 1, and then in verses 2, 6, 8, 10, and 11. In the next few verses he will speak of "preaching Christ" three times (1:15, 17, 18). Soon thereafter he will affirm, "To me to live is Christ, and to die is gain" (1:21). Later he will report that he has gladly discarded all his past accomplishments to gain one great treasure: "the surpassing worth of knowing Christ Jesus my Lord" (3:8). Now his only desire is to "gain Christ and be found in him, not having a righteousness of my own that comes from the law, but that which comes through faith in Christ" (3:8–9).

Jesus is the avenue through whom God will answer Paul's prayer for his friends. Jesus is the conduit through whom God pours overflowing love, with discerning wisdom, into their thirsty hearts. Jesus is the wellspring of life from whom they are absorbing nutrients that enable them to bear the fruit of peaceable righteousness. In Philippians 2 Paul will retell the story of Christ, who was and is God himself, yet who became a human being and a servant, and then became obedient to death, death on a cross—Rome's grotesque implement of torture and shame and the emblem of divine curse (Gal. 3:13, quoting Deut. 21:23). Then Christ rose from the dead to live and rule over all with God his Father in heaven.

When we trust and rest in Jesus, God declares that—as to our *relationship* with him—we are forgiven, reconciled, accepted, well pleasing. He credits us with all that Jesus has done for us. That is why Paul was no longer interested in "a righteousness of my own," but instead relied on the righteousness that God gives through faith in Christ (Phil. 3:9). But God does even more "through Jesus Christ." Not only does he restore our relationship with himself, but also he begins a lifelong project of transforming our character, making us eager to love others as we have been loved by God in Christ. Elsewhere Paul writes, "Walk in love, as Christ loved us and gave himself up for us" (Eph. 5:2). Jesus is not only the *example* who shows what love looks like, but also the *enlivener* who makes love spring from our grateful hearts like ripe figs from well-watered trees. The more we discover how much we are loved by the Lord of the universe, the more we love him and want to show his love to others.

So how has Paul's prayer for his friends expanded your own understanding of prayer? Do you want for yourself what Paul wanted for the Philippians: overflowing, heart-stretching love? Do you want love blended with knowledge and discernment, as committed to truth as it is compassionate toward people? Where can you find the strength to love when the love's price is high and its returns are low? Go to God your Father and ask him, as Paul did for the Philippians, to make your love abound to those you now find hard to love, to fill you with "the fruit of righteousness that comes through Jesus Christ," for his own glory.

4

Prison Is Great—Wish You Were Here!

Philippians 1:12–18a

I want you to know, brothers, that what has happened to me has really served to advance the gospel, so that it has become known throughout the whole imperial guard and to all the rest that my imprisonment is for Christ. (Phil. 1:12–13)

Her name was Joy, and it fit her beautifully. I was a young pastor when I was called to shepherd the church to which Joy belonged. Just a few years out of seminary, I treasured my hours in the study, interpreting the Bible and preparing sermons. I was far less eager to visit our aging congregation's shut-ins. It was not that I lacked sympathy for their pain, loneliness, and frustration. It was just that I felt overwhelmed and unable to offer words that would bring them comfort. I did what I could: I listened attentively, read the Word, and prayed for my suffering brothers and sisters. But often I left with a heavy heart, questioning whether I had brought them the solace that they needed. My visits to Joy, on the other hand, were different.

Joy was not elderly. But by the time I met her, rheumatoid arthritis had curled her hands into tiny fists, confined her to a wheelchair, and filled her days and nights with pain. At the start of my pastorate, she had the strength to attend worship services now and then. Later came years of home confinement and repeated hospitalizations. In her own family she alone trusted Christ. If anyone had reason to complain and pity herself, it was Joy. Yet every visit to Joy brought bright encouragement to her visitors. She took the initiative in our conversations, and the questions were all about our family, our children, or others in the church. Rarely would she mention her physical pain and personal trials. Gratitude to her God and concern for others overflowed from her heart, so that we all felt that we gained far more than we gave in every contact with this tiny champion of faith and hope.

Hearing my description of Joy, you may be inclined to think, "Some people are just born cheerful. I envy them." That may be true of some, but I can assure you that joy did not come naturally to Joy. She struggled often with loneliness, and at times with discouragement and confusion over God's promises and purposes. She would have been the first to insist that the resource that made her so refreshing to others came from outside herself. Because Joy's inner reservoir was filled by streams flowing from Jesus her Lord, her joy gave the rest of us reason to trust that her Lord and ours could and would sustain our own spirits in times of trial.

That is precisely the message that the apostle Paul intends to convey to his Philippian friends (and to us) as he turns the corner from his prayer for them (Phil. 1:9–11) to an update on his own situation in Roman custody, as he awaits a legal verdict from the emperor himself.

As Paul writes this letter, he is chained to a Roman soldier every hour of every day, confined to living quarters for which he has to cover the rent (Phil. 1:13; see Acts 28:20, 30). En route to Rome he had endured an extended imprisonment in Rome's provincial capital on the Judean coast (Acts 24–26), and then an arduous and dangerous voyage that included storm, shipwreck on the island of Malta, and a venomous snakebite from which God miraculously healed him (Acts 27–28). Now Paul awaits the emperor's decision on his appeal of unclear accusations filed against him by the Jewish leadership in Jerusalem. The spectrum of possible outcomes ranges from vindication and release, on the one hand, to condemnation and execution, on the other.

To say the least, Paul is not writing a vacation postcard from a comfortable beachfront villa on the Italian Riviera.

And yet, as Paul updates his concerned friends in Philippi on his current situation,[1] his tone radiates sheer joy. His high spirits are only one of several surprising features that emerge from Paul's report on how things are going for him in Rome. He also reveals that his own confinement, the result of his commitment to make Christ known, has not inhibited the spread of this good news of Christ, as we might have expected. Instead, Paul's chains themselves are preaching, in a way, to the soldiers who find themselves bound, literally, for hours on end to this man who loves to talk about Jesus. Moreover, others are stepping forward to carry Jesus' message where Paul cannot go. Instead of being intimidated by the threat of joining Paul in captivity, they find their zeal inflamed by his chains and spread the good news all the more boldly. Something counterintuitive is happening in Rome, not only in Paul but also in other Christians. Finally, there is Paul's striking reaction to the fact that some are preaching Christ from malicious motives, seeing themselves as Paul's rivals. Contrary to their expectations, Paul applauds their success, finding joy whenever and by whomever Christ is proclaimed.

Paul's Philippian friends need to know about these unexpected developments in Rome for two reasons. First, because they care deeply about Paul and are concerned about his difficulties, he wants to set their minds at rest by sharing the news of his circumstances and his response to them. Second, and more importantly, Paul invites his friends to glimpse the perspective through which he is processing his imprisonment because they, too, are suffering for Jesus' sake (Phil. 1:28–30). They need to see in Paul, their father in the faith, how to handle the pain of persecution in a way that brings honor to their Lord and joy to their own hearts. Just as Paul opened his heart in his prayer report (1:9–11) to show the Philippians how to pray for

1. The ESV's "what has happened to me" reflects a Greek expression (*ta kat' eme*) that would be rendered more literally "the things concerning me" (Phil. 1:12). This expression and the corresponding construction, "the things concerning you" (*ta peri hymōn*, 1:27; 2:19), signal shifts of focus from news about Paul's situation (1:12–26) to the situation of the Philippian believers (1:27–2:18) to a discussion of travel plans that include both the Philippians and Paul himself (2:19–30). Gordon D. Fee, *Paul's Letter to the Philippians*, NICNT (Grand Rapids: Eerdmans, 1995), 2–14, 37–39, 54–55, and others have observed that Paul's letter to the Philippians resembles other Greco-Roman letters of friendship in these transitions between news concerning the author's situation and discussion of the readers' issues or experience.

one another, so his optimistic assessment of his situation models how they should evaluate their own difficulties.

It is unlikely (though not unthinkable) that your suffering for Jesus' sake has reached the level of chains or the threat of death, as Paul's had. Yet that is all the more reason for you to join the Philippians as the apostle invites them and you to view his trials through his eyes, and then to apply to the challenges you face the same approach that Paul used for evaluating how well things were going in his life. Paul's shocking, counterintuitive, unnatural perspective on his imprisonment provides answers to questions that confront each of us: (1) How can I be set free from the petty pursuit of my own comfort, to devote my life to a cause bigger than I am, the cause of Jesus the King? (2) How can I be set free from the fear of what others may do to me, to lay my life on the line for Jesus the King? (3) How can I be set free from a competitive attitude, to surrender my reputation to the glory of Jesus the King? As we explore the answers that Paul's experience offers to these questions, the question in the back of our minds will be: (4) Can what worked for Paul work for me? Can Paul's secret liberate me from myself, my fears, and my competitive streak, to find freedom through serving Jesus and to find his abundant life in losing mine?

These are our questions. How does this portion of God's Word answer them?

Celebration in the "Cell" for the Gospel's Advance

Paul mentioned his "imprisonment" in his thanksgiving for the Philippians' partnership (Phil. 1:7). Now in reporting on his current situation, he refers twice more to this "imprisonment" (1:13, 17). Behind the ESV's "imprisonment" in all three verses, Paul's Greek[2] speaks more concretely of "chains." We should not envision Paul behind bars in a jail cell, but rather under a restrictive house arrest, chained to a guard at all times. Apparently he was free to receive visitors (Acts 28:17–18, 20–21), to keep current about events in the outside world (Phil. 1:14–18), and even to have contact with believers in Caesar's household (Phil. 4:22). Nevertheless, though Paul was not rotting in a rat-infested dungeon, we should not underestimate how such confinement might have frustrated such a man of action. Earlier he

2. Plural forms of *desmos* (*BAGD*, 176: "*bond, fetter*"), a cognate of the verbs *deō* ("*bind, tie*") and *desmeuō* ("*bind, tie up*"). The NIV rightly renders *desmoi* as "chains" in Philippians 1.

had written to the church at Rome that he was eager to reach the imperial capital, "that I may impart to you some spiritual gift to strengthen you" and "that I may reap some harvest among you as well as among the rest of the Gentiles" (Rom. 1:11, 13). Since his calling "to preach the gospel, not where Christ has already been named," had been completed in the eastern Mediterranean (15:19–20), he was restless to push westward to the frontier of gospel expansion, stopping at Rome en route to Spain (15:23–24). Paul's plans were always subject to his sovereign Savior's will, of course. He had prayed to reach Rome "by God's will" (1:10), and by God's means he did—though not for the brief stay that he had expected. Now he planned "in the Lord" to send Timothy to Philippi and was persuaded "in the Lord" that he himself would follow (Phil. 2:19, 24). So Paul was a submissive planner, but he was ambitious—not for his own fame but for Christ's. So the chain on his ankle and the soldier at his side, keeping him from the marketplace or the synagogues in the Jewish quarter across the Tiber River, could not have been welcome encumbrances.

Yet the first piece of news that Paul shares about his circumstances has nothing to do with the aggravation of his chafing chains. He views his situation from one perspective only: it is serving "to advance the gospel" (Phil. 1:12). The key term, *advance*,[3] by which Paul sets the tone at the start of his update, reappears as *progress* at the conclusion, where he expresses his anticipation that he will "continue with you all, for your *progress* and joy in the faith" (1:25). His news update falls into two phases, the first describing current developments (1:12–18a) and the second weighing future possibilities (1:18b–30). And Paul surrounds both his current situation and his future destiny with the motif of *advancement*, deflecting attention away from his problems toward the progress of Christ's gospel and others' faith. Since God is furthering these goals through Paul's chains, Paul himself rejoices in the present (1:18a) and "will rejoice" in the future (1:18b), whether his pending legal process results in further ministry or a martyr's death. Later he will say that if his captivity ends in martyrdom for the sake of the faith of the Philippians, he will rejoice and will insist that they share his joy (2:17–18). This prisoner turns his "cell," as it were, into celebration, his heart spilling over in exultation.

3. The Greek noun *prokopē* appears in the New Testament only in Philippians 1:12, 25 and again in 1 Timothy 4:15. The cognate verb, *prokoptō*, expresses Jesus' progress in wisdom and stature (Luke 2:52), and the young Saul's progress in Judaism (Gal. 1:14).

Paul reports that the gospel is advancing on two fronts. On the first front, Paul himself is capitalizing on the opportunity that his chains provide to speak about Christ to Roman military personnel. The result is that "the whole imperial guard," as well as many others, has learned that Paul's imprisonment is "for Christ." The imperial guard, also known as the Praetorian Guard,[4] constituted nine thousand elite soldiers selected for their military skill and loyalty to serve as the emperor's personal security force in the imperial palace. We are not explicitly told whether the soldiers chained to Paul in six-hour shifts belong to these crack troops, but it makes sense to conclude that the news about Paul and the Christ for whom he endured chains has been disseminated to Rome's military elite through those guards who have become their captive's "captive audience." No doubt he has told them all about Jesus of Nazareth, a Messiah whose execution was authorized by Roman authorities but who came to life again. Jesus now offers forgiveness and eternal life to everyone, from every race, who believes in him. Probably each guard can also recall the prisoner's looking him in the eye and challenging him, then and there, to trust in Jesus, the Lord far greater than Caesar. So word spreads through the barracks about this strangely cheerful prisoner and the faith that buoys up his spirits. From outward appearances, the months and years of Roman custody, first in Judea and now in Rome, seem like a setback to Paul's agenda to carry the gospel to places where Jesus' name has not been heard. In fact, however, this restriction of Paul's mobility has opened doors for the gospel into the halls of power to which he could never have gained access as a free agent! So Paul brims over with joy, because through his own containment he sees God breaking through barriers to bring good news to new ears and hearts.

Paul has been set free from a petty preoccupation with his own comfort. He has been liberated by the power of a message—a gospel, a piece of unimaginably good news—that has captured his heart for an infinitely bigger cause than himself. Now his personal circumstances don't matter so much, except as they provide a platform for getting the good news of Christ out to people everywhere.

Have you experienced that liberation from yourself, from treasuring your own comfort and convenience and amusement? In her disturbing novel *Chil-*

4. Paul's Greek term, *praitōrion*, reflects the Latin *praetorium*, and can refer either to a commander's residence (as in Matt. 27:27; Mark 15:16; John 18:28, 33; 19:9; Acts 23:35) or to the troops who guard it. Here Paul groups *praitōrion* with "the rest," indicating that the reference is to persons, not a building.

dren of Men, P. D. James portrayed a futuristic society uncomfortably like our own, in which people's dominant values and driving desires are "security, comfort, and pleasure."[5] Perhaps these have been your goals. Perhaps you have reached a certain degree of relief from danger, pain, and boredom—or maybe those goals still lie tantalizingly out of reach. Yet when you overhear Paul's excitement in a situation that was surely short on security, comfort, and pleasure, your imaginary picture of what the good life would look like begins to seem pale and hollow. Doesn't the glimpse of Paul's joy, drawn from wellsprings deeper than his circumstances and deeper than himself, whet your appetite to taste Paul's secret of surprising joy? Don't you long to be captured by something bigger than yourself, a cause of eternal significance, and a cause worth suffering and even dying for?

If you are a believer who has been taken captive by God's grace in Christ, how has the mercy that you have received influenced the way that you view life's frustrations and setbacks? When your agenda is thwarted as Paul's plans to reach Spain were, do you follow his lead in looking around in the situation that you would never have chosen for yourself, eager to see and seize unexpected opportunities to show Jesus' kindness and speak about his grace to people whom you would not have otherwise met?

OTHERS TAKE COURAGE FROM PAUL'S CHAINS

Not only has the gospel been invading the Praetorian Guard, but it has also been advancing on a second front in Rome. The chains that have opened access to Caesar's premier troops prevent Paul's movement around the city, but they embolden other believers "to speak the word without fear" all over the imperial capital. Many Christians see Paul shackled as a criminal for testifying about Jesus and his grace, and for some strange reason they want to imitate the behavior that got him arrested.

One of the purposes for which governments inflict severe punishments on lawbreakers is to deter others from following their example. Heavy sentences and harsh treatment in custody are supposed to send the signal that criminal behavior has unpleasant consequences. Sensible people, who know what is good for them, will not imitate such conduct. This holds

5. P. D. James, *The Children of Men* (New York: Knopf, 1993), 60.

true whether the civil law itself is right (forbidding robbery or murder, for instance) or wrong (such as banning true worship or the spread of the gospel). It is just not natural for self-protective people to engage in conduct for which we see others being punished, rightly or wrongly. So we might have expected the Christians in Rome, seeing Paul's high-visibility and high-risk legal troubles, to keep a low profile, flying "under the radar" and perhaps rationalizing that it would be better to let their lives be a subtle or silent witness rather than to offend people and attract attention by making too much noise about Christ.

Yet Paul reports that his Roman brothers and sisters have the opposite reaction from the predictable one: Seeing him in chains, they are *all the more eager* to commit the same "crime" that put him in those chains—the crime of spreading the news about Jesus the supreme Lord! Paul's captivity is setting other people free from fear, free to put their own lives on the line for the message of Christ. In verse 14 the apostle heaps up words to stress the courage that his problems have evoked in others: "having become confident in the Lord . . . much more bold . . . without fear." The chains that keep Paul confined are propelling the word about Jesus throughout the city, carried by many messengers to many more people than Paul himself could have reached.

Paul mentions the many brothers and sisters who are being emboldened by his chains in order to keep the Philippians (and the rest of us) from thinking that his joyful spin on suffering for the gospel's advance is somehow unique to him. We who knew our sister Joy could not attribute her radiant Christian hope to her innate optimism. Likewise, as we read Paul's response to his chains, we cannot simply dismiss his example, rationalizing to ourselves, "Of course *Paul* rejoices in miseries, but then he is one of a kind. He is a super-spiritual, workaholic masochist, a Christian Rambo who takes on all comers single-handedly, a lonely hero whose strength of will sets him apart from the rest of us." Paul does not let his Philippian friends imagine that his single-minded devotion to Christ, which makes his circumstances seem insignificant, is his personal achievement. Rather, he is implying, "Others share my joy and boldness, and so can you."

Paul's friends in faraway Philippi need this reassurance that Paul's joy is contagious. They are already enduring hardship akin to Paul's suffering. He will soon mention the opponents who threaten to intimidate them (Phil. 1:28) and tell them that God's grace has given them not only faith in Christ but also the privilege of suffering "for his sake, engaged in the same conflict that you saw I had and now hear that I still have" (1:29–30). The specifics of their suffering might differ from Paul's, but that does not diminish the solidarity that they share with him, as well as with their fellow believers in Rome. What the Philippians do need, it seems, is the boldness that many ordinary Christians in Rome had "caught" through Paul's chains, so that they would stand firm in one spirit, "striving side by side for the faith of the gospel" (1:27).

The reality that has set Paul's heart free to find joy in jail is not for him alone. Others in Rome are experiencing it, and fearlessly risking reputation and personal safety to make the message of Christ known. That confidence can also rescue the Philippians from fright in the face of intimidation in their city. It is God's answer to the threats that may tempt you to play it safe and keep a low profile regarding your allegiance to Jesus as you talk with coworkers, fellow students, neighbors, or family members.

REJOICING IN RIVALS' SUCCESS

This passage holds a third surprise. It is strange for a prisoner to celebrate in his cell. It is counterintuitive for others to notice his chains and then to imitate the actions that led to his imprisonment. Finally, it is unexpected for a strong leader to rejoice in his rivals' success.

Anyone who carefully reads the New Testament knows that Paul was a strong leader, a dynamo driven by unwavering commitment and tireless motivation. When he wrote to the church at Rome five years earlier, he expressed his eagerness to come to that city and strengthen the church there, and then to move on to Spain, a new frontier where Christ's name was not yet known (Rom. 15:20–24). By the time he writes to the Philippians, Paul has reached Rome, though not under the conditions that he had anticipated a half-decade before. Though he has some contact with believers in Rome and leaders of the Jewish community (Acts 28:15–28), in comparison with the mobility that Paul enjoyed in the eastern Mediterranean, his house arrest has "clipped his wings" and constricted his circle of influence. Now, with Paul confined to the sidelines, some see an opportunity to overtake and surpass him in influence and prestige.

This is the sober truth that Paul addresses in Philippians 1:15–18: some of the gospel preachers whose eagerness has been aroused by Paul's chains have mixed motives at best. Many spread the word about Christ because they love Paul. They recognize God's approval of his ministry, and they know that he has been "put here"—confined by chains through God's sovereign appointment[6]—"for the defense of the gospel." But some are preaching Christ to exalt themselves and humiliate Paul. They see themselves as Paul's rivals for eminence in the Christian community in the imperial capital. Since Paul seems to be sidelined by his legal troubles, in their "envy and rivalry" they swoop into the vacuum and strive to rack up a convert count that will put Paul's to shame. Surely, they imagine, their success will make Paul's chains feel even heavier!

But Paul does not play their game. Instead, he applauds their efforts and their successes. As long as they are truly preaching Christ, he really does not care who gets the credit, or who gets the converts. The privilege of preaching Christ is not a race for spiritual scalps or a contest to recruit delegates for a political convention. "As long as Christ is preached," writes Paul, "I am rejoicing and will keep on rejoicing!" (Phil. 1:18).

Now, Paul could celebrate his competitors' successes on only one condition: that they were *truly preaching Christ*, accurately setting forth the good news of who Jesus is and what he has done—in his perfect obedience, sacrificial death, and triumphant resurrection from the dead—and explaining that sinners are united to Jesus and his saving benefits by trusting him alone, and abandoning all trust in their own efforts. The fidelity of preachers' message to the truth of the gospel makes all the difference to Paul. He wrote his letter to the Galatians to protect his spiritual children from a band of spiritual predators, Judaizers, whose motives were self-serving (Gal. 4:17; 6:12–13), like the motives of the evangelists in Rome who saw themselves as Paul's competitors. Paul bluntly identified the selfish hypocrisy that animated the Judaizers in Galatia, but what really ignited his outrage was the fact that their message subtly contradicted the grace of God in the gospel. Under the banner of Christ, they promoted a semi-"good news" formula that blended humans' best efforts with a supplement of God's mercy—a "gospel" that, Paul insisted, did not deserve the name "good news" (Gal. 1:6–7). Paul, who

6. "The Greek *keimai* is a theological term, emphasizing that his appointment is a divine commission." Ralph P. Martin, *Philippians*, NCBC (Grand Rapids: Eerdmans, 1980), 74.

had been raised in such a system, knew that its outcome was not holiness pleasing to God but a self-satisfaction that disgusted God (Phil. 3:3–11). So Paul was harsh toward people who peddled a false gospel instead of the good message that calls us to rest in Jesus' blood and righteousness alone. He pronounced purveyors of pseudo-gospels defiled and damned (Gal. 1:8–9), Satan's servants disguised as apostles of Christ (2 Cor. 11:13–15).

In this passage, by contrast, Paul states repeatedly (Phil. 1:15, 17, 18) that the evangelists in Rome, though animated by envy rather than love, were indeed *preaching Christ*.[7] Their motives were shamefully selfish, but their message truly presented the divine-human person of Christ, his redemptive work, and the gracious means by which sinful people receive his saving benefits. If their proclamation had diverged on any of those key elements, Paul could never have rejoiced in their success.[8] As long as the real good news was going out, Paul would applaud his rivals' work, even if their hearts were tainted by competition or envy.

Here again, Paul's reaction surprises us. No doubt it surprised those who saw themselves as his rivals, who expected that their successes would irritate the ambitious and hyperactive apostle, now almost immobilized by Roman chains. You know, don't you, that twinge of envy or resentment that you have felt when others excel at your specialty, or when they receive recognition while your faithfulness and accomplishments are overlooked? And it is especially hard to celebrate their achievement when you suspect that behind it is naked ambition that craves the spotlight. So how could so energetic and effective a leader as Paul rejoice when others surpassed his

7. "They are labeled Christian preachers, and Paul does not condemn the substance of their message. They are presenting Christ." Peter T. O'Brien, *The Epistle to the Philippians*, NIGTC (Grand Rapids: Eerdmans, 1991), 103. "They are not false teachers—such as Judaizers—because Paul does not object to the content of their message: they proclaim Christ." G. Walter Hansen, *The Letter to the Philippians*, PNTC (Grand Rapids: Eerdmans, 2009), 74.

8. J. B. Lightfoot theorized that Paul's rivals in Rome had theological affinities with the Judaizers whom he resisted in Galatia, and whom he would call dogs, evildoers, and mutilators later in Philippians (3:2). J. B. Lightfoot, *Saint Paul's Epistle to the Philippians* (1913; repr., Grand Rapids: Zondervan, 1953), 88–89. Moisés Silva does not endorse Lightfoot's identification of the rivals as Judaizers, but believes that they contradicted Paul doctrinally and sought "to subvert the apostle's authority and to establish a form of Gentile Christianity that was friendlier to Judaizing influences." Moisés Silva, *Philippians*, 2nd ed., BECNT (Grand Rapids: Baker, 2005), 65. Similarly, Gordon Fee and Ralph Martin speculate that Paul's Roman rivals (as they saw themselves) shared the infatuation with power that characterized the "super-apostles" at Corinth (2 Cor. 11–12). Fee, *Philippians*, 121–23; Martin, *Philippians*, 73–74.

achievements? The answer to this puzzle, like the solutions to the mysteries of Paul's celebration in the "cell" and his true friends' finding courage in Paul's chains, is the grace of God in Christ.

The King Who Won by Losing

Paul knows that the whole gospel-spreading enterprise is not about his freedom to move or minister, his personal comfort, or his influence as a religious leader. It is all about Jesus: the grace he displays and the glory he deserves. Did you notice that in every sentence of this text Paul mentions a *message* or a *person*—or the *person who is the message*? He celebrates the advance of the gospel (Phil. 1:12), his chains are for Christ (v. 13), his brothers boldly speak the Word (v. 14). Some preach Christ out of envy (v. 15) but others do so out of love, knowing that Paul is where he is for the gospel (v. 16). Even when rivals preach Christ out of unworthy motives (v. 17), Paul rejoices that Christ is proclaimed (v. 18).

Paul is not a Stoic. The source of his emotional equilibrium in adversity is not the Stoic theory that taught adherents to steel themselves against life's disappointments through aloof indifference or emotional disengagement. No, Paul's source of joy is a person: the eternal Son of God who has always been equal with God the Father but did not use his equality for his own comfort and convenience. Rather, this majestic person, Jesus Christ, humbled himself, became our human brother, obeyed even to the point of the torture and shame of a Roman cross, and then was raised to life and exalted above everyone everywhere (Phil. 2:5–11).

This is an astounding message—a message so unlikely that no human mind would have invented it. This good news of amazing grace has set Paul free from himself and his previous pursuit of self-focused religious and moral achievement. It enables Paul to say, as we will hear a few sentences later: "It is my eager expectation and hope that I will not be at all ashamed, but that with full courage now as always Christ will be honored in my body, whether by life or by death. For to me to live is Christ, and to die is gain" (Phil. 1:20–21). The Christ who is the hero of Paul's message has become the center of his very life.

But can it work for you? Still we wonder, "Why should I think that what worked for Paul would work for me?" To be sure, it is encouraging that Paul could report that other followers of Jesus, not apostolic superstars but ordinary folk like us, had found courage through Paul's chains to put their own freedom and safety on the line as they filled Rome with the message about Jesus. Moreover, it is clear that Paul expects his friends in Philippi to "catch" his joy over the gospel's advance and to "catch" the courage of Roman believers. He invites them to take a new perspective on his trials, from the shackles on his ankles to the competitors on the streets. He urges them, and us with them, to see his sufferings from a Jesus-focused point of view, so that they can view their own opponents and troubles through the same lenses.

Then again, you think to yourself, that was long ago and far away. Does it still work today? Yes—for the same Christ who stooped to conquer, who descended from heaven's heights to the cross's lowest shame and rose again to supreme glory, still lives and rules today at God's right hand. This great King, who gave his life to win life for his people, who defeated death by dying, makes his presence known among us through the personal power of his Holy Spirit. Yes, what worked for Paul and the Jesus-followers in Rome and Philippi will still work for you in the twenty-first century.

But sometimes, like Paul's Philippian friends, we need to see Christ's promises enfleshed in people we know. That is why I am so grateful that, so early in my pastoral ministry, Christ introduced me to Joy, whose name fit her so beautifully. Despite Joy's pain and confinement, her compassion overflowed to refresh those who met her, including her young, sometimes less-than-joyful pastor. What gave Joy, and Paul before her, such freedom to rejoice in circumstances that were anything but pleasant? What reservoir of hope and contentment kept their hearts vibrant in hope and in concern for others during seasons of their lives that, to the outside observer, looked more like parched wastelands than like verdant gardens? Like mature trees with long taproots, they were drawing life-sustaining nourishment from a source far deeper than life's surface circumstances. The prophet Jeremiah painted a vivid picture of the secret supply that keeps people flourishing in the midst of suffering:

> Blessed is the man who trusts in the LORD,
>   whose trust is the LORD.
> He is like a tree planted by water,
>   that sends out its roots by the stream,

> and does not fear when heat comes,
> for its leaves remain green,
> and is not anxious in the year of drought,
> for it does not cease to bear fruit. (Jer. 17:7–8)

The Lord himself is the deep source of courage and contentment in your dry seasons. The faithful encouragement of his Holy Spirit's constant presence replenishes joy "in the year of drought."

If you are anything like me, your mood often flourishes or withers depending on your circumstances. If your health is good and others appreciate you and there is some money in the bank, those happy circumstances bear fruit in your sunny outlook on life. But when things go wrong—whether from a cold or cancer, others' indifference or outright persecution—your spirits plummet and your inner landscape becomes dry and brittle. For Paul, his friends in Rome, and my dear sister Joy, on the other hand, the taproot of their hearts drew life-giving sustenance from the wellspring of God's Son, Jesus, through the Holy Spirit. That is why Paul is about to affirm his confidence that "the help of the Spirit of Jesus Christ" will give him courage to stand strong for Christ's honor, whether ongoing life or violent death awaits Paul himself. Because the waters of God's grace never run dry, the refreshment provided by Christ's Spirit will set your heart free, inexplicably but truly, from the tyranny of fluctuating circumstances—and from fixation on your own comfort or reputation. This is why Paul judged his situation not by how it served his convenience, but rather by how it served to advance the gospel. He found such freedom from his personal aspirations and apprehensions because he knew that he belonged to Jesus, the Lord of glory who won by losing, who conquered death by dying, and that in Christ he had everything he would ever need.

Are you attached to your pleasant standard of living, afraid to let go for fear that Jesus might ask you to "hurt" for the sake of his global kingdom? Well, he just might. But when he does, he will "index" your joy to his constant faithfulness, rather than to the fluctuations that surround you in the stock market, real estate prices, employment opportunities, or retirement portfolios.

Do you need to be set free from the feverish quest to be best, to be first, or to achieve and gain recognition? Are you frustrated and bitter when it doesn't happen, either because you aren't as great as you think you are or because others fail to recognize your worth? Only one person can set your heart free from the heavy burden of your own reputation, free to sing in a cell (as Paul and Silas did at Philippi), to rejoice in the success of your rivals, and to put your life on the line for a cause bigger than yourself.

That person is not Paul—far from it! That person is the Lord Jesus whom Paul served, in whom Paul rested his heart—the Lord of lords who became Servant of all, who laid down his life to ransom many, who has taken up his life again, who now calls you to surrender to his grace, and to find in him the joy for which your heart was made, a joy that springs from fountains deeper than the fluctuating circumstances of our lives, from the infinite joy of God the Father in God the Son through God the Holy Spirit.

5

A Tough Choice, but One Supreme Goal

Philippians 1:18b–26

It is my eager expectation and hope that I will not be at all ashamed, but that with full courage now as always Christ will be honored in my body, whether by life or by death. For to me to live is Christ, and to die is gain. (Phil. 1:20–21)

Hamlet, prince of Denmark, faced an excruciating dilemma. His uncle had murdered his father, seduced and married his mother, and seized Denmark's throne. Because Hamlet could see no way to avenge his father and restore justice, he contemplated ending his inner torment through suicide. Taking his own life might offer escape from "the slings and arrows of outrageous fortune . . . a sea of troubles . . . the heart-ache and the thousand natural shocks that flesh is heir to." So, he reasoned, "'tis a consummation devoutly to be wished." Yet he hesitated, fearing the unknown future that lay beyond the grave:

> To die, to sleep—
> To sleep, perchance to dream. Ay, there's the rub,

For in that sleep of death what dreams may come,
When we have shuffled off this mortal coil,
Must give us pause. . . .
For who would bear the whips and scorns of time,
Th' oppressor's wrong . . . the law's delay,
The insolence of office . . .
But that the dread of something after death,
The undiscovered country from whose bourn [realm]
No traveler returns, puzzles the will
And makes us rather bear those ills we have
Than fly to others that we know not of?
Thus conscience does make cowards of us all . . . .[1]

The words that Shakespeare put into Hamlet's mouth show a mind at the end of its tether. Most of us rarely live in that dark place. Nevertheless, Shakespeare voices a fear that we instinctively feel when we face the unknown future. It is not only the prospect of God's judgment beyond the grave and our own uneasy consciences that "make cowards of us all." We cannot even see the next thing to come in this life, and so we dread "the slings and arrows of outrageous fortune" that (we suspect) lie in wait to ambush us. As miserable as our current situation may be, the ominous possibility of worse to come does indeed make us "rather bear those ills we *have* than fly to others that we *know not of*." You were laid off last month, and money is tight today. Will tomorrow bring bankruptcy and foreclosure? Yesterday's surgery left a legacy of discomfort, and today chemotherapy saps your strength and takes your hair. Tomorrow, will the cancer take your life, in excruciating pain, anyway?

As the apostle Paul writes from Rome to his Christian friends in Philippi, his current circumstances are "a sea of troubles." He is shackled to a soldier and confined to house arrest. And on the near horizon looms a more daunting threat that he "knows not of." He has been accused of fomenting civil unrest—a charge that always got the attention of Roman authorities. Invoking his right as a Roman citizen, he has appealed to Caesar to decide his case. Now he awaits the emperor's verdict. As his ruminations in our text imply, the possible outcomes are extreme: either vindication and release, or condemnation and death. In Philippians 1:18 Paul turns his attention from

1. William Shakespeare, *The Tragedy of Hamlet, Prince of Denmark*, act 3, scene 1, lines 56–83.

his inconvenient present to his unseen but imminent future, tying present and future together with the theme of joy. Since the effect of his current imprisonment is that Christ is being proclaimed, "in that I rejoice" (1:12–18a). And as he looks ahead to the pending outcome of his legal appeal, he can still predict, "I will rejoice" (1:18b–26).

Paul, like Hamlet, ponders the pros and cons of ongoing life in this world of woes, on the one hand, and an imminent death that would end earthly suffering, on the other. Yet Hamlet's and Paul's soliloquies on the "to be or not to be?" conundrum are radically different in perspective and tone. Hamlet was paralyzed by indecision, unable to choose the lesser of two evils: miserable life or worse misery beyond the grave. Paul, on the other hand, is "hard pressed" to decide between the greater of two goods: ongoing life to serve Christ's people, or a martyr's death that would usher him into Christ's presence. Hamlet was Shakespeare's fictional invention, whereas Paul was a real historical person. Yet for many people, Hamlet's words of dismay sound more credible than Paul's words of hope. What explains Hamlet's and Paul's radically different reactions to the uncertainties of the future? More importantly, how can we grasp the truth that set Paul free from fear as he faced an unseen but imminent life-or-death future?

It is in order to show others the truth that had freed him from fear of the future that Paul opens the windows of his heart. He is inviting his Philippian friends (and us) to observe his inner wrestling with the life-or-death possibilities in his future not only to calm their concerns for his well-being, but also—more importantly—to show them how being captivated by Christ's preeminence colors a person's reaction to suffering and relationship to others. The Philippian believers are suffering at the hands of opponents whose intimidation tempts them to be frightened, putting their joy at risk (Phil. 1:28–30). Moreover, those outside pressures have revealed fissures in the believers' bonds with one another. Paul wants them to stand "firm in one spirit, with one mind striving side by side for the faith of the gospel" (1:27). Paul shows how he himself is processing possibilities looming in the future to guide both their joy in reaction to suffering and their humility in relation to each other.

Paul's joy has its source in his commitment to one supreme goal and in his confidence that he would reach that goal, by whatever path God chose to lead him there. His certainty of reaching his destination casts a distinctive light on the alternative routes that lie before him—ongoing life or impending death. The advantages of each option compound his dilemma in choosing which of them he should request from his sovereign Savior. In the end, though, Paul's supreme goal so transforms his deepest desire that it "tips the scale" of his own preference in the direction of ongoing life for the sake of promoting others' progress and joy in trusting Jesus.

PAUL'S ONE SUPREME GOAL: THE GLORY OF CHRIST

Paul's discussion of what might await him in the future begins with an announcement of his confident expectation that his supreme desire will be fulfilled (Phil. 1:18b–20). Paul's opening declaration, that he will go on rejoicing because "I know that . . . this will turn out for my deliverance," is attention grabbing, even surprising. As Paul shows in what follows, the verdict that Caesar would hand down is by no means a foregone conclusion. By the end of this text Paul's inner debate will reach a conclusion about the *probable* outcome, but Paul knows that the coming weeks or months could bring him either life . . . or death.

How, then, can Paul assert so confidently that his current imprisonment will definitely result in his "deliverance"? The answer lies in the meaning of *deliverance*. The Greek term that Paul uses, *sōtēria*, was sometimes applied to rescue from physical threats and harm. In Acts 7:25 it refers to the Israelites' liberation from slavery in Egypt. Some New Testament scholars, therefore, have concluded that in Philippians 1:19 Paul claims to know that Caesar would exonerate him and order his "deliverance" from both execution and chains.[2] There are stronger reasons, however, to believe that the *sōtēria* "deliverance" that Paul anticipates so confidently is a far greater salvation than escape from Roman chains and sword.[3] When Paul uses the word *sōtēria*, he means comprehensive salvation from sin's power, from

2. For example, Gerald F. Hawthorne and Ralph P. Martin, *Philippians*, rev. ed., WBC 43 (Nashville: Thomas Nelson, 2004), 49–50.

3. J. B. Lightfoot, *Saint Paul's Epistle to the Philippians* (1913; repr., Grand Rapids: Zondervan, 1953), 91; Peter T. O'Brien, *The Epistle to the Philippians*, NIGTC (Grand Rapids: Eerdmans, 1991), 109–10; Gordon D. Fee, *Paul's Letter to the Philippians*, NICNT (Grand Rapids: Eerdmans, 1995), 130–32; Moisés Silva, *Philippians*, 2nd ed., BECNT (Grand Rapids: Baker, 2005), 69–72; John Reumann, *Philippians: A New Translation with Introduction and Commentary*, AYB (New Haven, CT: Yale University Press, 2008), 242–44; G. Walter Hansen, *The Letter to the Philippians*, PNTC (Grand Rapids: Eerdmans, 2009), 77–79.

condemnation under God's wrath, and ultimately from physical and eternal death at the end of history.[4] Sometimes he looks back to the salvation from spiritual death that the Holy Spirit brought about at the beginning of the Christian life: "by grace you have been saved" (Eph. 2:5, 8). Sometimes he looks forward to the consummation of salvation at Christ's return: "Since, therefore, we have now been justified by his blood, much more shall we be saved by him from the wrath of God" (Rom. 5:9). Here, Paul's focus is on both his current and his future experience of God's saving power in Christ.

Paul explains the salvation that would result from his imprisonment and trial in the short term in the following verse (Phil. 1:20), when he announces his "eager expectation and hope" that before Caesar's tribunal he would have *the courage to bring honor to his Lord*, whatever the legal outcome, "whether by life or by death." Paul craves and expects "salvation" from anything that would tempt him to cringe back in shame from bearing a bold witness to the glory and grace of Jesus. To have Christ "honored"—Paul's Greek says "shown to be great"[5]—is Paul's one supreme goal. The "salvation" that Paul expects is deliverance from the temptation to be ashamed of Christ. By God's saving power, Paul hopes to serve one transcendent purpose—displaying the greatness of Jesus—whether God had determined that Paul's "chief end"[6] would be better accomplished in that venue by the death of his body or its ongoing life on earth.

Paul can have such confidence because he knows that his ultimate destiny is not determined by human opinions—not even the verdict of the mighty emperor over Rome's far-flung domain. Paul blends the words of Job, the ancient sufferer, into his own expression of confidence. Despite the accusations of his friends, Job had affirmed his innocence and his trust in God: "Though he slay me, I will hope in him; yet I will argue my ways to his face. This will turn out[7] for my salvation" (Job 13:15–16). Like Job, Paul faces an ordeal that could well end in death; and like Job, Paul is confident

4. See Phil. 1:28; 2:12; Rom. 1:16; 10:10; Eph. 1:13; 1 Thess. 5:9; etc.

5. Greek *megalynō*, used only once elsewhere by Paul (2 Cor. 10:15) with geographic connotations. "Magnify" as descriptive of ascribing greatness to God in praise appears in Luke 1:46; Acts 10:46; 19:17; and frequently in the Psalms, such as Pss. 34:3 (33:4 LXX); 35:26 (34:27 LXX); 40:16 (39:17 LXX); 70:4 (69:5 LXX).

6. WSC, answer 1: "Man's chief end is to glorify God, and to enjoy Him for ever."

7. I have modified the ESV wording of Job 13:16 to reflect the Greek Septuagint, *touto moi apobēsetai eis sōtērian*, which Paul echoes word for word in Philippians 1:19.

that, whatever the earthly outcome of his trial, his salvation before God's heavenly tribunal is secure.

Paul also weaves another Old Testament theme into his confession of confidence in God's salvation. In Psalm 34:3–6, the psalmist invited his fellow worshipers to "magnify[8] the LORD with me," promising that those who look in faith to the Lord "shall never be ashamed"[9] and confirming this promise from his own experience: "This poor man cried, and the LORD heard him and saved[10] him out of all his troubles." Like the psalmist, Paul sees the connection between salvation, freedom from shame, and the privilege of magnifying the Lord. The "shame" that Paul wants to escape is embarrassment over confessing his faith. He had written to the Christians at Rome that he was not ashamed of the gospel (Rom. 1:16). Now his hearing before Caesar would offer opportunity to prove that stance in practice.

For the psalmist and for Paul, to "not be ashamed" not only entails standing courageously rather than cringing in timid silence. It also expresses confidence that their hope in God will not prove misplaced, that he will not let them down by failing to keep his word. Other sources on which people rely might embarrass those who rest on them for security, collapsing in the moment of crisis. But David and Paul had staked their lives on the Lord's faithfulness and power, and he never fails—never shames—those who trust him. Both Paul and Peter applied Isaiah 28:16 to Christ, the cornerstone that God would set in Zion, promising that "whoever believes in him will not be put to shame" (Rom. 9:33; 10:11; 1 Peter 2:6). The "salvation" from shame that Paul anticipates is not only deliverance from cowardice before Rome's emperor. It is also rescue from shattered hopes at the last judgment before the Lord of all creation.

8. Greek *megalynō* (Ps. 33:4 LXX), the same term that Paul uses in Philippians 1:20, rendered "will be honored" in the ESV.

9. Greek *ou mē kataischynthē* (Ps. 33:6 LXX), roughly equivalent to Paul's "I will not be at all ashamed" (or, more woodenly, "in nothing I will be ashamed," *en oudeni aischynthēsomai*).

10. In the LXX, the Greek verb is *sōzō*, cognate to the noun *sōtēria*, used by Paul here. In the Greek Scriptures (LXX and New Testament), as in other ancient literature, the verb *sōzō* connotes rescue from a wide variety of threats and dangers, both temporal and eternal, physical and spiritual. Hints in Psalm 34 of its historical setting suggest that David had been delivered from violence at the hands of human enemies. If Paul is intentionally alluding to the confluence of the themes—"magnify," "save," and "shame"—in Psalm 34, he brings into view another dimension of God's saving power: enabling suffering saints to stand boldly, whether or not physical rescue is forthcoming.

Paul knew that God would grant him this great salvation, but he also knew that the sovereign Creator uses creaturely means to fulfill his Word. God would supply[11] "the Spirit of Jesus Christ" to him, imparting courage, and God would flood Paul's heart with the Spirit's power "through your prayers" (Phil. 1:19). As Paul had prayed for the Philippians to grow in love, so he needed their prayers to conduct himself so as to bring credit to the Lord. Paul never underestimated the formidable spiritual forces of evil arrayed against believers. Nor did he overestimate his own intrinsic strength for the battle. During this same imprisonment he wrote to the church at Ephesus, "We do not wrestle against flesh and blood, but against the rulers, against the authorities, against the cosmic powers over this present darkness, against the spiritual forces of evil in the heavenly places" (Eph. 6:12). Therefore, he begged them to support him in prayer, "that words may be given to me . . . boldly to proclaim the mystery of the gospel, for which I am an ambassador in chains, that I may declare it boldly, as I ought to speak" (6:19–20).

What is your supreme goal in life, your "eager expectation and hope" (Phil. 1:20)? Paul expressed his supreme goal—to promote Jesus' glory, whatever the cost or benefit to Paul himself—in order to whet the Philippians' and our appetites for the same heart-satisfying aim. He was not setting himself apart from the rest of us as an otherworldly ascetic, to be admired from a distance by people whose devotion could not match his. Rather, he wanted to make us all to feel his thrill at the privilege of magnifying Christ. This privilege, and nothing less, is what you and I were made for. The author of Psalm 73 once envied those who enjoy attractive but ephemeral rewards. But when God brought him to his senses, he realized that nothing could compare to the priceless treasure he already possessed:

> Whom have I in heaven but you?
> And there is nothing on earth that I desire besides you.
> My flesh and my heart may fail,
> but God is the strength of my heart and my portion forever. (Ps. 73:25–26)

11. "Supply of the Spirit of Jesus Christ" is a more accurate and vivid translation than "help of the Spirit." In Galatians 3:5 Paul described God as "he who *supplies* the Spirit to you," using the cognate verb (*epichorēgeō*) to the noun (*epichorēgia*) used in Philippians 1:19. Without denying the divine personality of the Spirit, Paul spoke of the Spirit as an empowering gift that God supplies to believers.

Just as "the heavens declare the glory of God" (Ps. 19:1) and the trees clap their hands and "sing for joy before the LORD" at his coming (Isa. 55:12; Ps. 96:12–13), so human beings who bear the very image of the Creator are designed to be "to the praise of his glory" (Eph. 1:12, 14). Even a glimpse of that goal, the honor of extolling God's honor, confronts you with a heart-searching question: What passion fills your thoughts in your waking hours and sleepless nights? Are you pursuing academic achievement, career success, health and fitness, a fulfilling marriage, respectful and accomplished children, financial stability, popularity, or community recognition? These are good goals, but none is big enough to be your supreme goal, the goal for which your Creator designed you. They *might* be consistent with his all-wise design for you, but if your sights are set no higher than these earthbound accomplishments, sooner or later your hopes will be dashed.

Or, to use Paul's other way of speaking, from what evil, above all, do you need salvation? Do you long to escape poverty, illness, abuse, injustice, violence, loneliness, failure, obscurity, or shame? Any sane person would want to escape such miseries. Yet no sane person would expect a life free of pain and adversity in this world. God has not promised complete deliverance from the world's woes short of Christ's return at the end of history. On that great day, when the Savior for whom we wait appears from heaven, not only will he transform our lowly body to be like his glorious body (Phil. 3:20–21), but he will also create a new heavens and earth, from which every form of evil and misery will be excluded (Rev. 21:1–4). In the meanwhile, the salvation that God does give now, which Paul *knew* that he would receive through the saints' prayers and the Spirit's presence, is deliverance from fear and protective self-centeredness. And God's grace will free you, too, enabling you to embrace the supreme goal that was Paul's "eager expectation and hope" and that gave him confident joy in the face of an uncertain future: the goal of seeing Christ glorified through you, whether through life or death, plenty or want, health or disease, admiration or rejection. Don't settle for less than the best!

PAUL'S DIFFICULT DILEMMA: TWO DESIRABLE ALTERNATIVES

Paul's mention of the life-or-death alternatives that await him in the near but unknown future prompts him to reflect on the pros and cons of each.

Paul's supreme goal is the glory of the Christ who had redeemed him and who controls his every circumstance. That goal casts a distinctive light on each fork in the road ahead in Paul's pilgrimage of faith. The supremacy of Jesus and Paul's passionate commitment to make his Master's majesty known explain two factors that might surprise us in Paul's ruminations on his options (Phil. 1:21–24).

The first surprise is that Paul, the prisoner, seems to assume that he, not the emperor, has the authority to decide whether he will be executed or released. As he weighs the alternatives, he expresses his quandary: "Yet which *I shall choose* I cannot tell" (Phil. 1:22). Here Paul sounds much like Hamlet, though their situations are so different. For Hamlet, to choose death would mean suicide by his own hand. For Paul, suicide is not an option. Years earlier Philippi's jailer was about to commit suicide, preferring death by his own hand to public shame and torture for allowing prisoners to escape. But Paul stopped him: "Do not harm yourself, for we are all here" (Acts 16:28). Knowing that God's Law demands, "You shall not murder" (Rom. 13:9), Paul would not have taken the power of life and death, which belongs exclusively to the Creator who gives life, into his own hands. From a human viewpoint, therefore, the decision between death and life did not belong to Paul the prisoner but to Caesar the judge. How could the defendant presume to speak as though he could choose his verdict and sentence?

The explanation of Paul's audacious assumption is the sovereignty of the Christ to whom Paul belongs. Christ, not Caesar, is the true Lord who controls Paul's fate. Paul will soon speak of Christ as the One who is and always has been equal with God (Phil. 2:6). He will go on to affirm that this divine person became human and was brought low in suffering obedience, but was then highly exalted by God as Lord of all, bearing the name that excels every name, the title that outranks every creaturely authority in heaven or on earth (2:9–11; see Eph. 1:21–23). Christ is the second person of the triune God who "works all things according to the counsel of his will" (Eph. 1:11). Therefore, when Paul made plans, they were always "in the Lord Jesus" and subject to his Lord's revision (Phil. 2:19, 24). And when Paul prayed, his requests were subject to God's will (Rom. 1:10). Although Caesar was unaware of it, his decisions, too, were subject to the sovereign control of the Lord Christ. Therefore, Paul the prisoner, through prayer, could go "over the head" of the Roman emperor; Paul had access to the

King of all kings. God Almighty controlled the emperor's every decision, so that the upcoming verdict on Paul's appeal, whatever it turned out to be, would "work together for good, for those who are called according to [God's] purpose" (Rom. 8:28). Paul's dilemma—his "choice," as he calls it—concerns which legal outcome he should request as he approaches his Lord in prayer, knowing that Christ, to whom he makes his petition, holds all authority in heaven and on earth and therefore controls Caesar's decision.

The second surprising feature of Paul's difficult dilemma is that he hesitates between two desirable alternatives: ongoing life or impending death. Some people have a love affair with death, a fascination with the macabre, even a depression that tempts them to suicide. Some find their current lives so intolerable and their future, as far as they can see it, so hopeless that they feel like the despairing Hamlet, torn between the lesser of two evils: life in an anguish-filled world or suicide. Other people—probably most people, most of the time—have a love affair with this life, with its opportunities for affection or achievement, its relationships, pleasures, and amusements. For them, the obvious choice would be "life," particularly since the "life" that Paul anticipated was not perpetual incarceration but release from custody and a return to freedom. To people who doubt or deny life after death, Paul sounds like a madman when claiming that leaving this world is better by far than a "get out of jail free" card that would free him to consume the stuff and satisfaction that this earth has to offer (as they would use their freedom). For Paul, however, both ongoing life in this world and sudden death have almost irresistible advantages, though they are not advantages that would occur either to those who are suicidal or to those who cling desperately to this life.

Both of the alternatives that Paul finds so appealing are mentioned in his confession, "To me to live is Christ, and to die is gain" (Phil. 1:21). Both his life and his death are defined by the fact that he belongs to Christ. In fact, because his "living is Christ," Paul can also affirm that "dying is gain." Paul wrote to the Galatians: "I have been crucified with Christ. It is no longer I who live, but Christ who lives in me. And the life I now live in the flesh I live by faith in the Son of God, who loved me and gave himself for me" (Gal. 2:20). When Christ died in his place, Paul himself had been executed under the just curse of God's Law. Paul's individual, independent standing before God came suddenly and mercifully to an end; he had entered into a new life, in which his identity, his motivation, and his capacity to love were all defined by the fact that Paul was now united to Jesus Christ. Paul had gladly forfeited the advantages of his former life, "that I may gain Christ and be found in him, not having a righteousness of my own that comes from the law, but that which comes through faith in Christ" (Phil. 3:8–9). Paul's current life in service to the gospel is sustained by his Savior's presence through "the Spirit of Jesus Christ" (1:19). Paul's heart goes out to the Philippians with "the affection of Christ Jesus" (1:8), in a selfless concern for their welfare. He urges them to show the same attitude in their treatment of each other, since this blend of humility and love expresses the mind-set that is his and theirs "in Christ Jesus" (2:3–5).

Because Paul's union with Christ now defines who Paul is, Paul can assess the alternative outcomes of his legal appeal—ongoing life or imminent death—not as competing evils, but as competing goods. In verses 22 and 24 Paul explores the benefit that would result if God's plan is for Paul to honor Christ through ongoing life on earth. Continuing to "live in the flesh" would result in "fruitful labor for me" (Phil. 1:22). This fruitful labor that Paul envisions is not for his personal advantage. For Paul, "fruit" pictures the transformation that takes place in people's lives when the Holy Spirit mercifully and mightily brings "home" to them the good news of Jesus, which Paul preached. Paul prays that the Philippians would be "filled with the fruit of righteousness that come through Jesus Christ" (1:11). He rejoices in their generosity not because he personally needs their gifts but because he seeks "the fruit that increases to your credit" (4:17). Years before he came to Rome, Paul wanted to reach the imperial capital "in order that I may reap some harvest [fruit[12]] among you as well as among the rest of the Gentiles" (Rom. 1:13). Paul has labored to plant the seed of the Word in the soil of human hearts, others have watered it, and God has made it grow, flourish, and produce fruit (1 Cor. 3:6–9). Therefore, Paul concludes, for him to "remain in the flesh is more necessary on your account" (Phil. 1:24). If his legal appeal results in his vindication and release, he would return to Philippi, to strengthen his friends' faith in person and to participate in their gospel witness to their fellow citizens. His arrival would further their "progress and joy in the faith," and enrich their confidence in Christ Jesus (1:25–26).

12. The Greek word that the ESV renders "harvest" is *karpos*, translated "fruit" in Philippians 1:11, 22; 4:17.

On the other hand, for Paul *personally*, death is an even more attractive option. Death would be "gain" not because Paul finds his current circumstances intolerable, nor because it offers a quick escape from threats looming on the horizon. The one thing that makes a speedy death better—in fact, "far better" (actually, to convey Paul's deliberately over-the-top Greek: "much more better")—is that to die is "to depart and be with Christ" (Phil. 1:23). Paul already enjoys, in a profound way, "the surpassing worth of knowing Christ Jesus my Lord" (3:8), but he is still straining ahead to lay hold of all that is involved in "gaining Christ." Paul will later remind his friends that Christians are awaiting "a Savior, the Lord Jesus Christ, who will transform our lowly body to be like his glorious body" (3:20–21). Christ's glorious return from heaven is believers' "blessed hope" (Titus 2:13). Until that great day of the Lord, followers of Jesus live in an uncomfortable tension, as Paul told the Corinthians: to be "at home in the body" is to be "away from the Lord," to "walk by faith, not by sight" (2 Cor. 5:6–7). Christ's presence with his struggling people on earth through his Holy Spirit is real, but not visible. At death, even as we await the resurrection of our bodies, those who belong to Christ enjoy his presence in a more intimate way than we now do on this earth. So, Paul wrote, "we would rather be away from the body and at home with the Lord" (5:8). What makes death "gain" is not the earthly misery that it puts behind us but the heavenly delight into which it will usher us, the delight of being with the Savior who loved us and gave himself for us. Paul's desire to depart from life on this earth is ignited by longing to be as near to Christ as possible. If his personal longing were the deciding factor between life and death, Paul would gladly choose the martyrdom that would bring him swiftly to his King of grace.

Paul's hesitation over the pros and cons of life and death poses a question to each of us. The question is not merely "*Which* would you choose, life or death?" It is rather "*Why* would you opt for ongoing life or for imminent death?" Perhaps you would choose life over death because life, despite its inconveniences, keeps you entertained with a modicum of comfort and pleasure. You view death, on the other hand, as a realm shrouded in mystery. You do not know whether it will bring endless unconsciousness or, as Hamlet feared, terrors to be dreaded more than the ills that you have known in life. Or perhaps you find death appealing precisely because you think (or hope) that nothing lies beyond it. Although you cannot know that it is so, you imagine death to be an endless anesthesia that finally numbs life's agonies of body or mind. But can you hear how these reasons for preferring life to death, or death to life, reveal hearts enslaved by earthbound dreams and fears, hearts bent in on themselves? Paul the prisoner was so much freer! Because he had been seized and saved by Christ, he would joyfully follow whichever path Christ chose for him, knowing that in either case the Lord who had loved him and rescued him from himself would grant his heart's chief desire: "that with full courage now as always Christ will be honored in my body, whether by life or by death" (Phil. 1:20).

Paul's Pastoral Preference: To Stay and Serve

By the end of this text Paul has reached a tentative conclusion (Phil. 1:25–26). As he has weighed the life-or-death alternatives before him and their consequences for himself and for his fellow believers, he has discerned which option would better enable him to glorify Christ at that moment. He introduces his conclusion with the carefully chosen word "Convinced"[13] to show that his confidence is not based on a special revelation granted to him uniquely as an apostle, but rather results from a process of reflection to which his friends in Philippi also have access: he has applied biblical norms to observable circumstances. Paul's purpose in opening the windows of his heart, so that they (and we) can watch him wrestle with the dilemma between desirable alternatives, is to present himself as a "case study" in how to assess our sufferings and how to find our deep desires transformed by the reality that defines our lives, as it did Paul's: "To me to live is Christ, and to die is gain" (1:21).

At this time in his ministry, Paul concludes that it is "more necessary" for the Philippians (and no doubt for other believers) that he "remain in the flesh" (Phil. 1:24). He expects not only to stay alive on earth but also to be released from custody, to return to Philippi and to "continue with you all, for your progress and joy in the faith" (1:25). Paul's arrival will give his friends even more occasion to "glory in Christ Jesus" (1:26). They

13. Paul's Greek word is *peithō*, which in the second perfect active (in Phil. 1:6, 25; 2:24) means "*be convinced, be sure, certain*" (*BAGD*, 639). In the active voice, *peithō* often refers to the process of persuasion through the presentation of evidence and argumentation, leading to firm conviction (for example, Acts 18:4; 2 Cor. 5:11).

will praise God not only because he will have answered their prayers for Paul's release and return, but also because, in response to their petitions, Christ's Spirit will have emboldened Paul to honor Christ throughout his legal proceedings. Although persuaded that God would restore him to his friends, Paul could not rule out the possibility that God planned for him to glorify Christ in a martyr's death in the near future. He will still write: "Even if I am to be poured out as a drink offering upon the sacrificial offering of your faith, I am glad and rejoice with you all" (2:17). Yet as he evaluates the Philippians' spiritual need and the opportunities for "fruitful labor" still available to him throughout the empire, Paul expects exoneration and release. He plans to send Timothy to Philippi as soon as the verdict is delivered (2:19–23), and he is "convinced[14] in the Lord that shortly I myself will come also" (2:24). From a later imprisonment, on the other hand, Paul would express a very different assessment of where God's providence was about to lead him: "I am already being poured out as a drink offering, and the time of my departure has come. I have fought the good fight, I have finished the race, I have kept the faith" (2 Tim. 4:6–7). At that later point, with a sense of "mission accomplished," Paul would joyfully embrace that "far better" alternative, "to depart and be with Christ" (Phil. 1:23). In the meanwhile, he will joyfully remain in this sin-stained, sorrow-pained world to draw others into the depths of Christ's redeeming grace.

In view of the confidence and joy with which Paul reached the resolution of his dilemma, we may still wonder why he led us through his inner turmoil in the first place. As he dictated the epistle to a copyist, did he not sense where his reflection would lead when his mixed feelings burst forth: "Which I shall choose I cannot tell. I am hard pressed between the two" (Phil. 1:22–23)? I suspect that from the outset, Paul glimpsed the conclusion to which his Christ-centered contemplation would lead. But the Spirit of Christ gave Paul words to lead us, his readers, through the process that brought Paul to the conviction that ongoing life in service lay in his immediate future. As we walk alongside Paul through the quandary and as we feel with him the intense apppeal of each option, Christ's Spirit is showing

14. The Greek *pepoitha*, second perfect of *peithō* (see note 13 above), is rendered "trust" by the ESV in Philippians 2:24. As in 1:25, *peithō* implies a process of inference by which Paul reached his confident conclusion.

us how to assess and address the issues of suffering and self-centeredness that confront us.

Paul is about to turn the topic of conversation from his situation in Rome to his friends' situation in Philippi (Phil. 1:27–2:18). He will speak of opponents whose threats must not intimidate them (1:28) and urge them to see their suffering for Christ's sake as a gift of God's grace (1:29–30). He will also urge them to strive to maintain a deep unity of mind and heart (1:27). They must resist the self-centered tendencies, rivalry and conceit, that are undermining their oneness in Christ's love, and they must replace divisive individualism with a Christlike humility (2:3–4).

Now through his own experience Paul has shown them how to view and bear suffering. He sees his chains as servants of Christ, carrying the gospel into the barracks of the imperial guard and multiplying the numbers of evangelists who proclaim Christ where Paul cannot go. Moreover, Paul sees death itself, even violent death, as Christ's servant and Paul's, bringing about their face-to-face reunion in glory. What is the worst that the Philippians' opponents could throw at them? Chains, flogging, sword? As God sustains them by his Spirit, every threat and pain could only further their one supreme goal: to see Christ honored in their bodies.

In his experience Paul has also shown them how to repudiate self-centeredness. Why does he stress that his personal preference is departure to Christ, when his final choice would be to stay here for the sake of the Philippians? He is not engaging in self-pity or venting resentment, nor is he loading false guilt on their consciences for keeping him from heaven's joys. True, Paul's staying to serve them would cost him. It would postpone his full enjoyment of his heart's chief treasure: face-to-face communion with his Lord. But the deferring of Paul's delight could not compare with the price Christ paid to make Paul and his Philippian friends—and you, if you trust Jesus—into citizens of heaven. Paul is merely a miniature replica of Jesus. Jesus is the Son who is equal to God, who left heaven and came to earth, became a man, and assumed the slave's subservience, ultimately paying an infinitely dearer price—laying down his life on the cursed cross—to serve and to save his wayward creatures. The more that we learn to say, with Paul, "To me to live is Christ," the more that our hearts are set free from self-interest by Christ's others-serving compassion, of which we are unworthy but most grateful beneficiaries. The gospel logic that calls us to love others as God in Christ loved us (Eph. 5:2) will take deeper root in our hearts, so that increasingly our "eager expectation and hope" is that, whatever happens to us, through us Christ will receive glory and others' faith in him will be advanced.

A pastor from a bygone era embraced Paul's one supreme goal for himself in prayer:

> Sovereign God, thy cause, not my own, engages my heart, and I appeal to thee with greatest freedom to set up thy kingdom in every place where Satan reigns; glorify thyself and I shall rejoice, for to bring honour to thy name is my sole desire . . . . Lord, use me as thou wilt, do with me what thou wilt; but, O, promote thy cause, let thy kingdom come, let thy blessed interest be advanced in this world! . . . Let me be willing to die to that end; and while I live let me labour for thee to the utmost of my strength.[15]

May you find the joy and freedom of serving the great cause of Christ's glory, so that you can make this pastor's prayer your own heart's cry, and Paul's affirmation the theme of your life: "To me to live is Christ, and to die is gain."

15. "God's Cause," in *The Valley of Vision: A Collection of Puritan Prayers & Devotions*, ed. Arthur Bennett (Edinburgh: Banner of Truth, 1975), 320–21.

6

Solidarity in Suffering

Philippians 1:27–30

Only let your manner of life be worthy of the gospel of Christ, so that whether I come and see you or am absent, I may hear of you that you are standing firm in one spirit, with one mind striving side by side for the faith of the gospel. (Phil. 1:27)

What do the Christians of first-century Philippi have in common with Christians in the twenty-first century? Both belong to marginalized minorities in societies driven by values that contradict our deepest commitments as followers of Jesus. I am not instinctively pessimistic, but sometimes it is hard to pay attention to the drift of things in our world and our society without an uneasy sense of approaching doom, of things falling apart that were more "together" (or seemed so) when I was growing up.

From a global perspective, to be sure, we see hopeful signs that the church is growing numerically in much of the developing world. In nations once thought completely closed to the gospel of Christ, tiny pockets of Christian believers are gathering and drawing others to encounter the Word of God. Yet despite this growth—and sometimes because of it—followers of Jesus in societies dominated by Islam, Hinduism, Buddhism, and Marxism experience social rejection, political oppression, economic hardship, and even violent persecution.

Even in countries and cultures that were once molded by the convictions and commitments of Christ's followers, we now see signs of societal decay that seem virtually impossible to reverse and repair. I am troubled by questions such as these:

- Do Europe and North America have the self-discipline and courage to resist international terrorism that is driven by religious conviction, while we in the West live for pleasure and entertainment?
- Will we recover the will to say "No" to escalating violence and explicit sensuality in our own entertainment media, while violence invades our schools, shopping malls, city streets, and suburban neighborhoods, and while sexual exploitation invades our homes through the Internet?
- Will the social atmosphere become even more coercively controlled by the idols of moral and ideological relativism and tolerance, so that in time the only intolerable minority will be the one that affirms our divine Creator and his norms for human desires and conduct?

For all these reasons, it is easy to start thinking that people who are serious about following Jesus are now, or will soon find themselves to be, a beleaguered, misunderstood, despised minority surrounded by a society that is indifferent at best and hostile at worst to the Lord whom we trust and serve. Then again, indifference may be worse than hostility.

It may seem cold comfort to be reminded that this is not the first time in history that people who take the lordship of Jesus Christ seriously have been a misunderstood minority in the midst of a hostile culture. Yet it is true. Consider, for example, Philippi in Macedonia, in the eastern Roman Empire during the reign of Nero Caesar. Philippians prided themselves on the fact that their city was a Roman colony, having received this honor from Octavian (later to become Caesar Augustus) and Marc Antony a century before Paul wrote his epistle. This status conferred Roman citizenship on every citizen of Philippi. Unlike the United States (since 1868, when the Fourteenth Amendment to the Constitution was adopted), Rome did not automatically confer citizenship upon everyone born within its territories.

In Rome, citizenship was first a privilege reserved for the city's leading families. As the Roman republic expanded and became an empire, citizenship status was conveyed to others through sponsorship by citizens. Many Philippians were retired soldiers on whom citizenship had been conferred in recognition of distinguished military service. For most subjects living under Rome's dominion around the Mediterranean Sea, however, citizenship's rights and responsibilities were out of reach. Paul himself, though a Jew, was a Roman citizen because his parents had been citizens. That news shocked a tribune whom he met in Jerusalem, because the officer had purchased his own citizenship at a hefty price (Acts 22:25–28).

Philippi further cultivated its Roman identity through its architecture and the use of Latin (setting it apart from the Greek-speaking cities of Macedonia). Politically, Philippi had special reasons for loyalty toward Rome and its emperors. Its allegiance, even reverence, was shown in its temple dedicated to Caesar Augustus and his Empress Livia and the recently completed shrine to the divine Claudius. Along with the worship of Rome's emperors, Philippians worshiped Greek deities (Apollo Comaeus, Dionysos), Asian deities (Cybele and her consort Attis), and local deities such as the Thracian hero-horseman. In other words, Philippi practiced paganism in a wide variety of forms. But when Paul and Silas's team arrived, they found no synagogue dedicated to the one true God who had spoken to Israel. Philippi was a mission field rife with religious weeds and undergrowth, its spiritually barren soil not even pierced by the plow of Israel's ancient Scriptures.

Not surprisingly, the introduction of the gospel into this proud Roman city of Macedonia met resistance. In Philippians 1:30, when Paul mentions the "conflict that you saw I had," he is alluding to the rocky start that this church experienced when he and his team first arrived on the scene (Acts 16:11–40). When Paul exorcised a "spirit of divination" from a young slave girl who was reputed to be clairvoyant, the girl's owners retaliated for their loss of future profits by accusing Paul and Silas of advocating "customs that are not lawful for us as Romans"—note the civic pride—"to accept or practice" (Acts 16:21). A mob gathered, a riot loomed, and to quell the disturbance the magistrates ordered Paul and Silas beaten and imprisoned, without a trial or opportunity to answer their accusers. That was definitely "conflict," as Paul says!

After Paul left, the opposition did not diminish. As he writes his epistle over a decade later, Christian believers at Philippi are still facing "the same conflict that you saw I had and now hear that I still have" (Phil. 1:30). The followers of Christ in Philippi still have influential opponents whose threats frighten and intimidate them (1:28). We do not know the exact form of the Philippians' suffering for the sake of Christ, but Paul compares their sufferings with his own: both those that they saw him experience in Philippi and those that he now endures in Rome, as they have heard. So it is safe to say that their suffering had already been more severe than the mild ways in which Christians are currently marginalized in the West, and was more akin to the oppression endured today by believers in the developing world.

When people find themselves surrounded and outnumbered by an unsympathetic or even hostile dominant culture, they tend to react in various ways. They might get feisty and lash out like a cornered wolf. Or they might get timid and pull back like a turtle into its shell. Then again, they might get sneaky and blend into their surroundings like a chameleon. Or they might get frustrated and turn on each other like a pack of pit bulls, venting hostility internally that they dare not show to their oppressors. To put it another way, when the whole world is against us, some of us become bold but abrasive, whereas others become humble but timid. Then again, we might display an unhealthy combination, cowering timidly toward those outside and arguing abrasively among ourselves. The mind-set of those who become bold but abrasive is: "*They* are the enemy; *we* are in the right, pure, and therefore persecuted. Someday they will get what's coming to them." The humble but timid, whether they withdraw or blend in, meekly try to persuade the dominant majority, "Please just leave us alone; we won't make waves—we promise!" And sometimes those who bow deferentially toward the unsympathetic outside world then turn around and assault fellow believers, whom they should treasure in times of trial.

But the apostle Paul, writing from the chains of imprisonment, calls his fellow followers of Jesus in Philippi to respond to the surrounding society's hostility in a way that does not come naturally. He does not underestimate the gravity of the opponents' evil. In fact, he states soberly that their aggression toward Christ's people is a signal that those opponents are on their way to eternal destruction. But Paul summons the Philippian believers and us to respond to those who despise our faith and our Savior with a distinctive blend of boldness and humility, neither intimidated nor belligerent, neither fearful of those outside nor frustrated with those inside the church. We must show courage that does not blink when opponents confront us, along with concern for fellow Christians with whom we stand, shoulder to shoulder. This new way of responding to the pressures of a society that has no sympathy for our faith is grounded in a deepening appreciation for the privileged status that Christ has conferred upon us by his grace.

Live in Your Earthly City as Citizens of Heaven

Paul has reassured his Philippian friends that he could rejoice in "what has happened to me," since it promoted the advance of the gospel (Phil. 1:12–18a). He has weighed the pros and cons of the potential life-or-death outcome of his legal appeal to Caesar, concluding that his continuing service on earth will advance the Philippians' faith (1:18b–26). In reporting on his current situation and future prospects, Paul has as his purpose not merely to calm his friends' concerns. More importantly, he has modeled how they themselves should react in their own situation of suffering (1:30). Having shared his news, he now turns to address what is happening in Philippi.[1] His opening "Only" underscores Paul's unwavering expectation for his Philippian friends, whatever the outcome of his own case. Whether he returns to see their faith in action, remains at a distance and hears reports, or even dies by Caesar's sword, *their* focus must be on serving as citizen-soldiers in courageous unity, so as to bring credit to the distant and majestic capital that defines their privileged status—not Rome to the west, but heaven on high (Phil. 3:20–21).

In his call to arms, Paul draws upon two pictures that were near to the heart of everyone in Philippi: *citizen* and *soldier*. Behind our English version's "let your manner of life be" (ESV) is a single Greek verb that has *citizen* at its

1. English versions, including the ESV, have trouble reflecting the parallel Greek expressions in Philippians 1:12 and 27 by which Paul demarcates the subjects that he addresses, first "the things concerning me" (*ta kat' eme*) (ESV: "what has happened to me"), and then "the things concerning you" (*ta peri hymōn*) (ESV: "of you"). See also Phil. 2:19 (*ta peri hymōn*, aptly rendered "news of you" by the ESV) and 23 (*ta peri eme*, ESV: "how it will go with me"). Gordon Fee demonstrates the importance of these phrases as structural signals of shifts of topic (from Paul's situation to the Philippians' and back again) in this "letter of friendship" joining a pastor and his people across the miles of their separation. Gordon D. Fee, *Paul's Letter to the Philippians*, NICNT (Grand Rapids: Eerdmans, 1995), 2–4, 54–55.

core. The Greek word for *city* is *polis*, from which we get the word *politics*. In writing to the Philippians—and only to them—Paul bypasses the *walking* metaphor that he uses so often to portray the pattern of behavior that befits our faith,[2] replacing *walk* with the citizenship-laden verb *politeuomai*: to maintain a standard of conduct befitting a citizen, to behave in a way that enhances the reputation of one's city.[3]

Besides being a colony of Rome that valued the citizenship privileges entailed in that status, Philippi was also a military town, heavily populated both by active-duty troops and by retirees from service in the Roman legions. Philippi's active and retired soldiers would associate vivid combat memories with the terms that Paul uses to rally Christian believers to keep "*standing firm* in one spirit, with one mind *striving side by side* . . . , and not frightened in anything by your opponents" (Phil. 1:27–28). They knew that steadfast courage and unity were crucial to victory.

Say "citizen" in Philippi, therefore, and you could see chests swell with pride. Say "soldier," and many of those citizens would stand even straighter, recalling the sweetness of triumph or the bitter loss of fallen comrades. In the first decade of the twenty-first century, as the Western allies fought costly wars in Afghanistan and Iraq, a new generation of Americans came to recognize the connection between the honor of citizenship and the duty of soldiering. In 2008 the rock group 3 Doors Down produced "Citizen Soldier," a stirring music video to support the recruiting efforts of the United States National Guard. The lyrics of "Citizen Soldier" extol everyday heroes who forgo their own comfort and safety to "be right there" for those who need protection, food, shelter, and consolation in the midst of catastrophes. The video moves from the Minutemen at Lexington and Concord in 1775 to current wars and disaster-relief efforts. The aim is to enlist patriotic

2. For example, Phil. 3:17–18; Rom. 8:4; 13:13; Eph. 2:2, 10; 4:1, 17. See Deut. 5:33; Pss. 1:1; 15:2; 26:3, 11; 119:1.

3. Among the interpretations of *politeuomai* offered by recent commentators are these: "as citizens of heaven live in a manner" (Peter T. O'Brien, *The Epistle to the Philippians*, NIGTC [Grand Rapids: Eerdmans, 1991], 144; G. Walter Hansen, *The Letter to the Philippians*, PNTC [Grand Rapids: Eerdmans, 2009], 93), "live out your 'citizenship'" (Fee, *Philippians*, 159, 161–62), "show yourselves to be good citizens" (Gerald F. Hawthorne and Ralph P. Martin, *Philippians*, rev. ed., WBC 43 [Nashville: Thomas Nelson, 2004], 66), "behave as citizens of heaven" (Moisés Silva, *Philippians*, 2nd ed., BECNT [Grand Rapids: Baker, 2005], 80), and "exercise your citizenship" (John Reumann, *Philippians: A New Translation with Introduction and Commentary*, AYB [New Haven, CT: Yale University Press, 2008], 261, 285–86).

participation by appealing to America's older tradition of defense through voluntary militias of ordinary citizens, rather than a standing army of professional warriers. Recruiting videos produced by the Army or Navy Reserves also show citizens combining civilian life and labor with their readiness to serve at the nation's call.

We do not know whether members of the Philippian church held citizenship status in Philippi, and therefore in Rome itself. If the jailer had served in Rome's legions, he may well have been rewarded with citizenship.[4] Whether citizens or not, no doubt all residents of a colony understood what Roman citizenship entailed. Citizenship conferred privileges such as exemption from some forms of taxation.[5] Even more relevant to the experience of Paul and Silas, who were Roman citizens, at the foundation of the church in the Roman colony Philippi, citizenship conferred the right of due process before courts in the Roman legal system. Citizenship also involved duties, particularly the responsibility to uphold the dignity of the empire and the emperor in the way one lived and related to others. Although Rome, the political, military, and commercial center of the empire, lay hundreds of miles to the west, that distant metropolis defined the identity of every proud citizen of Philippi.

So Paul opens his treatment of the Philippians' situation with a politically loaded directive that the Philippians know well. Yet Paul is not thinking of their citizenship in Rome, the capital of Caesar's far-flung empire, a status that only some in the congregation would have enjoyed. Roman citizenship conferred a status that paled in importance when compared to the citizenship that Paul has in mind. In Philippians 3:20 he identifies the city that confers privilege and responsibility on every believer in Philippi: "Our citizenship

4. Everett Ferguson, *Backgrounds of Early Christianity* (Grand Rapids: Eerdmans, 1987), 38: "The [Roman] legions were composed of citizen soldiers. With the extension of the citizenship the provinces came to provide many of the legionaries. . . . Probably many recruits obtained citizenship on induction."

5. A. N. Sherwin-White, *The Roman Citizenship*, 2nd ed. (Oxford: Clarendon Press, 1973), 317, cites an edict of Octavian (Augustus), "granting the citizenship to veterans resident in the eastern provinces," using "the formula 'let them be Roman citizens of the highest grade and condition exempt from taxation.'" He also offers evidence that under later emperors, the "Italian right," *ius Italicum*—the legal right granted to a colonial city outside Italy to be regarded as an extension of Rome, the imperial capital itself—included immunity from certain taxes: "the *ius Italicum* brought with it a great practical advantage, freedom from the land-tax, *tributum soli*, and apparently also from the poll-tax, *tributum capitis*" (276). Ferguson, *Backgrounds*, 32, lists Philippi among the cities privileged to be "*coloniae civium romanorum* (colonies of Roman citizens). These were mostly towns in which military veterans were settled. They were sometimes granted partial or complete immunity from taxation."

is in heaven, and from it we await a Savior, the Lord Jesus Christ." If some of these believers had been proud of their status as citizens of Caesar's imperial capital, all of them could now marvel that they are citizens of an infinitely more majestic celestial city, where Christ their Lord and Savior reigns at God's right hand.

In effect, Paul is saying to his dear Christian friends at Philippi, "Some of you have felt patriotic pride that your status as citizens of Philippi gives you dignity and rights as citizens of Rome itself. You recognize that your behavior brings credit or dishonor not only to the city in which you live but also to the distant city in which the emperor reigns. But I am telling you that you have a connection with an infinitely more glorious city than Rome, where the Ruler of the whole universe sits enthroned. Your behavior had better show it!"

What does behavior befitting a citizen of heaven actually look like? Paul invokes the "soldiering" imagery to introduce the two key qualities that distinguish heaven's citizens. On the battlefield, survival and victory depend on *unity* with one's fellow soldiers and *courage* when confronting the enemy. Likewise, the main features of heavenly citizen-soldiers are (1) a selfless humility that fosters unity, and (2) a courageous confidence that withstands suffering. Paul introduces the importance of unity in Philippians 1:27, and he will spell out the details in 2:1–4. In most of the text before us (1:28–30), his stress falls on how the Philippians must respond to their opponents with unflinching confidence, grounded in their union with Christ by faith. Cultivating this odd combination of humility and courage, says Paul, will enable his friends at Philippi to stand in solidarity against the pressure of the surrounding society.

STAND ARM IN ARM IN SOLIDARITY THROUGH SELFLESS HUMILITY

The first quality that must characterize the citizens of heaven is unity. Paul hammers this point home in the successive phrases in Philippians 1:27: "in one spirit," "with one mind," and "striving side by side." Most English versions and commentators interpret "in one spirit [Greek *pneuma*]" and "with one mind [or soul] [Greek *psychē*]" as essentially synonymous expressions, calling us to a unified perspective and loyalty to one another—as the French say, *esprit de corps*. There are sound reasons, however, to capitalize the word *Spirit* as a reference to God's Holy Spirit, the divine Creator and Preserver of Christians' oneness of soul with one another.[6] Later in this same discussion, Paul will refer to believers' "participation in the Spirit," along with encouragement in Christ and love (from the Father) as the rationale provided by the triune God to enable and motivate us to have the same mind and love for one another (2:1–2). Paul's other uses of the Greek word *pneuma*, "Spirit," in Philippians (1:19; 3:3) are clear references to the Holy Spirit. In other letters, Paul's use of the construction "one Spirit" (*hen[i] pneuma[ti]*) refers to God's Holy Spirit as well (1 Cor. 12:13; Eph. 2:18; 4:4). God's Spirit is the powerful Protector in whom Christ's soldiers stand strong, and the Spirit is the divine Guardian of our unity of soul and mind.

Because God's one Spirit subdues our instinctive independence and individualism, Christians can and must "with one mind striv[e] side by side for the faith of the gospel" (Phil. 1:27). "Striv[e] side by side"[7] is derived from a root that sometimes refers to athletic competition. (The birthplace of the original Olympics lay some distance south of Philippi in Achaia.) Yet ancient Greek athletics developed out of military training for combat, as we still see today in sports such as the javelin, the hammer, the discus, and wrestling. Here, Paul is thinking in terms of mortal combat. He paints the picture of an advancing line of Roman legionnaires, their long shields forming both a seamless wall before them and a "roof" over their heads against the enemy's arrows and spears. His point is: "Don't let the opposition divide you! Instead, let the pressures from your opponents draw you together in a deeper and stronger unity!" He will develop this theme more fully in 2:1–4, even repeating the terms from 1:27 to make the point: "participation in the Spirit . . . being in full accord." Then he will make clear that only self-denying humility can maintain the deep unity that the triune God creates among believers: "Do nothing from rivalry or conceit, but in humility count others more significant than yourselves" (2:3). The pressures of social isolation, discrimination, ridicule, and rejection may threaten to undermine Christians' unity, as each person is preoccupied with private concerns or tempted to evade persecution by avoiding other believers (see Heb. 10:24–25). But

6. See especially Fee, *Philippians*, 163–66, many of whose arguments are summarized here. Ralph Martin also understands *heni pneumati* in Philippians 1:27 as a reference to the one (Holy) Spirit. Ralph P. Martin, *Philippians*, NCBC (Grand Rapids: Eerdmans, 1980), 83.

7. Greek *synathleō*, used again by Paul in Philippians 4:3 and nowhere else in the New Testament. Paul's one use of the uncompounded verb *athleō* invokes athletic imagery (2 Tim. 2:5).

citizen-soldiers who serve heaven's King, Jesus, will not compete as rivals or withdraw in introspective self-pity, each licking his or her own wounds and ignoring others' pain. Rather, their costly compassion for others, their humble honoring of others, will reflect the humility and compassion of the Lord and Savior who reigns in their true city. *Selfless solidarity!*

Paul himself is an object lesson in selfless solidarity. He rejoices in his chains because they have given the gospel access into Rome's military elite, the Praetorian Guard. Others are stirred, whether by love or by rivalry, to proclaim Christ throughout the imperial capital, where Paul could not go. This, too, gives Paul joy. Anticipating his hearing before the emperor, Paul looks forward to enhancing the honor of Christ, whether the outcome proves to be death or life. Though death would usher Paul immediately into his Savior's presence, life would enable Paul to help others. So Paul prefers the painful path of further service, deferring his face-to-face reunion with King Jesus in order to bolster others' faith.

Having shown them what it looks like to behave as a citizen of heaven in a hard spot on earth, Paul now says: "You do the same. Stay together; keep united by putting your personal agenda on the back burner. Better yet, take your safety and comfort, your reputation and self-respect out of the picture altogether."

Putting other people's needs and agendas before our own does not come naturally to any of us. Yet we can see how necessary it is, if the followers of Christ are going to maintain a united front amid an unsympathetic or hostile environment. When the pressures mount, whether because the world is out of whack or because others take offense at your commitment to Christ, it is easy to turn in on yourself in self-pity. Our self-protective instinct is to focus in on our own hardships and to feel that we just have no emotional energy to care for others' problems. Since, as Paul says, the Philippians were facing sufferings similar to his, was it insensitive of Paul to lay upon them so superhuman an obligation: "Care more for each other than each of you cares (so naturally, so instinctively) for himself"? The deep and self-forgetful love that genuinely puts others first is just not the way we're wired at birth. So where could we get it?

Such love is found in the gospel message, through which heaven's Lord enrolled us as heaven's citizens. That is Paul's point in describing the behavior befitting our citizenship as conduct that is "worthy of the gospel of Christ"

(Phil. 1:27). As we'll see, the Lord who stooped to be our Servant (2:5–11) can turn hearts inside out, enabling us to care for others even when we ourselves are under attack. In response to Christ's good news, and in the strength of Christ's one Spirit, heaven's citizen-soldiers can maintain solidarity with one another through humility and heartfelt compassion.

But this selfless humility is by no means spineless timidity or quivering fear!

Stand Face to Face with Your Opponents in Calm Endurance

As an officer in Christ's army, Paul wants to see or hear that the believers of Philippi are "standing firm," an order that he will issue again in Philippians 4:1, surrounding it with warm affection: "Therefore, my brothers, whom I love and long for, my joy and crown, stand firm thus in the Lord, my beloved." The martial overtones of this term's summons to steadfast resistance can be heard in Paul's bullet-point orders in 1 Corinthians 16:13: "Be watchful, stand firm in the faith, act like men, be strong." To the Philippians Paul bolsters his call to courage by issuing a preemptive strike against their fears: they must not be "frightened in anything by your opponents"[8] (Phil. 1:28). Paul bypasses his customary word for *fear*[9] to use a term that evokes a vivid scene of terror, like spooked horses heedlessly stampeding,[10] or an army breaking ranks and fleeing pell-mell in retreat. The Philippian believers must not succumb to the intimidation posed by the opponents' aggression, scattering in panic or silenced by fear.

Paul has the credibility to issue this call to stand fast. In verse 30 he reminds them that they and he are truly comrades-in-arms, since they are "engaged in the same conflict that you saw I had and now hear that I still

8. The identity of these opponents is debated by scholars. Some (Hawthorne and Martin, *Philippians*, 72; tentatively Silva, *Philippians*, 82–83) identify them with aggressive advocates of Judaism, against whom Paul issues a sharp warning in Philippians 3:2, calling them dogs, evildoers, and mutilators. Yet the comparison that Paul draws in Philippians 1:30 between his own conflict and the Philippians' suffering suggests that the opponents in view in chapter 1 threatened physical harm and coercion, whereas the danger counteracted in chapter 3 was theological.

9. Greek *phobeomai*, which he uses both in the positive sense of reverence toward God (Col. 3:22) and authorities appointed by God (Eph. 5:33) and in the negative sense of feeling apprehension about threats that do not warrant our fear (Gal. 2:12).

10. Paul's term here, rendered "frightened," is *pturō*, which "could denote the uncontrollable stampede of startled horses." O'Brien, *Philippians*, 152. See also J. B. Lightfoot, *Saint Paul's Epistle to the Philippians* (1913; repr., Grand Rapids: Zondervan, 1953), 106; Hawthorne and Martin, *Philippians*, 72.

have." He is a veteran in the spiritual conflict in which every follower of Christ is engaged, and he bears the scars of battle. He closed his letter to the Galatians with the reminder, "I bear on my body the marks of Jesus" (Gal. 6:17). These marks no doubt include the scars of beatings such as the one that he and Silas endured before the Philippians' eyes (Acts 16:23, 33; see 2 Cor. 11:23–25). He bolsters their courage with a stirring pep talk, saying, in effect, "Yes, you have opponents. So do I. But I assure you that you do not need to be intimidated by anything your opponents throw at you, figuratively or literally!"

American Christians, on the whole, are such wimps! We whimper over inconveniences and complain when we are slightly disrespected, forgetting that our brothers and sisters throughout history and across the globe even today experience far greater suffering for the faith on an ongoing basis. Since Paul called his Philippian friends to stand fast and fearlessly, surely Christ's Spirit says to us, "No sniveling. No quivering. No cringing. Look your enemies in the eye and tell them calmly, 'Do your worst to my body, but you will not win the real battle. I will *not* succumb to fear.'"

How can we gain such unflinching courage? Knowing that we are citizens of heaven gives us a glimpse of victory still ahead. Heaven's citizen-soldiers "await a Savior" who will return from heaven, the Lord Jesus Christ; and when he comes, he "will transform our lowly body"—the body that current enemies can now harm and destroy—"to be like his glorious body" (Phil. 3:20–21). Although it seems now that the church's opponents have the upper hand, their current persecution of believers, whether severe or mild, is a harbinger of a radical reversal to come. It is actually "a clear sign" of the enemies' own impending destruction and of believers' eventual rescue by God himself (1:28). That preview of the end of the battle must not make us smug, however. It makes us calm, but not arrogant. The very fact that our salvation is "from God" (1:28) reminds us that our eventual victory is nothing that we can take credit for: "He who began a good work in you will carry it on to completion until the day of Christ Jesus" (1:6 NIV). Paul reminded Titus that "we ourselves were once foolish, disobedient, led astray, slaves to various passions and pleasures, passing our days in malice and envy, hated by others and hating one another." Because God saved us "according to his own mercy," we must "show perfect courtesy toward all people," even those who aggressively oppose or arrogantly marginalize us (Titus 3:2–5).

The rage of the gospel's enemies marks them as people heading for eternal destruction. That sober reality must move our hearts not to thoughts of revenge, but to patience and to pity.

This blend of boldness and humility, confidence and selflessness is a rare combination. How can we behave in ways befitting our heavenly citizenship, in ways that bring credit to our heavenly Emperor, when that way of reacting to hostility and trouble is so alien to the way that we are naturally wired? By grace!

Two Gifts of Grace

The basic thrust of this paragraph is a command, a duty that falls to those who are citizen-soldiers of heaven. Yet grace saturates Paul's call to courageous solidarity. Twice in verse 27 Paul mentions "the gospel." There is a pattern of conduct that must characterize heaven's citizens because it is "worthy" of the gospel (Phil. 1:27). But the gospel is not a set of responsibilities to be fulfilled. It is good news to be believed, the joyful report of a mission accomplished on our behalf despite our unworthiness and helplessness. Therefore, Paul speaks of "the faith of the gospel," and concludes his charge by mentioning the ability to believe in Christ as one of two gifts of grace that come from God (1:29–30). These two gifts—faith and suffering—are Paul's explanation of his concise but momentous assertion that our final salvation will be "from God" (1:28).[11]

Paul's concluding reassurance in verse 29 is truly amazing: "For it has been granted to you that for the sake of Christ you should not only believe in him but also suffer for his sake." The opening "For" shows that this twofold gift reveals Paul's meaning in saying that the Philippians' salvation would be "from God." The gifts granted to them have been bestowed by God himself, and this truth instills both gratitude and confidence. Moreover, out of several Greek words available to express the idea of "giving" or "granting," Paul has chosen the one that has *grace* at its heart. The Greek word for *grace*

11. Of course, it is not only our eventual and complete salvation that finds its source and guarantee in God, though that future consummation is Paul's focus of attention here. Our salvation began by grace through faith, not generated from within ourselves but imparted by God's gift (Eph. 2:8–9). We are currently working out our salvation—persisting in faith and practicing love—in the confidence that God is at work in us, imparting both the desire and the power to live lives worthy of the gospel, pleasing him (Phil. 2:12–13).

is *charis*, and you can hear it in the word *charizomai*, which enters English as "it has been granted." Grace, divine favor *in spite of* what we deserve, is the source of these gifts. Our wonder at divine grace is further amplified as we are reminded that the gifts we have received are "for the sake of Christ." God's twofold gift comes into our lives not because we have done anything to earn or deserve it, but because God loves his Son Jesus and *for his sake* lavishes grace on people like us, who belong to his Son.

The first aspect of God's twofold gift is *faith*: "it has been granted to you that . . . you should believe in [Christ]" (Phil. 1:29). This faith that brings salvation involves transferring our trust away from our own good intentions, resources, and achievements. It is relying instead on Jesus and what he has done. Paul will soon narrate the mission of Christ, who, though equal with God, humbled himself, taking human nature and the slave's role, offering flawless obedience and enduring the cursed death of the cross, subsequently to be exalted to the heights by God the Father (2:5–11). Paul will go on to affirm that Christians rely no longer on our own righteousness but on "that which comes through faith in Christ, the righteousness from God that depends on faith" (3:9). The Westminster Confession aptly says that "the principal acts of saving faith are accepting, receiving, and resting upon Christ alone for justification, sanctification, and eternal life."[12] Moreover, we claim no credit even for this faith, for the salvation that comes by grace through faith "is not your own doing; it is the gift of God" (Eph. 2:8).

Yet Paul's emphasis in our text is not on the first aspect of God's gracious gift, but on the second: "but also *suffer* for his sake" (Phil. 1:29). Here Paul's perspective takes us aback. It is humbling but also heartening to be told that our *faith* is not our independent contribution to our salvation but is rather God's gracious gift. To discover that my faith comes from God encourages my assurance, my astonished love, and my overflowing thanks. But can we believe that our *suffering* is a gift of God's grace? The whip's lash, shackles on ankles, shipwreck, exposure, hunger, and the threat of the sword—can these really be received as tokens of God's grace?

Yes, exactly. As you realize more deeply how God used Jesus' suffering to bring you everlasting joy, you begin to have surprising reactions to your own suffering. You see suffering as providing opportunities to bring Jesus

12. WCF 14.2.

even more glory, by the comfort that his Holy Spirit brings. In the church's earliest days, the apostles were imprisoned, interrogated, threatened, and beaten by the high priest's colleagues in Jerusalem. Upon their release, they were actually "rejoicing that they were counted worthy to suffer dishonor for the name" of Jesus (Acts 5:41). Paul, too, has exhibited this very mind-set in describing both his current imprisonment and his impending trial, which may bring either swift death or release to continue his ministry. Whatever the outcome, Paul eagerly anticipates the Spirit's strength to save him from shame and enable him to bring glory to Jesus—*amazing grace*!

Because Christ's death has reconciled us to God, his grace sets us free from the fear of death. Death remains the last enemy, to be eliminated altogether only when Jesus returns in glory. But death's sting has been nullified in Christ's cross, so death can do no more than reunite believers with the Lord who loves us: to depart and be with Christ is "far better" (Phil. 1:23). Because God the supreme Judge has inflicted the condemnation that we deserve on his beloved Son, his grace frees us from fear of rejection or ridicule from mere human beings. The only Evaluator whose opinion actually counts has welcomed us, not because we are intrinsically worthy but because we are robed in Jesus' righteousness.

Yet the very grace that silences our fears at the same time shatters our pride, our edgy competitiveness, our bristly impatience with our fellow soldiers, with whom we stand, shoulder to shoulder, in the King's army. Their citizenship and ours in the Lord's heavenly city are free gifts of his grace, not awards for our valor. We can and must stand strong in solidarity, not because we have something to prove about ourselves but because we have a Champion who is worthy of our wholehearted allegiance. When you find your Christian brother or sister frustrating—perhaps his agenda for the church conflicts with yours, or her thoughtless comment grieved your heart—then you will be tempted to see that fellow Christian as the enemy and to withdraw from him or her. That is precisely when you need to remember two truths: first, that your real Enemy, Satan, is intent on disrupting believers' unity with each other (Eph. 4:26–27; 6:10–12); and second, that you both stand together *in need of* God's redemptive grace, and *as recipients of* that grace through Jesus. Humbled by grace, then, take the painful initiative to speak truth in love to the one who has hurt you (or to go in humble repentance to the one whom you have hurt), striving to keep the unity of the Spirit in the bond of peace (Eph. 4:3).

Surrendering to God's grace in Christ produces in us an unusual reaction to opposition and our marginalized position in society. Grace makes us confident and humble at the same time. In *The Reason for God*, Pastor Timothy Keller of Redeemer Presbyterian Church in New York confesses:

> When my own personal grasp of the gospel was very weak, my self-view swung wildly between two poles. When I was performing up to my standards—in academic work, professional achievement, or relationships—I felt confident but not humble. I was likely to be proud and unsympathetic to failing people. When I was not living up to standards, I felt humble but not confident, a failure. I discovered, however, that the gospel contained the resources to build a unique identity. . . . The Christian gospel is that I am so flawed that Jesus had to die for me, yet I am so loved and valued that Jesus was glad to die for me. This leads to deep humility and deep confidence at the same time. It undermines both swaggering and sniveling.[13]

Here Paul applies this power of the gospel to a group of people who feel the pressure of being marginalized by the society around them, people who might be tempted to retreat in fear, or to lash out in retaliation, or to vent frustration on each other. But Jesus offers another avenue of response: confidence grounded in God's grace that frees you to react to opponents with calm kindness, and to failing fellow believers with humility and forgiveness.

Citizens of heaven, behave in ways befitting the character of your King, who rules now and will return in glory. By his transforming grace, show courageous humility, bold gentleness, and selfless solidarity, calmly enduring all that this decomposing culture can throw at you. All the while, invite the very people who would intimidate you to share in the gifts of the King's grace, to join you in believing in him and in suffering for his sake.

13. Timothy Keller, *The Reason for God: Belief in an Age of Skepticism* (New York: Dutton, 2008), 180–81.

7

Hearts Turned Inside Out

Philippians 2:1–4

So if there is any encouragement in Christ, any comfort from love, any participation in the Spirit, any affection and sympathy, complete my joy by being of the same mind, having the same love, being in full accord and of one mind. (Phil. 2:1–2)

For such a brief epistle, Paul's letter to the Philippians contains many majestic mountaintops of Christian devotion and truth. His cheerful assessment of his imprisonment and possible execution reaches such a summit when he proclaims, "To me to live is Christ, and to die is gain" (Phil. 1:21). In Philippians 2:5–11, the apostle guides our climb to the heights of mystery in the incarnation, humiliation, and glorification of Christ. The third chapter displays a Mount Everest of gospel grace, as Paul recounts his trajectory out of self-produced, Torah-defined righteousness and into glad surrender to God's gift of righteousness by faith (3:3–11). Other memorable pinnacles rise above the clouds: "I press on toward the goal for the prize of the upward call of God in Christ Jesus" (3:14), "rejoice in the Lord always" (4:4), "I can do all things through him who strengthens me" (4:13), and others.

By contrast, Philippians 2:1–4 may seem, at first glance, like dusty flatlands surrounded by snow-covered ranges of "purple mountain majesties." Southern Californians see just such a view as they drive north from Los Angeles on Interstate 5, over the Tejon Pass, and then down into the San Joaquin Valley. There are green forests on the coastal range to the west, but they are hidden by the valley's haze. To the east, now and then, we may glimpse the faint shapes of the Sierra Nevada's snow-capped peaks. Straight ahead, however, lies a long, straight ribbon of multilane asphalt, surrounded by flat acres of furrows and furrows of crops soaking in the blazing sunlight. The central valley that runs through California like an out-of-kilter backbone is one of America's most fertile breadbaskets, but few (except farmers) would call it a scenic drive. Likewise this text, which lies between "To me to live is Christ, and to die is gain," on the one hand, and "At the name of Jesus every knee should bow," on the other, seems to bring us down from soul-stirring heights to the flatlands of everyday life. Here we read instructions such as "get along with each other, stop being selfish, care as much about others' concerns as about your own." So mundane, so down to earth.

I suggest, however, that the mountain ranges on either side of our text are just as down to earth, just as grounded in the flatlands of everyday experience with its heat and its hassles. After all, Paul writes momentous words such as "To me to live is Christ, and to die is gain" while he is chained to a succession of Roman guards. In fact, he has mentioned his real-world "conflict" in Philippians 1:30 just before our passage. Likewise, Paul is about to show that the Jesus whom every creature will acclaim as Lord became "obedient to the point of death, even death on a cross" (2:8). Nothing could be more ugly and down to earth than a cross. The very word *cross* brought shudders to Rome's subjects, evoking images of bloodied beams and the stench of death. The Christ who is exalted above all knows by experience the dirty, ugly, painful reality of life in this fallen world. The apostle Paul is being strategic when he places his own experience in prison on one side of our text and the Lord Jesus' experience of suffering on the other. Paul's chains and Christ's cross set the context in which Paul's friends in Philippi should process their own situation.

After all, those Philippian followers of Jesus were not savoring a mountaintop experience. They were not basking in fresh air and clear skies above the smog line of their society. They were a marginalized minority, pushed to the edges of Philippi's interpersonal, economic, and political networks. They had opponents whose antagonism tempted them to terror, so Paul urged them to stand fast and not to succumb to intimidation (Phil. 1:27–30). Paul also urged the Philippians to pursue and maintain their unity: "in one spirit," "with one mind striving side by side" (1:27), "of the same mind, having the same love, being in full accord and of one mind" (2:2). Not only did he stress this repeatedly at this point in the letter, but he also returned to the theme of unity later, in a personal word of appeal to two treasured coworkers (4:2). Of course, we should not exaggerate the threats to unity in the Philippian church. There is no evidence that this congregation suffered from the competition and sharp schism that had troubled the congregation in Corinth, to the south (1 Cor. 1:10–17). Nevertheless, it does seem that Paul's prescription—his strong appeal for unity—implies his diagnosis of their spiritual malady: these Christians were at least tempted to turn in on themselves, each licking his or her own wounds, and some might even have been turning on each other in competition and conceit. Just as wise physicians do not prescribe antibiotics for nonexistent maladies, so Christ's apostle would not say, "Do not be frightened," if fear and intimidation posed no danger to his friends. He would hardly waste his breath forbidding "rivalry or conceit" (Phil. 2:3) and urging each to care for others' interests unless some, at least, were focusing on their own concerns and agendas, neglecting those in need, and jeopardizing the congregation's unity. Paul would not amass reasons for staying unified in heart and mind, as he does in Philippians 2:1, unless the oneness of conviction and affection in this church at Philippi, whom he loved so dearly, faced a real threat.

We have seen that in Philippians 1:27 Paul turns the topic of conversation from his situation in Rome to his friends' situation in Philippi. He summons them to conduct befitting citizens, to display values and behavior that reflect the city that defines their identity and gives them significance. When he states that their conduct as citizens must be "worthy of the gospel of Christ," he implies—as he will make explicit in 3:20—that the city that defines their dignity and duty is not Rome but heaven, where their Savior and Lord Jesus Christ reigns and from which he will return. With the privileges of citizenship Paul introduces the related responsibility of military service in such terms as "standing firm" and "striving side by side for the faith of the gospel." Christians are citizen-soldiers of heaven, charged to *stand fast* in the face of threats from the gospel's enemies, but also to *stand together*, shoulder to shoulder, in mutual compassion and support.

Followers of Jesus, the Lord who stooped to servitude in order to conquer our Enemy, must therefore exhibit a rare combination of qualities. We are called to be bold and humble, fearless and selfless. Heaven's citizen-soldiers must be unflinching in the face of opponents' aggression but gentle and openhearted toward our Christian comrades-in-arms. In Philippians 1:27–30, Paul has emphasized the need for fearless courage in the face of opposition and suffering. Now he focuses on Christians' relationships to each other, showing how imperative it is that we "fight for" unity of conviction and affection through humility of heart. Both themes—courage toward hostile non-Christians and compassion toward annoying fellow Christians—show us what it means to fulfill our calling as citizens of heaven.

Paul leads up to his summons to oneness of mind and heart by rehearsing the abundant reasons that the triune God has provided to motivate and empower believers to stand together in solidarity with one another: the encouragement of Christ, the love of the Father, and the participation in the Spirit, expressing the triune God's affection and sympathy for us (Phil. 2:1). He then describes the unity that will complete his joy, as his dear friends approach that goal (2:2). Finally, he profiles the mind-set of humility and compassion for others that will turn the Philippians' hearts inside out, and so enable them to stand in solidarity (2:3–4). In our consideration of Paul's summons to oneness in mind and heart, we will slightly rearrange Paul's order, focusing first on the goal and then on the mind-set that leads to the goal, and concluding with the irresistible rationale that God's grace provides to sustain our persistent pursuit of fearless, selfless love.[1]

1. In discussing the "indicatives" of Philippians 2:1 last, closing with these truths that provide the rationale and motivation for Paul's "imperatives"—the exhortation to unity and to the others-focused humility that promotes unity—I am following the interpretive strategy of Moisés Silva, *Philippians*, 2nd ed., BECNT (Grand Rapids: Baker, 2005), 85–88. Although Philippians 2:1–4 follows Paul's customary epistolary structure, laying "indicative" foundations first and then showing the "imperative" implications that flow from the gospel, in this case it makes homiletical sense to begin by clarifying Paul's expectation of unity through humility, which so radically contradicts our instinctive self-centeredness, and then to apply the heart-changing force of the grace of the triune God to inflame both our desire and our hope for exhibiting this mind-set, which is ours in Christ. In Ephesians 4:1–6, another text advocating Christian unity, Paul moves from the imperative (4:1–3) to the supportive indicatives (also structured in terms of the persons of the Trinity!).

Unity of Conviction and Affection

Although our English versions find it necessary, in the interests of clarity, to subdivide this four-verse passage into several sentences, in Greek the entire text is a single sentence. Its only imperative verb is "complete my joy," and the verbs in the subordinate clauses and phrases that follow ("being of the same mind, having the same love, being in full accord . . . count others . . . look") serve to elaborate how the Philippians can complete Paul's joy through harmonious interpersonal relationships and the humble mind-set that would preserve their unity (Phil. 2:3–4).

At first hearing, "complete my joy" might sound strangely self-centered as a lead-in to Paul's directive that his friends should *not* think first about *their own* joy or happiness or comfort or convenience, but rather care more about others than about themselves. We would be misreading Paul, however, if we took him to be saying, hypocritically, "You need to be selfless and sacrificial, but I do not. Forgo your happiness so that I can be happy." In fact, he is hinting that his own experience and example show that there is a joy that is richer, deeper, and sweeter than the satisfaction of self-centered desire, whether that desire is for comfort, pleasure, reputation, or achievement. Christ's grace has turned Paul's heart inside out, so that his joy is now bound up in seeing his Christian friends grow more like Jesus. In Philippians 4:1, he will actually call them "my joy and crown," brothers and sisters whom he loves and longs for. We have already heard him say that as long as Christ is preached, he doesn't care who gets the credit: "I rejoice . . . , and I will continue to rejoice" (1:18 NIV). Writing from the "comforts" of captivity, he has announced that his most passionate desire is that "Christ will be honored in my body, whether by life or by death" (1:20). It is in that down-to-earth situation of discomfort and risk that he issues the ringing affirmation, "To me to live is Christ, and to die is gain" (1:21). Paul's "mountaintop" is not on a snow-covered peak of the Sierras; it's down in the muggy humidity of a Mediterranean metropolis.

So Paul sincerely and credibly calls his friends away from a mind-set dominated by "rivalry [and] conceit" (Phil. 2:3). Another version amplifies our understanding of these terms by rendering them "selfish ambition or vain conceit" (NIV). The gospel of Jesus has impressed on Paul the counterintuitive truth that the pursuit of happiness, when fueled by selfish ambition, is bound to end in bitter disappointment, whereas the highest, strongest joy surprises and overtakes those who find their hearts so drawn to others' well-being that their personal comfort and pleasure slip from their view. Surely the opinion-makers of the ancient world could not see Paul's point. After planting churches in Philippi and elsewhere in Macedonia, the apostle traveled south to Athens and encountered its urbane intelligentsia, notably Epicurean and Stoic philosophers.[2] Ancient Epicureans were not mere addicts to sensual indulgence; rather, they sought a sensible moderation that would maximize pleasure while minimizing its painful consequences.[3] Still, the aim was all about the shrewdest strategy for achieving one's own contentment. Stoicism offered the ideal of a self-sufficient contentment in which the man of superior intellect protects himself from emotional or physical distress by cultivating *apatheia*, aloofness from feeling and therefore from suffering. In our day, too, the consensus of those who mold minds—whether in the rarefied atmosphere of the academy, in the grasping ambition of Wall Street and Washington, or in the appetites aroused through entertainment and marketing—is that happiness comes to those who pursue it for themselves at all costs, no matter who else gets hurt or left behind in the hunt.

For Paul, on the other hand, his personal escape from the miseries of being marginalized and abused by a hostile world did not matter so much, when compared to the reputation of Christ and the needs of Christ's church. So Paul's single direct imperative in the long, complex sentence that makes up Philippians 2:1–4 could be paraphrased: "Because my heart embraces yours, my fullness of joy depends on your growing unity with one another. So it pains me to hear that you have tensions among you, and to realize that those tensions are the result of selfish ambition and vain conceit. Even if the way you are wounding each other does not bother you, if you care about me, swallow your pride, break out of your prison cell of self-centeredness, and humbly treat each other as more significant than yourselves. *This* is what will complete my joy!"

Paul's surprisingly selfless "complete my joy" poses a challenge both to pastors and to church members. Do we who are pastors find our hearts so passionately invested in those whom we serve that we can honestly say that

2. "The Epicureans and Stoics were the chief rivals for the allegiance of educated men in the Hellenistic Age." Everett Ferguson, *Backgrounds of Early Christianity* (Grand Rapids: Eerdmans, 1987), 301.

3. Ibid., 299–300.

our joy finds completion as we see them maturing in unity with each other by the grace of the gospel? And do all of us who have pastors and elders who care for our spiritual well-being hear in Paul's words our own shepherds' heart cries? Has it occurred to us that pastors lose sleep over the friction between Christians? If we love our shepherds, we must complete their joy by loving each other, relinquishing our rights, caring compassionately for brothers and sisters with whom we have differences, and humbly honoring each other.

The heaven-befitting behavior that brings full joy to Paul's heart—to Christ's heart—is the goal that Paul profiles in four terms in verse 2: "being of the same mind, having the same love, being in full accord and of one mind." These four descriptions focus on two dimensions of unity: unity of conviction and unity of affection.

One crucial dimension of the unity for which we should strive is oneness of *conviction*. The first and the last descriptions of Christian unity here focus on what we think and believe: "being of the same mind" and "of one mind." These contain the same Greek verb, *phroneō*, which appears no fewer than ten times in Philippians, more frequently than in any other epistle of Paul. Although it is hard to translate into English, our word *think* or the idea "have a certain mind, mind-set, or attitude" is a close equivalent. A wooden translation of these phrases in Philippians 2:2 would be "that you think the same thing . . . thinking the one thing," or, more interpretively, "that you adopt the same viewpoint . . . adopting one viewpoint." Paul will use this word again in verse 5 when he says that our "mind" or attitude must reflect the perspective that is ours because we are "in Christ Jesus." Jesus, who by God's grace now defines our identity, is the fullest display of that "mind" that does not cling to rights and privileges, but is willing to pour out oneself utterly to meet others' needs (2:6–8).

As we grow in our grasp of God's Word, increasing clarity in our understanding will also draw us together. We will see the truth more clearly, and therefore we will see eye to eye more often. When Paul writes to the Ephesian church, he says that God's goal for us is that we will "all attain to the unity of the faith and of the knowledge of the Son of God, to mature manhood, to the measure of the stature of the fullness of Christ" (Eph. 4:13). True doctrine, far from dividing believers into rival theological camps, actually promotes and preserves true Christian unity! We are not there yet, but we long for that complete unity of conviction and confession.

Because none of us has yet reached full and clear knowledge of Christ, we need not and must not make our oneness contingent on 100 percent agreement on every issue, large and small. Later in this letter, Paul himself will admit that he has not yet arrived at the final goal for which he is striving, and he urges all who are mature to "think [*phroneō* again] this way." Then, demonstrating how this humble recognition that we are all "in process" applies to the differences of conviction that remain among us, he comments, "If in anything you think otherwise, God will reveal that also to you" (Phil. 3:12–15). United in the essentials of the gospel, we can wait patiently and expectantly for God himself to bring our minds into unity on other matters, as we continue to search the Scriptures together, humbly and prayerfully.

The unity among his Philippian friends that will fill up Paul's joy includes growing theological agreement, but it goes beyond such consensus of conviction. Between the two "thinking"-oriented expressions in Philippians 2:2, Paul inserts two terms that focus on believers' *attachment in affection*: "having the same love, being in full accord." Actually, "being in full accord" represents a single, rare Greek word, *sympsychoi*, which speaks of souls in harmony with one another. Our English expression *soulmates* captures this wonderful word well. It's as if Paul is saying, "It is not enough to agree with each other theologically: God actually calls you to *care for* each other deeply, in a love that binds your souls together so strongly that differences of perspective cannot pull you apart." This strong bond of affection, grounded in the truth of the gospel, stabilizes believers' relationships with each other so that they can address their differences—whether doctrinal or interpersonal—in patience, humility, and love.

Have you noticed that at times Christians may agree on almost every detail of doctrine, but still not get along interpersonally? Theologically, there may be no more than a hair's breadth of difference between them. But personally, they grate on each other, or simply ignore each other. Was that what was going on with Euodia and Syntyche at Philippi (Phil. 4:2)? Paul commends their faithful service to the gospel, and mentions no doctrinal disagreement that needs his apostolic adjudication. Yet he must entreat each of them, in the same terms that he addresses to the whole church in our present text, "to agree in the Lord"—literally, "to *think the same thing* in the

Lord" (*to auto phronein*, virtually replicating the wording rendered "being of the same mind" in 2:2). Do you ever have such points of friction with folks with whom you agree in all the big points of doctrine and almost all the small points, too? The unity that fills Paul's heart and, more importantly, Jesus' heart with joy certainly encompasses our common conviction and aims for our full agreement in God's truth, but it goes beyond conviction to bind our souls together in deep affection and sacrificial love.

Paul has summoned heaven's citizen-soldiers to fight hard for our unity, standing shoulder to shoulder to resist forces that would divide us (Phil. 1:27). Where are those hostile, divisive forces to be found? Paul has called for courage in the face of outside opponents (1:28–30). No doubt the pressure that external persecution exerts can widen the cracks in believers' relationships with each other, yet the more insidious enemy to Christian unity does not lie outside the church or its members, but in each of our hearts. That is why he moves on from his general exhortation to unity (2:2) to contrast the heart attitudes that either damage or deepen our unity in Christ (2:3–4).

Replacing Self-Centered Pride with Others-Centered Humility

In Philippians 2:3–4, Paul moves disconcertingly from comfortable theorizing to annoying meddling. In the abstract, we all admire unity of conviction and affection in any community or society. But Christ's apostle is not content to articulate a pleasant ideal that wins merely sentimental and superficial acceptance. He unpacks his command "complete my joy" by probing the motives of our hearts, the deep and driving forces that either undermine or reinforce harmony with others in the body of Christ. And he starts with the negative because he knows that this congregation, which he loves so dearly, actually exhibits the symptoms that he now calls by name: "rivalry or conceit" (ESV), or "selfish ambition or vain conceit" (NIV). Paul would not waste papyrus and ink to name these unpleasant motives if they were not a real threat to the Philippian Christians.

Paul used the first term, "rivalry" or "selfish ambition," in Philippians 1:17, when he described the motives of some believers who were spreading the gospel in Rome, as he sat in chains. The word appears only seven times in the New Testament, five of which are in Paul's letters. Its contexts suggest that it encompasses primarily a deep-seated desire for personal preeminence and then, secondarily, the interpersonal competitiveness that results from such self-centeredness.[4] Although many became bolder in expressing their faith out of love for Paul and a desire to extend his gospel witness during his own captivity, there were others whose evangelistic zeal was driven by a desire to garner attention for themselves, hoping to make Paul envious of their freedom of movement and effectiveness in ministry. No doubt the Philippian believers, who had demonstrated their deep affection for Paul in past gifts and Epaphroditus's current service, were shocked to hear of the shameful competitiveness, the "selfish ambition," of those who sought to compound the pain of Paul's chains. Did they marvel that those speaking of Christ and his grace could exploit the gospel to further their personal delusions of grandeur and place their beloved Paul in the shadows?

Now, however, Paul turns the spotlight on the Philippian Christians themselves and uses the same term to diagnose the cause of their own interpersonal friction or indifference. He challenges each of them—and us—to face the question: "*Why* do I do what I am doing in service to God in the church, in my home, in my workplace? Am I driven by self-centered motives, even when I am supposedly helping others? Am I self-serving even while serving others, wanting and hoping to be noticed, so that I receive the appreciation and recognition that I think I deserve? Whether I express it outwardly or not, do I nurse resentment when my hard work is ignored or my brilliant ideas are not followed?"

There are more blatant forms of self-centeredness, of course: greed for more stuff and indifference to the financial needs of others; demanding one's own way rather than being open to compromise; expecting people to jump at one's bidding instead of looking for ways to serve them. But Paul's linking of "selfish ambition" (NIV) with "conceit" (ESV) or "vain conceit" (NIV) shows that he is most concerned about the self-centeredness that seeks credit and praise for what one has done. The Greek word was

4. The Greek term is *eritheia*, which appears among other strife-related terms in lists of vices in 2 Corinthians 12:20 (ESV: "hostility"); Galatians 5:20 (ESV: "rivalries"); and James 3:14, 16 (ESV: "selfish ambition"). See also Romans 2:8, where the ESV translates "those from *eritheia*" as "those who are self-seeking." *BAGD*, 309, notes that the meaning of *eritheia* in Paul's usage is "a matter of conjecture." The meaning "*strife, contentiousness* . . . cannot be excluded. . . . But *selfishness, selfish ambition* . . . in all cases gives a sense that is just as good, and perh[aps] better."

compounded of *empty* and *glory*, and its components unlock its meaning.[5] Believers who eagerly await Christ's return from heaven anticipate the joy when he "will transform our lowly body to be like his *glorious* body" (Phil. 3:21), so we "rejoice in hope of the *glory* of God" (Rom. 5:2). On the other hand, the mind-set that is unworthy of heaven's citizens is "*empty* glory," the pursuit of honor and admiration that is *void, hollow, and baseless* because it is self-focused. Paul is about to weave the two pieces of "vainglory" into his narrative of Christ's redemptive mission, first tracing the condescension by which the eternal Son "made himself nothing [*kenoō*]"—through incarnation, humiliation, and death—and then recounting the Servant-King's exaltation "to the glory [*doxa*] of God the Father" (2:7, 11). The life of the Savior shows that self-*emptying* service is the route to real *glory*, whereas self-centered ambition yields only "empty glory," a hollow shell. We need no graduate degree in psychology or the social sciences to realize that when our hearts cherish longings such as these—when we crave credit for all the good and sacrificial things we have done—such appetites breed competition, and competition undermines our unity of conviction and affection with one another.

So what must replace selfish ambition and vain conceit as the driving force of our desire? The alternative is the most unnatural attitude imaginable: "in *humility* count others more significant than yourselves. Let each of you look not only to his own interests, but also to the interests of others." The Greek word rendered "humility" is, literally, a "lowly mind-set" (*tapeinophrosynē*). It is not an unrealistically dour inferiority complex, which at bottom is as self-centered as glib pride is. It is, rather, a readiness to forget oneself and to exalt others both with respect and with concern. It is to have the grace of the Holy Spirit so turn our hearts inside out that we eagerly honor and care for others, as we instinctively do for ourselves.

This is not the way we are naturally wired.[6] Greco-Roman society was frank to admit this. The pagan authors of Paul's era used the term *humility* derisively, to describe "weakness, lack of freedom, servility, and subjection,"

5. Greek *kenodoxia*, which appears only here in the New Testament and a handful of times in the LXX in the apocryphal books Wisdom and 4 Maccabees. Paul uses a related adjective, *kenodoxos*, once, in Galatians 5:26: "Let us not become conceited."

6. Writing to the Ephesians, Paul appeals to our instinctive self-centeredness to show husbands what loving their wives really means: "No one ever hated his own flesh, but nourishes and cherishes it" (Eph. 5:29).

as one scholar put it.[7] The term "conveyed the ideas of being base, unfit, shabby, mean, of no account."[8] Masters might look for such obsequious self-deprecation in slaves, but no self-respecting Roman citizen should exhibit such a low view of his own significance! Yet Paul, building on the Old Testament promise of God's grace to the humble (*tapeinos*, Prov. 3:34) and captivated by the Lord who described himself as "lowly in heart" (*tapeinos*, Matt. 11:29), unashamedly commends "lowly-mindedness" as the hallmark of heavenly citizenship. Paul will soon say that Christ Jesus, heaven's King, "humbled [*tapeinoō*] himself by becoming obedient to the point of death, even death on a cross" (Phil. 2:8). This gracious and surprising descent of the Lord of glory to the shame of the cross has upended Rome's whole scale of social values, and those who have been rescued from sin and death by the servitude of God's Son have the honor of expressing his lowliness in selfless service to others.

The others-embracing, others-serving mind-set of Christ is so unnatural to our self-preserving instincts. Yet when God's grace grasps us deeply, it begins to develop into our deepest, strongest desire. We begin to care for all our brothers and sisters in Christ with the same passionate intensity that we so automatically and easily lavish on our own comforts and concerns. We learn to give as much weight to their opinions as we do to our own. We start feeling their disappointments and pains and grief as intensely as we do our own. We put their needs before our own.

Such counterintuitive humility is the prerequisite to a profound Christian unity that can weather the storms of external opposition and internal disagreements. It is the fountain from which flows a oneness of conviction and affection that gives joy to Paul's heart and gives glory to Jesus. But how can our hearts be turned inside out, to love selflessly like this? It is not a matter of teeth-gritted discipline, ruthlessly suppressing our every selfish thought. That only breeds further resentment against those whom we are called to love, and resentment toward the God who demands such unnatural affection. The only solution is to have our hearts overwhelmed with wonder at the fact that we have received such unnatural, supernatural, selfless love from the Creator of the universe, the triune God who pours out

7. G. Walter Hansen, *The Letter to the Philippians*, PNTC (Grand Rapids: Eerdmans, 2009), 115.

8. Gerald F. Hawthorne and Ralph P. Martin, *Philippians*, rev. ed., WBC 43 (Nashville: Thomas Nelson, 2004), 88.

his manifold grace on us in encouragement, love, comfort, partnership, and tender compassion. Therefore, Paul opens his summons to unity through humility with this irresistible, irrefutable rationale (Phil. 2:1).

The Triune God's Investment in Your Unity

The strongest reasons for us to love and honor others selflessly are the encouragement, love, and partnership graciously lavished on us by Christ, God the Father, and the Holy Spirit. Paul's Greek style in verse 1 is terse. English versions have to supply a verb such as "there is" (ESV) or "you have" (NIV), but Paul's original has none. Paul's Greek says simply: "if any encouragement in Christ, if any comfort of love, if any partnership of Spirit, if any affection and mercies . . . ." Paul grounds his summons to selfless unity in our relationship to each member of the Trinity. The mercy that the Son and the Father and the Holy Spirit have shown us by drawing us to trust the gospel gives us the strongest motive to put others before ourselves and the strongest reason to hope that we can grow in such selfless humility.

Not all scholars see a Trinitarian reference in Philippians 2:1. Some believe that the second clause, which refers to "comfort from love," refers to Christ's love, Paul's love, or even Christians' love for each other.[9] Paul elsewhere speaks of the love of Christ the Son (Gal. 2:20) and the love of God the Father (Rom. 5:5, 8), sometimes in close succession (Rom. 8:35, 39). Since Christ was just mentioned in connection with "encouragement," it would be plausible to see him as the source of the love that brings comfort, as the NIV's insertion of a pronoun implies: "if any comfort from his love." There are, nevertheless, good reasons to believe that Paul may be thinking of *the Father's* comforting love, though he does not mention God the Father by name.[10] First, two of

9. Seeing the love as Christ's love are Ralph P. Martin, *Philippians*, NCBC (Grand Rapids: Eerdmans, 1980), 86; Peter T. O'Brien, *The Epistle to the Philippians*, NIGTC (Grand Rapids: Eerdmans, 1991), 172; and others. Hansen, *Philippians*, 108, views this love as including "Christ's love for the church, Paul's love for the church, and the love of believers for one another." In the first edition of WBC, Hawthorne said that the love was Paul's, but the second edition (revised by Martin) favors "God's love or Christ's love for his people." Hawthorne and Martin, *Philippians*, 83. See NIV: "encouragement from being united with Christ, if any comfort from his love." Gordon D. Fee, *Paul's Letter to the Philippians*, NICNT (Grand Rapids: Eerdmans, 1995), 179–82, has persuaded me of the Trinitarian structure of Philippians 2:1abc. John Reumann, *Philippians: A New Translation with Introduction and Commentary*, AYB (New Haven, CT: Yale University Press, 2008), 321–23, concurs.

10. Fee, *Philippians*, 179–82, argues for the Trinitarian structure of Philippians 2:1abc on the basis of linguistic similarities to the Trinitarian benediction in 2 Corinthians 13:14 and conceptual parallels

the three persons of the Trinity are explicitly mentioned: "encouragement in *Christ*" and "participation in the *Spirit*." This in itself raises the possibility that there is a reference to God the Father here as well. Second, in his letter to the Ephesians, written from the same Roman imprisonment, Paul made a strong appeal to Christian unity through humility—the very issue he is addressing in Philippians 2:1–4—by citing the investment of each member of the Trinity in the church's oneness: "with all humility . . . eager to maintain the unity of the Spirit in the bond of peace. There is one body and one Spirit—. . . one hope that belongs to your call—one Lord [Jesus], one faith, one baptism, one God and Father of all, who is over all and through all and in all" (Eph. 4:2–6). As in Ephesians, so in our present text it seems likely that Paul grounds his call to unity in the personal involvement of each of the persons of the Trinity. Finally, the only other place in which Paul uses the expression "participation in the Spirit" (*koinōnia pneumatos*) is in the Trinitarian benediction with which he closes 2 Corinthians: "The grace of the Lord Jesus Christ and the love of God and the fellowship of the Holy Spirit [*hē koinōnia tou hagiou pneumatos*] be with you all" (2 Cor. 13:14). In that passage, Paul mentions Christ first as the Mediator of our redemption, then the Father's love as the source of redemption, and finally the fellowship in which the Spirit applies redemption to us.

Paul's point is this: despite our sinful self-centeredness—the preoccupation with our own ideas, our own preferences, our own convenience and comfort that is so instinctive to every fallen child of Adam—the triune God has encouraged us, comforted and loved us, and established his partnership with us. The last two elements in verse 1, which the ESV renders "affection and sympathy," drive home the reality that the benefits of encouragement, consolation, and companionship that we receive from Christ, his Father, and his Spirit are expressions of our God's passionate and compassionate attachment to us. "Affection" (*splanchna*) is the gut-wrenching expression of intense inner yearning that we heard in Philippians 1:8, where Paul described his longing for his Philippian friends with "the affection of Christ." Alongside the Trinity's passionate affection for us Paul puts God's overflowing mercies (*oiktirmoi*, here rendered "sympathy" by the ESV). Although this term is

---

to Romans 5:1–5, which speaks of God's (the Father's) love poured out in believers' hearts by the Spirit (5:5) and demonstrated in history by Christ's death (5:8). He does not cite the Trinitarian "case" for unity in Ephesians 4:1–6, but it seems an apt parallel, originating in the same period of Paul's ministry.

applied to humans' mercy in the New Testament (Col. 3:12; Heb. 10:28), the present context directs our attention to God as the source of mercies. God is "the Father of mercies [*oiktirmoi*] and God of all comfort," as Paul calls him in 2 Corinthians 1:3. In response to God's mercies (*oiktirmoi*), Paul exhorts his Roman readers to present their bodies as living sacrifices (Rom. 12:1). Christ's encouragement, God's comforting love, and the Spirit's partnership toward suffering and beleaguered believers in Philippi and elsewhere are not merely reinforcements sent by a distant monarch to troops at the front of the battle. They are concrete expressions of the deep affection and mercies that tie God's heart to ours and show that he is with us in the valleys of life.

Lest we doubt his personal engagement with us, we must remember the price God paid to make us his beloved partners. In the next paragraph (Phil. 2:5–11), Paul will describe what it has cost the Son, the Father, and the Spirit to draw us into the circle of their love: the Son, though he was in very nature God, became human, became a slave, and became obedient to death, even death on a cross—the death that you and I deserve because we have turned our hearts in on ourselves, rather than turning our hearts upward to God in adoration and outward to others in compassion. Elsewhere Paul shows that the cross also displays the love of the Father, who did not withhold his beloved Son but graciously gave him over for us all (Rom. 5:8; 8:32). In divine love the Spirit is given to us to draw us to Christ and to impart life in him.

How, then, can we fail to respond to such divinely imparted encouragement, comfort, love, and companionship? How can we refuse to lay down selfish ambition and vain conceit, our hurt feelings and competitive urges? Having received such love, let us beg our Savior to turn our hearts inside out, to treasure others as more important than ourselves, to care about their needs even more than we care for ourselves, to "fight for unity" by cultivating the humility that we see in Jesus, the King who stooped to serve us.

What might it look like, in practice, to count others more significant than yourself, to include their interests instead of focusing narrowly on your own? It might mean pausing in the midst of a busy day to offer a listening ear to someone who needs a hearing and a word of encouragement—and doing so without glancing at your watch or stewing internally about a waste of your precious time. It could involve allowing others to bask in the spotlight of admiration, knowing that you bore the brunt of the task. It might entail listening carefully and respectfully, even sympathetically, to another person's viewpoint, rather than spending that person's "speaking time" formulating your own counterargument. When full agreement on an issue seems out of reach and no biblical principle is jeopardized, such others-focused empathy could even lead you to defer to your brother's or sister's preference or conviction, and to do so willingly, not grudgingly or resentfully.

Steve Fry captured well the humility-producing, unity-promoting power of the gospel of grace in his prayer-song:

> Let it be said of us that the Lord was our passion,
> That with gladness we bore ev'ry cross we were given,
> That we fought the good fight, that we finished the course,
> Knowing within us the pow'r of the risen Lord. . . .
> Let it be said of us: We were marked by forgiveness,
> We were known by our love and delighted in meekness,
> We were ruled by His peace, heeding unity's call,
> Joined as one body that Christ would be seen by all.
>
> Let the cross be our glory
> And the Lord be our song,
> By mercy made holy,
> By the Spirit made strong
> Let the cross be our glory
> And the Lord be our song,
> 'Til the likeness of Jesus
> Be through us made known.
> Let the cross be our glory
> And the Lord be our song.[11]

Let *this* be said of us!

11. Steve Fry, "Let It Be Said of Us," Copyright © 1994 Universal Music—Brentwood Benson Publishing (ASCAP) / Word Music, LLC. All Rights Reserved. Used By Permission.

8

The King Who Stooped to Conquer

Philippians 2:5–8

Christ Jesus, . . . though he was in the form of God, did not count equality with God a thing to be grasped, but made himself nothing, taking the form of a servant, being born in the likeness of men. And being found in human form, he humbled himself by becoming obedient to the point of death, even death on a cross. (Phil. 2:5–8)

We have heard the story again and again in various forms. The details differ from one telling of the tale to the next, but the premise of the plot always intrigues us. It is the story of the incognito king.

J. R. R. Tolkien portrayed the motif memorably in his trilogy *The Lord of the Rings*. Early in volume 1, *The Fellowship of the Ring*, the hobbits Frodo, Sam, Pippin, and Merry set off from their home in Hobbiton on a quest related to the mysterious ring that Frodo's Uncle Bilbo had entrusted to him. When they reached the Prancing Pony Inn in the town of Bree, their paths crossed with a mysterious, menacing stranger, a Ranger whose rootless wanderings caused settled folks to view him with suspicion. Folks called him Strider, though that was not his real name. Only as the drama unfolds over the three books do the hobbits discover not only that the ominous Strider is an ally, not an enemy, but also that he is in fact the king, Aragorn, long exiled from his rightful throne! Long before Tolkien, English folklore contained the theme of a disguised monarch moving unrecognized among his subjects. Legends of Robin Hood spoke of Richard the Lionhearted returning from the Crusades costumed as an abbot, and in his classic *Ivanhoe* Sir Walter Scott portrayed Richard as a mysterious Black Knight.

Yet the drama of the unrecognized ruler is far older than English legend, and far truer than legend's blending of fact and fiction. God's Word itself shows us ordinary people encountering the Creator-King of the universe in the routines of ordinary history, only to be shocked in retrospect by the discovery that the stranger whom they had hosted, resisted, or rebuked was in fact the Lord of glory. As Abraham entertained a group of nomads, his wife chuckled in the tent at the absurdity of their guests' outlandish promise. But Abraham's guest was the God of promise himself, who always keeps his word (Gen. 18:1–15). Though fearful of facing his bitter brother Esau, Jacob found the stamina to wrestle through the night with an anonymous adversary. As the sun rose, the truth dawned on Jacob that his Opponent was divine. He had seen God face to face, and yet survived (Gen. 32:30). When, at Jesus' direction, the fisherman Simon dropped his nets back into the lake, he may have wondered how much the carpenter-turned-teacher knew about fishing. Only when fish filled the nets to the breaking point did Simon recognize and fall in awe before the Lord of creation, who sat calmly in Simon's skiff (Luke 5:4–11). Through tear-filled eyes Mary caught a blurry glimpse of a gardener and begged him to reveal where her Lord's body had been moved. Only when he spoke her name, "Mary," did her sorrow and confusion clear, so that she could report, "I have seen the Lord" (John 20:11–18). Heartbroken travelers were trudging to Emmaus in the aftermath of a Passover that had turned tragic. An unknown companion joined them and showed them from the Scriptures that Messiah must pass through suffering to glory. Later, as he broke bread, they discovered that their teacher was the King himself, who had walked beside them all along (Luke 24:13–32).

When the King comes among us incognito, he seems so ordinary, so much like the rest of us. Then, in a flash of recognition, we discover that the person whom we had taken for granted or contradicted turns out to be Someone of such majestic dignity that we are overwhelmed with awe. This is the familiar but amazing true story that Paul rehearses for his friends in Philippians 2:5–11. In heightened language and matchless eloquence—probably a poem, or even a hymn sung by the early church in praise of Christ the Lord[1]—Paul tells the story of the King who stooped to serve, and who, by serving, conquered.

Paul's call to unity of mind and soul in the previous passage (Phil. 2:1–4) followed an indicative-imperative order, first cataloguing the Trinitarian indicatives—Christ's encouragement, God's love, the Spirit's partnership—that motivate and enable us to treasure others above ourselves (2:1), and then defining the attitude of unity and humility that is believers' right response to grace (2:2–4). Now Paul reverses that order to imperative-indicative: he first exhorts us to embrace and exhibit the mind of Christ (2:5), and then he shows how Christ's mind-set of selfless love found expression in his incarnation, humiliation, and death, leading to his exaltation to unparalleled heights of glory (2:6–11).

1. Gerald F. Hawthorne and Ralph P. Martin, *Philippians*, rev. ed., WBC 43 (Nashville: Thomas Nelson, 2004), 99–100, seem correct when they observe that in this significant and challenging text "there is at least one thing that calls forth almost universal agreement. It is that Philippians 2:6–11 constitutes a signal example of a very early 'hymn' of the Christian church." Since the analysis of Ernst Lohmeyer in 1928 (*Kyrios Jesus: Ein Untersuchung zu Phil. 2,5–11*), most scholars have regarded this text as a sample of the early church's poetic-musical adoration of Christ, sung in worship, and then embedded by Paul in his epistle. They note such poetic features as parallelism of lines, alliteration, chiasm, and hints of meter; the appearance of words unparalleled elsewhere in Paul's letters; and the shift in tone from pastoral second-person admonition (2:1–4, with resumption at 2:12) to third-person doxological narration. For a discussion of the text's poetic features, see Peter T. O'Brien, *The Epistle to the Philippians*, NIGTC (Grand Rapids: Eerdmans, 1991), 188–93; Hawthorne and Martin, *Philippians*, 99–104; G. Walter Hansen, *The Letter to the Philippians*, PNTC (Grand Rapids: Eerdmans, 2009), 122–33; and especially Ralph P. Martin, *Carmen Christi: Philippians ii.5–11 in Recent Interpretation and the Setting of Early Christian Worship*, rev. ed. (Grand Rapids: Eerdmans, 1983) (the third edition is entitled *A Hymn of Christ*, 1997). Among recent commentators Gordon D. Fee, *Paul's Letter to the Philippians*, NICNT (Grand Rapids: Eerdmans, 1995), 40–43, 192–97, demurs from this consensus. Fee contends that disagreements among scholars' attempts to reconstruct the "hymn's" original poetic structure (often requiring the supposition that Paul inserted key phrases that do not fit a conjectured parallelism of lines), the passage's narrative flow, the reappearance of important terms elsewhere in the epistle, and other factors weaken the case for interpreting this text as an early Christian hymn about Christ. Rather, Fee contends that this Christological masterpiece is elevated prose freshly composed by Paul under the Holy Spirit's inspiration. Yet Fee also reports singing Francis Bland Turner's hymn based on Philippians 2:5–11 on the Sunday after completing his commentary on this text, commenting, "The passage obviously sings, even if it was not originally a hymn!" Ibid., 226n42.

The King Who Stooped to Conquer Shares His Mind of Humility with Us

This passage is, as we have observed, a majestic mountain peak, towering over the surrounding countryside. It is a pinnacle of theological truth, piercing the heavens and probing the mystery of the incarnation. Its dramatic movement traces the inverted arc of Christ's redemptive mission from divine glory down into humiliation and death, and then up again to heaven's heights in resurrection splendor. These seven verses may have generated more scholarly comment and theological reflection than the other ninety-seven verses of Philippians put together, and for good reason. This brief and beautiful text is one of the fullest, most explicit descriptions in the New Testament of the identity of our Redeemer, Jesus Christ.

Yet Paul retells this drama of costly love and rescue not merely to fill our minds with correct doctrine, nor to move our imaginations with wonder. Paul does not sit down to compose or incorporate this marvelous hymn about Christ because he thinks, "In a hundred years Docetists will arise to deny the humanity of Jesus!" or "In a few hundred years Arius will appear to deny the divinity of Jesus!" or "In a couple thousand years (1977) John Hick will edit *The Myth of God Incarnate*, urbanely dismissing the surprising truth of God-become-man, which is humanity's only hope!" Of course the Holy Spirit knew that those heresies would come down the road, and that we would need passages such as this to stabilize us when theological attacks are launched against our Savior. But Paul writes this rich Christological feast in the context of the trials confronting his Philippian friends. He seeks to give them (and us) a new way of looking at the challenges that threaten to thwart Christians' efforts to stand fast and stand together in the midst of an unsympathetic society. Paul is concerned about the threats to Christian unity that arise when our preferences and priorities bump up against one another, when such tensions bring to the surface pockets of "selfish ambition" or "vain conceit" that lie hidden in our hearts.

So having summoned his friends to have "the same mind" and "one mind" (Phil. 2:2), Paul clarifies that "this mind" is the attitude that "is yours in Christ Jesus" (2:5).[2] Then he recounts the story of the King's mission of

2. Behind the ESV's "being of [the same] mind" and "of [one] mind" in 2:2 and "Have this mind" is the Greek verb *phroneō*, which in this context means "*have thoughts* or *(an) attitude(s), be minded* or *disposed*" (*BAGD*, 866).

self-sacrifice, in order to illustrate what that humble mind-set looked like in its purest, costliest form, when the divine Lord of glory clothed himself in a slave's shame and died the cross's cursed death for others' sake.

There has been considerable debate over the meaning of Paul's transitional directive in verse 5, and therefore over how we should respond to this narrative of Christ's humble descent and glorious ascent. The familiar wording of the KJV, "Let this mind be in you, which was also in Christ Jesus," implies that Paul is presenting Jesus' humble mind-set as an *example* to the Philippians and to later generations. Down through history, most students of Scripture have read the text in this way.[3] The verb "think" or "have a certain mind-set" that ties our text to the exhortation that precedes it (Phil. 2:2, 5) gives us reason to conclude that the humble mind-set that was evident in Jesus must be seen in us who follow him. Likewise, at the end of the hymn Paul links Christ's costly obedience (2:8) to the following exhortation that the Philippians must continue to obey (2:12), as they always have.[4] Jesus' obedience is the pattern that theirs must follow.

Yet Paul's Greek wording is more terse and ambiguous than the KJV rendering suggests. Paul wrote, literally, "This think in you which [*] also in Christ Jesus." Paul expects his readers to mentally "fill in the blank," providing a verb where I have placed the asterisk (*). But which verb? A form of *to be*, as the KJV translators proposed? "Let this mind be in you, which *was* also in Christ Jesus"—that is, imitate the attitude of humility that Jesus showed? Or should we supply "[you] think" from the first part of the sentence, expressing a meaning such as: "Have the mind-set among yourselves that expresses *the mind-set you have* in your union with Christ"? The ESV implies that this is Paul's point: "Have this mind among yourselves, which is yours in Christ Jesus"—that is, "as you think about each other, let your perspective be formed by your shared identity in Christ." On this reading, Paul's compact Greek still points to Jesus' self-sacrificing humility as an example to be emulated (as the KJV and other versions suggest), but it also reminds

3. See, for instance, ACCS NT 8:236; John Calvin, *The Epistles of Paul the Apostle to the Galatians, Ephesians, Philippians and Colossians*, ed. David W. Torrance and Thomas F. Torrance, trans. T. H. L. Parker (Grand Rapids: Eerdmans, 1965), 246–47; J. B. Lightfoot, *Saint Paul's Epistle to the Philippians* (1913; repr., Grand Rapids: Zondervan, 1953), 110.

4. The adjective *obedient* (Greek *hypēkoos*) in "becoming obedient to the point of death" (Phil. 2:8) is a cognate of the verb *obeyed* (*hypakouō*) in "you have always obeyed" (2:12).

us that unbreakable cords of grace bind believers to our Savior so tightly that *Christ conveys his mind-set to us* through his Holy Spirit. When Christ Jesus left the bliss of heaven for the miseries of earth for you, his purpose was *not only* to rescue you from your sin's just deserts (though it was that). It was *not only* to set you an example of humility (though it was that). It was *also* to reconfigure the inclinations of your heart, so that his mind-set (that is, his joy in selflessly serving others) is becoming your mind-set! As you hear the song of the King who stooped to serve in verses 6 through 11, let the drama of his descent remold your own affections and perceptions of yourself and others.

The drama is told in three acts—the song is sung in three stanzas: (1) the height from which the King stooped (Phil. 2:6), (2) the depths to which he stooped (2:7–8), and finally (3) the highest heights to which he soared (2:9–11). Because this passage so richly reveals Christ's divine majesty and redemptive mercy, in this chapter we can explore only the first two stanzas, reserving the finale—the exaltation of our humble King—for the next.

THE HEIGHT FROM WHICH THE KING STOOPED

The story began in the far reaches of eternity past, infinitely beyond the comprehension or imagination of time-bound creatures such as we are. To begin to glimpse the wonder of his condescension, we need to know the divine dignity and glory that forever characterized the One who became a human baby born in Bethlehem, given the name *Jesus* because his mission on earth would be to "save his people from their sins" (Matt. 1:21). The eternal Son of God existed eternally not only before Bethlehem, but even before the creation of the universe itself (John 1:1; Heb. 1:2–3). He was "in the form of God" and he had "equality with God" (Phil. 2:6). Both of these descriptions have been debated for centuries.

First, we consider the expression that Christ was "in the form of God." Paul uses the word *form* twice in this passage—"form of God" and "form of a servant"—and nowhere else in his epistles.[5] The ESV's "in the form of God" and the NIV's "in very nature God" may seem to be making different points. Although Greek philosophers had used the word behind *form* to

5. Greek *morphē*, which is found elsewhere in the New Testament only once, in later manuscripts of Mark (16:12).

refer to something's essence or nature, generally in ancient Greek the term focused on visible appearance.[6] Some scholars believe that "in the form of God" is an Old Testament allusion to Adam's and Eve's having been created "in the image of God" (Gen. 1:27). Paul does, in fact, call Jesus "the image of the invisible God" in Colossians 1:15, echoing the creation account in Genesis 1. But Paul does not use *image* (*eikon*) here, although he could have. It seems that Paul (or the poet whose hymn he now incorporates into inspired Scripture) selected *form* because it blended the ideas of *nature* and the *visible display of that true identity*. Paul is speaking of One who not only is and has eternally been truly God, but also displayed his glory as God in preincarnate radiance. One commentator concludes that the phrase "in the form of God"

> is best interpreted against the background of the glory of God, that shining light in which, according to the [Old Testament] and intertestamental literature, God was pictured. . . . The expression does not refer simply to external appearance but pictures the preexistent Christ as clothed in the garments of divine majesty and splendor.[7]

In Jesus the very nature of the invisible God is displayed in a form that we can access, as John testified that the Word who "was God . . . became flesh and dwelt among us, and we have seen his glory, glory as of the only Son from the Father, full of grace and truth" (John 1:1, 14). The epistle to the Hebrews opens on the same note, describing the Son as "the radiance of the glory of God and the exact imprint of his nature" (Heb. 1:3). In prayer to his Father, Jesus himself referred to "the glory that I had with you before the world existed" (John 17:5).

The term *form* perfectly served Paul's purposes for two further points he would make. First, we are about to learn that Christ, in addition to being in the glorious form of God, has condescended to assume another identity, visibly expressed in shame and weakness: "the form of a servant"—more exactly, the form of a *slave*.[8] The form of God is stunning radiance; the

6. For a recent summary of the ancient usage and meanings of *morphē*, see Joseph H. Hellerman, "MORPHE THEOU as a Signifier of Social Status in Philippians 2:6," *Journal of the Evangelical Theological Society* 52 (2009): 779–98.

7. O'Brien, *Philippians*, 210–11.

8. Greek *doulos*, which (unlike other terms such as *diakonos* and *pais*) connoted a worker who was the property of an owner/master. Paul applied *douloi* (plural of *doulos*) to himself and Timothy in Philippians 1:1 to emphasize that Christ Jesus was their Owner who could use them as he chose.

form of a slave entails subservience to others and subjection to their scorn. Christ now bears both. Second, although Paul uses the noun *form* only in this hymn, elsewhere he uses verbs in which this noun is embedded; he does so to highlight how Christ's adding "slave-form" to his "God-form" benefits us. In Philippians 3:10 Paul expresses his desire to experience both the power of Christ's resurrection and a partnership in his sufferings, "being con*formed*[9] to his death." A few sentences later, he looks ahead to the return from heaven of our "Savior, the Lord Jesus Christ, who will transform our lowly body to con*form*[10] to his glorious body" (3:21). As Paul wrote earlier to the church at Rome, the change of our identities to resemble Christ's holiness has already begun: "be trans*formed*[11] by the renewal of your mind" (Rom. 12:2). Christ assumed our human nature, and he did so not in the form of the "high and mighty" but as a scorned slave. His purpose was that we might be transformed now by the renewing of our minds, and when he returns by the resurrection of our bodies.

The height from which Christ the King stooped is also implied in Paul's second statement in verse 6: he "did not count equality with God a thing to be grasped." Again, Bible scholars have debated Paul's meaning at length. The heart of the puzzle is that our English expression "a thing to be grasped" represents a single Greek noun (*harpagmos*), a word that appears only here in the New Testament, never in the Greek translation of the Old Testament, and rarely in all other ancient Greek literature. The noun is derived from a verb (*harpazō*) meaning "to grasp or snatch," but that takes us only so far in understanding the noun. Basically, the meaning of this rare noun, and therefore of the clause in which it appears, has been understood in one of four ways:[12]

(1) Christ did *not* possess equality with God, but he did not regard that status as a prize that he should *try to seize* for himself. In other words, Christ was not like Adam and Eve, who succumbed to Satan's offer: "You will be like God, knowing good and evil" (Gen. 3:5). Instead he obeyed God's will, and was therefore exalted by God after his

9. Author's translation of the verb *summorphizō*, which the ESV renders "becoming like."

10. Author's translation of the adjective *summorphon*, which the ESV renders "to be like."

11. The Greek verb is *metamorphoō*, from which we get *metamorphosis* in English.

12. For a more detailed discussion, see N. T. Wright, "*Harpagmos* and the Meaning of Philippians 2:5–11," *Journal of Theological Studies* 37 (1986): 321–52.

suffering.[13] This explanation fits Paul's call to replace "conceit" with a mind-set of humility. But it conflicts with the affirmation that Christ was "in the form of God," sharing God's identity and displaying divine glory, as well as Paul's other statements of Christ's deity (for instance, Romans 9:5, where he calls Christ "God over all, blessed forever").

(2) Christ *did* have equality with God, but did not selfishly *cling to* that divine prerogative. Instead, he *gave up* his divine standing in order to become a human being. This interpretation again fits the emphasis in the context on selfless humility. But some who have held this view—such as the "kenotic" theologians of the nineteenth century—also embraced the idea that the Son, in letting go of his equality with the Father, relinquished not only his divine *status* but also his divine *nature*—the identity that set him apart as Creator from all his finite creatures. Such an understanding of the incarnation diverges from the Bible's teaching and the church's historic confessions that our triune God is unchangeable (Ps. 90:1; Mal. 3:6) and that the incarnate Christ who walked the earth, endured the cross, and rose again was and is fully God as well as fully human.

(3) The puzzling noun should not be translated as "a *thing* to be grasped" at all, but rather as "an *act* of grasping." It is beside the point to debate whether equality with God was at first within Christ's grasp or not. The issue is the action of seizing itself. (This is one way we could take the KJV's "thought it not robbery to be equal with God.") Unlike the assumptions about power and prestige that structured Greco-Roman social relations, Christ expressed his equality with God not by grabbing but by giving.[14] This view fits the context, since Paul's point is that Christlikeness, which is Godlikeness, is really a matter of giving, not getting. It also avoids the problems of the previous two views, since now "equality with God" is not viewed as a prize that Christ refused to grab or refused to retain. Yet I think a fourth alternative is even more persuasive.

13. Morna Hooker, "Philippians 2:6–11," in *Jesus und Paulus*, ed. E. Earle Ellis and Erich Grässer (Göttingen: Vandenhoek & Ruprecht, 1975), 160–62.

14. C. F. D. Moule, "Further Reflections on Philippians 2:5–11," in *Apostolic History and the Gospel*, ed. W. Ward Gasque and R. P. Martin (Grand Rapids: Eerdmans, 1970), 164–76.

(4) Christ was and remains equal with God, but he did not regard that equality as a "perk" to be exploited for his own advantage, "a windfall, a fortuitous springboard to be used for self-aggrandizement."[15] This approach does not try to interpret the rare noun ("grasping" or "a thing grasped") by itself, but in combination with the verb "consider" (ESV: "count"). This combination appears with some frequency in ancient Greek, referring to a situation (often unexpected) that a person sees as an opportunity to exploit for personal profit or advancement. Christ's equality with the Father was not a random or unexpected windfall, of course. But it was a status that gave him every right to demand that others serve him. Yet, as he said, "the Son of Man came not to be served but to serve, and to give his life as a ransom for many" (Mark 10:45).

Actually, the last two explanations make virtually the same point. Both affirm that Christ did not view or use his equality with God the Father as a pretext for self-serving or a platform from which to achieve his own interests (as the Philippians were prone to do, Phil. 2:3). In fact, Paul explains most clearly what he means when he says that Christ "did not count equality with God a thing to be grasped" in the next stanza of the hymn (vv. 7–8), which describes the King's descent from heaven's heights to the depths of messy, miserable life on this earth.

Now, before we move from Act 1 to Act 2, from eternity into history, we should reflect on two implications for ourselves that flow from Paul's portrait of the preincarnate glory of Christ. First, consider the vast—actually, infinite—distance in dignity that separates the Son who is equal with God the Creator himself and who radiates his glory, on the one hand, and the most exalted and admirable members of the human race. However we measure worth—character, intelligence, courage, strength, influence, or some other quality—the very best that our human

15. R. B. Strimple, "Philippians 2:5–11 in Recent Studies: Some Exegetical Conclusions," *Westminster Theological Journal* 41 (1979): 247- 68, citing prior research by Werner Jaeger and R. W. Hoover. (The quoted words are on 264.) O'Brien, *Philippians*, 214–16, concurs. Lightfoot, *Philippians*, 134, seems to have anticipated the findings of Jaeger and Hoover, interpreting the phrase "consider [equality with God] a thing to be grasped" as a single unit with the sense "He, though existing before the worlds in the form of God, did not treat His equality with God as a prize, a treasure to be greedily clutched and ostentatiously displayed."

family has to offer falls so far short of the majesty of the divine Son who is the theme of Paul's song. That discovery prompts uncomfortable questions, doesn't it? Just how great do you think you are? How much respect and honor do others owe you? When Christ's apostle told you to "count others more significant than yourselves" (Phil. 2:3), did you find his instruction surprising or demeaning—since, after all, you see yourself as more significant than many people (and they may see you that way, too)? Think long and hard about the magnificence of Christ. It will put your sense of your own importance into proper perspective—and bring it down to size. It will narrow the gap between your self-image and your appreciation for those whom you have viewed as less significant than yourself.

Second, consider the fact that the divine Son did not regard his equality with the Father as a pretext for grasping but as a platform for giving. Consider how his perspective on privilege and power upends the scale of values that we often assume to be fitting and proper. As early as Eden, Satan has spun the lie that the Creator's motives are suspect and selfish—that though God made us in his image, he is holding out on us, unwilling to share the best part of being like God, "knowing good and evil" (Gen. 3:5). Our first parents were told, and foolishly believed, that to be Godlike we have to grab what looks good to us, what we imagine will be good for us. Jesus reminded his disciples that societies still operate on the same self-serving assumption: "You know that the rulers of the Gentiles lord it over them, and their great ones exercise authority over them" (Matt. 20:25). That is the way the world works: people in power give orders and are waited on by subordinates who do their bidding. But Jesus says, "It shall not be so among you. But whoever would be great among you must be your servant" (20:26). Jesus has unique, unchallengeable authority to define leadership as servanthood because he, the Son of Man (destined to rule the whole world forever, Dan. 7:13–14), "came not to be served but to serve, and to give his life as a ransom for many" (Matt. 20:28). In your home, in your workplace, in the church, when you are tempted to throw your weight around, to "pull rank" in order to get your way, pause and ponder the wonder of the mind of Christ—the mind that is becoming yours as you rest in him—the mind-set that exhibits the incomparable glory of God not in self-seeking grasping but in self-sacrificing giving.

THE DEPTHS TO WHICH THE KING STOOPED

Paul traces the King's descent in two stages: first Christ "made himself nothing" by becoming a human being, with a slave's disenfranchised status (Phil. 2:7). Then he "humbled himself" further through obedience to the point of death, even the criminal's death on a cross (2:8).

The Greek verb *kenoō*, which the ESV (with the NIV) renders "made himself nothing," has also been translated as "emptied Himself" (NASB). Kenotic theologians of the nineteenth century, taking their cue from a literalist interpretation of this verb, alleged that in his incarnation Christ divested himself of divine attributes such as omniscience and omnipotence, confining himself completely to the creaturely limitations of his human nature. We already noted some of the problems of this view, which contends that One who had been God, infinite in his perfections, abandoned those divine perfections in order to become human. The kenotic view of Christ also misunderstands the verb itself. The verb is related to an adjective meaning "empty" or "devoid of contents" (Luke 1:53; 20:10–11). But Paul, the only New Testament author to use the verb, always uses it *metaphorically* in the sense of "render void, ineffective, powerless" (Rom. 4:14; 1 Cor. 1:17; 9:15; 2 Cor. 9:3). In other words, in none of these four other New Testament uses of the verb *kenoō* (all by Paul) does the context mention or imply that some type of "contents" have been "emptied" from some sort of receptacle: "the promise," "the cross," "my/our boasting."[16] Paul's use of the verb is not to invite the question "When Jesus 'emptied' himself, *of what* did he empty himself?" Paul's point, rather, is that Christ "emptied" *himself*—made *himself* powerless, ineffective. He embraced a role of insignificance and impotence, assuming "the form of a servant." He humbled himself in this way not by *abandoning* his divine might and knowledge but by *adding* to his divine nature a complete human nature, limited like ours, yet unstained by Adam's sin. Except on the Mount of Transfiguration, his divine glory was not visible to physical sight. (Even there, no doubt, his radiance was "filtered" to protect his closest disciples.)

16. The ESV translation of 1 Corinthians 1:17—"lest the cross of Christ be emptied of its power"—might suggest to our ears the metaphor of "the cross" as a receptacle and "power" as its contents. Here, however, the ESV borders on paraphrase, using the *whole* English phrase "be emptied of its power" to represent the *single* Greek verb *kenōthē* [a form of *kenoō*]. A translation that would be closer, formally, to the Greek is "lest the cross of Christ be rendered null and void."

As he introduces what it cost Christ to become human—"being born in the likeness of men," "being found in appearance as a man" (NASB[17])—Paul begins by describing the King's condescension as "taking the form of a servant." He repeats the word *form* (*morphē*) to set the visible display of the divine glory that was the Son's eternally in contrast to the lowly role, identity, and appearance that he undertook in assuming our human nature. Christ became human not as a monarch in a palace, nor as a wealthy noble in an estate house. He came as a servant—a slave, without rights—whose whole purpose in life was to meet the needs of others. Instead of demanding his rights to be served as God, he stooped to perform the slave's most menial task: washing the road filth from his friends' feet. Our version captures the force of the metaphor just right: he "made himself nothing."[18]

When Paul's hymn says that Christ became "in the likeness of men" and "in appearance as a man," there is no suggestion that there was any pretending or illusion going on when the Son appeared as a human being. He was not a "god in disguise," pretending to be human, as the Greek myths described. He really became a flesh-and-blood human being. He was "made like his brothers in every respect" (except for our sin, which is not intrinsic to our being human) (Heb. 2:17). But at the same time, he was unsettlingly, frighteningly different. From sheer physical exhaustion he slept through a storm (Mark 4:38), yet the next moment he spoke as the stormy sea's sovereign Creator, and stilled the waves to silence (4:39; see Ps. 107:26–32). The men whom he rescued from death by drowning were more fearful of him than they were of the storm itself!

We are so accustomed to the Bible's teaching that Jesus is God and man in one person that we may forget how utterly shocking it was for a first-

17. Here the NASB reflects the Greek text more accurately than the ESV.

18. Although at this point in the hymn the focus is on Christ's becoming our human brother, it is possible that the striking use of the term "made himself nothing" looks beyond Bethlehem to the purpose for which the Son came: the cross. Christ came as a servant not only to perform a slave's subservience in a Roman household but also to fulfill the suffering of the Servant of the Lord prophesied in Isaiah 53. If "[he] emptied himself" here walks the border between the literal sense it bears in extrabiblical Greek and Paul's metaphorical usage, it may allude to the description of the Servant's death, when he "poured out his soul [his life, himself!] to death," as he "bore the sin of many" (Isa. 53:12). Though the hymn does not explicitly state that Christ's death was to atone for others' sins, as Paul does elsewhere (1 Cor. 15:3; 2 Cor. 5:21; Gal. 2:20), this possible allusion to the song of the Suffering Servant hints at the redemptive purpose for which Christ became human: to serve and to give his life a ransom for many (Mark 10:45). For a defense of this interpretation of "he poured out himself," see Strimple, "Philippians 2:5–11 in Recent Studies."

century Jew to make this claim. The new-age paganism of our day loves to talk about the spark of the divine in all of us, in every living creature, in Mother Earth . . . and we all start humming Elton John's "Circle of Life" from *The Lion King.* But as an orthodox Jew, Saul of Tarsus knew better. He knew that humans were created in the image of God. But that is a far cry from saying that God, the Lord of Israel, had *become* one specific human being. The Creator must not be confused with his creatures, nor with any specific creature (Rom. 1:25)! The Gospels tell us that during his ministry Jesus was repeatedly accused by his fellow Jews of making a blasphemous claim to be equal with God (Matt. 26:63–66; Mark 14:61–64; John 5:18; 8:58–59; 10:30–33). Moses had directed that any Israelite who advocated allegiance to any god other than Israel's covenant Lord should be executed by stoning (Deut. 13:6–11; see 17:2–7). Evidently the crowds picked up stones and their leaders passed the death sentence on Jesus because they understood him to be promoting another god besides the Lord God—claiming that he himself was that god! Although the Judaism of Jesus' day sometimes envisioned various angelic and human figures sitting in God's presence in heaven, in the minds of the Jewish leaders Jesus' prediction that he would sit at "the right hand of Power" (Matt. 26:64) and would come as the Son of Man and Judge of all seemed to violate the sharp boundary between the Creator and his creatures.[19] Why would an orthodox rabbi such as Saul now say that Jesus is God incarnate? The third stanza of this hymn to Christ will show us. But first, we must trace the King's descent to the depths.

In Philippians 2:8 Paul makes explicit what his allusions to Isaiah 53 were hinting: This One who was in very nature God, equal with God—the ever-living God, the very source of all life—this One, having become human, having embraced his calling as Servant, "humbled himself by becoming obedient to the point of death, even death on a cross." Again, it is shocking for a Jew to suggest that the living God could *die*. The prophet Elijah confirmed

19. "Jesus' 'blasphemy' consisted not in a formal misuse of God's name but in claiming for himself a unique association with God, sitting at his right hand. While a claim to be the Messiah was not in itself blasphemous, what Jesus said in response to the high priest went far beyond that claim: he was not only Messiah and Son of God but also, as the Son of Man predicted in Daniel 7:13–14, he was now to share God's throne. Such outrageous claims must either be accepted, which was unthinkable, or repudiated as blasphemous." R. T. France, *The Gospel of Matthew*, NICNT (Grand Rapids: Eerdmans, 2007), 1020–21. See also Darrell L. Bock, *Blasphemy and Exaltation in Judaism: The Charge against Jesus in Mark 14:53–65* (Tübingen: J. C. B. Mohr [Paul Siebeck], 1998; Grand Rapids: Baker, 2000), 234–37.

the truth of his prediction with an invincible guarantee: "As the LORD, the God of Israel, *lives*, before whom I stand . . ." (1 Kings 17:1). If God could die, his prophets' words might fail. But that would not, could not happen!

Yet the Christ who is everlastingly, ever-livingly divine became a mortal man. He added to his infinite, unchangeable deity our limited and mutable humanity, mysteriously uniting these two natures in his one person. That was the very purpose for which God the Son became our human brother, veiling his divine glory behind a slave's rags and human flesh, torn by Rome's cruel lash. It was the Father's plan, the necessary route to the redemption of his wayward children. So Christ "humbled himself" and became "obedient to the point of death." His self-humbling defines the humility of mind that will enable the Philippians to count others more significant than themselves (Phil. 2:3). His obedience sets the pace for their ongoing obedience to the will of God (2:12). But his humility and obedience to death did more than set a noble example. As the Lamb of God, by his blood he "ransomed people for God from every tribe and language and people and nation" (Rev. 5:9). As Isaiah said in the famous Servant song, "upon him was the chastisement that brought us peace, and with his stripes we are healed" (Isa. 53:5).

Then Paul leads us to the final drop-off point in Christ's descent from the heights to the depths: "even death on a cross." This is the ultimate insult. Execution by crucifixion was reserved for slaves and terrorists. The cross was distasteful for Roman citizens even to mention in conversation. (Yet Paul the Roman citizen does it all the time in his letters—even boasting in that weapon of shame and torture, Gal. 6:14.) As a zealous Pharisee, Saul had considered it his duty, for the honor of the God of Israel, to do all he could to wipe out this cult that acclaimed a crucified criminal as Messiah and Lord. He knew well what being hanged, stripped, and subjected to public exposure really meant, for he had read Deuteronomy 21:23, which he would quote to the Christians in Galatia: "Christ redeemed us from the curse of the law by becoming a curse for us—for it is written, 'Cursed is everyone who is hanged on a tree' " (Gal. 3:13). To Gentiles, the cross meant scandalous and shameful impotence. To Jews, it meant death under God's curse.[20]

Nothing in the world could have opened the mind of Saul the Pharisee to the possibility that Jesus the Nazarene, executed on a cross, rejected by

20. F. F. Bruce, *Commentary on Galatians*, NIGTC (Grand Rapids: Eerdmans, 1982), 237.

his people and his God, a helpless victim of the Romans, could be anyone worth admiring, much less the promised Messiah. Nothing could have changed Saul's mind so radically, except the indisputable demonstration of Jesus' resurrection, exaltation, and divine splendor (the theme of the hymn's third stanza).

Saul of Tarsus, Paul the apostle, surrendered to Jesus as Lord and then proclaimed Jesus as Lord, simply because he could not help it. On the road to Damascus the overpowering, blinding divine glory of Jesus the Nazarene, whom Saul was trying to persecute, pulverized Saul's unbelief and pride in a moment. What else could Saul do but confess that "Jesus Christ is Lord, to the glory of God the Father"?

Conclusion

How, then, should we respond to this incredibly rich disclosure of the person and saving mission of the King of heaven, who walked incognito among us and died a slave's death in our place?

First, we can be confident in the great Savior whom God has given to us. Christ is building his church on the confession that Simon Peter offered to the all-important question "Who is Jesus?" By revelation from God the Father, Peter confessed, "You are the Christ, the Son of the living God" (Matt. 16:16). The hymn of Philippians 2:6–11 adds detail to that bedrock of our faith: Christ is eternally the Father's equal, arrayed in divine splendor; he became our human brother, the Suffering Servant; and (as we will soon see) he is risen and ascended as Lord of all, to whom every knee will bow and whom every tongue will confess. Though non-Christians find Jesus' claims offensive and inventive heretics try to tame them, God's Word shows him to be that unique person that the church has confessed for centuries: "one Lord Jesus Christ, the only-begotten son of God, begotten of his Father before all worlds, God of God, Light of Light, very God of very God, begotten, not made, being of one substance with the Father; by whom all things were made; who for us and for our salvation came down from heaven and was incarnate by the Holy Spirit of the virgin Mary, and was made man."[21]

21. Nicene Creed (A.D. 325), expanded by the Council of Constantinople (A.D. 381) and adopted by the Council of Chalcedon (A.D. 451).

Second, surely we must be astonished, again and again, that Jesus the eternal Son made this excruciating journey into death, and back to life and lordship again, for self-centered people such as we are! In grateful wonder, let our knees bow and our tongues confess now what every creature everywhere will one day acknowledge: Jesus Christ is Lord.

Finally, we must be humbled and transformed: Jesus, who blazed the trail down into selfless service in order to rescue us, now leads us up to joy with his heavenly Father. His Spirit, meanwhile, is teaching us the unexpected lesson that our highest privilege lies in reflecting the triune God's self-giving love in our treatment of one another. Whatever it may cost you now to put your own interests, comforts, conveniences in second place behind others' needs, in the end you will discover that you have lost nothing of value in having your heart turned inside out to serve others. In fact, through your giving away of your rights, God will have given you the most amazing gift: the joy of starting to look (a little) like himself, more and more, until you, too, at Christ's return and your resurrection, are so radically transformed that you find pure delight in bringing glory to his Father!

9

The Son Exalted for the Father's Glory

Philippians 2:9–11

Therefore God has highly exalted him and bestowed on him the name that is above every name, so that at the name of Jesus every knee should bow, in heaven and on earth and under the earth, and every tongue confess that Jesus Christ is Lord, to the glory of God the Father. (Phil. 2:9–11)

People are innately, instinctively, incurably purpose-driven. We do what we do for reasons. We are drawn toward certain options and away from others because we expect that our choices and our actions will produce outcomes that we want. For some, the target that spurs them to action may be modest. Those just getting by at subsistence level struggle day by day just to find some food and a little shelter from the elements. Others strive after dreams that fly higher: education, marriage and family, interesting employment that supplies more than life's bare necessities, a vocation that serves others and betters society. Our reasons may sometimes be unfounded, our purposes thwarted, and our hopes disappointed. Nonetheless, though our objectives sometimes elude us, most of us remain undeterred from setting goals, large and small, and undaunted from pursuing those goals. Ironically, even intellectuals who embrace a naturalistic worldview that repudiates the very idea that purposeful intentionality orders the cosmos cannot help but live their lives as purposeful persons—setting goals, laying strategies to achieve their aims, and investing effort to implement their plans. What is the source of this purposefulness that seems built into our personhood?

From its opening pages the Bible shows us a personal God who brings a universe into existence out of nothing by the power of his Word, who sets its contents all in order, and who pronounces the product of his creativity "good" and "very good"—meeting his criteria for approval, fitting his purpose and design. And this Creator designed one particular creature to bear his image and likeness—a creaturely replica of his personality and purposefulness and a creaturely representative of his authority over his handiwork. We are purpose-driven people because we are made in the image of our purposing Creator. Of course, God's Word soon shows us how quickly our own purposing—the aims and objectives that motivate our choices—became deflected and disoriented from the Creator's purposes for us, purposes that would have made our own plans and efforts flourish under his good pleasure. Still, our goal-setting and striving—even when reach exceeds our grasp—bear a quiet testimony that is hard to deny: we bear the image of a personal Creator who has and pursues and accomplishes his purposes.

What is the Creator's purpose for his cosmos? Worship! God's chief end is his own worship—the display of his unique magnificence to evoke the adoration of all his creatures, especially human beings, whom he designed to bear his image and enjoy his friendship. The last book of the Bible, the Revelation granted to John, shows us scene after scene, containing song after song, in which joyful worship is offered to God seated on the throne and to the Lamb, who has rescued people from all nations and transformed them into an entourage of priests who eagerly serve in the presence of their Creator.

Yet we live and we worship at cross-purposes with the Creator's cosmic purpose. In his epistle to the Romans, Paul traced the source of human wickedness to our exchange of God's truth for "a lie," as we worship the creation instead of its Creator (Rom. 1:25). We look in all the wrong places for the contentment we crave, the unbreakable love for which we long. We direct our affection and devotion to objects that do not deserve our wholehearted allegiance and adoration. We rest the full weight of our hopes and our hearts on fragile relationships and fleeting resources that will, sooner or later, collapse under such pressure, bringing us down in the process.

Perhaps the most tragic scene in the Oscar-winning film *Chariots of Fire* shows the great sprinter Harold Abrahams in the aftermath of his victory at the 1924 Olympics. He places his gold medal on his Jewish prayer shawl, latches his suitcase, and glumly slips away from his British teammates' locker-room celebration. When one calls him to join their joy, a more experienced competitor hushes him:

"Let him be. He's whacked."

"But he won!"

"Exactly. One of these days, you're going to win yourself. Then you will know it's pretty difficult to swallow."

The scene shifts to a close-up of Abrahams' grim face as he and his coach, Sam Mussabini, "celebrate" their triumph late that night by getting drunk in a Parisian bar. His race won and his goal achieved, Abrahams had proved himself and chastened the anti-Semitic prejudice of England's academic and athletic elites. Yet the viewer sees in the victor no jubilation but only a bleak numbness, perhaps from blending postcompetition depression with too much wine. Having grasped the object of his worship, he found his achievement small and unsatisfying. What purpose could give his life meaning tomorrow?[1] Whatever god or goal we choose to define ourselves, our hopes, and our happiness, the heartbreaking disappointment of misdirected worship always ensues.

Christ's redemptive mission as the Lamb was to reconquer, reclaim, and re-create us to be worshipers of the living God, who alone deserves worship. That gripping true story has been the theme of the song of the King who stooped to conquer (see Phil. 2:6–11), which we began to consider in the previous chapter. In the first two "stanzas" of the hymn, we heard of the divine and glorious height from which this King descended (2:6) and the depth of his humility in his incarnation, suffering as a servant, and death on the cursed cross (2:7–8). Now in the third stanza our hearts and minds are directed upward to the purpose and result of Christ's self-humbling and sacrificial suffering, in his exaltation by God his Father above all creation,

1. The film's postscript notes that Abrahams subsequently became a widely respected sports journalist and a statesman-advocate for amateur athletics.

to receive adoring worship from every creature everywhere. We glimpse God's purpose in creation and in redemption, reaching its divinely designed destination: his own worship by his creatures.

The Divine "Mutual Admiration Society"

Before we explore Paul's exposition of the awe-evoking exaltation that reversed Jesus' self-denying humiliation, we need to consider how this third stanza fits into Paul's purpose for introducing this *Carmen Christi*, this "song of Christ," in the first place. In Philippians 2:3–4 the apostle exhorted the Philippian Christians to replace their natural tendency toward self-centered ambition with a selfless readiness to "count others more significant than yourselves" and to pursue not only their own concerns but also those of others. He showed that the mind-set that should govern relationships among Christians is the mind-set of humble servanthood that Christ displayed in his incarnation and suffering, a mind-set that now belongs to believers because we have been united to this suffering Savior (2:5). Jesus is the great example of what it means to shun self-interest and to serve others selflessly.

To think of Jesus as an example to be followed may make you uneasy, if you are a Protestant who treasures the Reformation's rediscovery of the biblical truth that we are justified through faith alone in Christ alone, so that God's vindicating verdict and acceptance do not depend on our own efforts to obey. Centuries of church history have shown how easy it is for well-meaning Christians to reduce Jesus, in practice if not in theory, to a mere moral model to be imitated, urging us to make day-to-day decisions by asking and answering, "What would Jesus do?" Yet the abuses of the theme of the imitation of Christ should not blind us to the fact that the inspired biblical authors themselves call us to conform ourselves, by the power of God's Spirit, to the character of our Lord and Savior. In his ancient Law the Lord told Israel, "Be holy, for I am holy" (Lev. 11:44–45; 19:2; see 1 Peter 1:15). Paul insists that the forgiving love of the Father and the sacrificial love of the Son for us set the pace for our love for each other:

> Be kind to one another, tenderhearted, forgiving one another, as God in Christ forgave you. Therefore be imitators of God, as beloved children. And

> walk in love, as Christ loved us and gave himself up for us, a fragrant offering and sacrifice to God. (Eph. 4:32–5:2)

Peter likewise expects us to follow Jesus' lead in accepting undeserved suffering with patience rather than retaliation: "For to this you have been called, because Christ also suffered for you, leaving you an example, so that you might follow in his steps" (1 Peter 2:21). If we are serious about the Reformers' principle that Scripture alone—and the whole of the Scripture—directs our beliefs and our behavior, we cannot turn a deaf ear to the Bible's pervasive summons to follow our Lord's example as he renews us and restores us into the image of God (Eph. 4:24; Col. 3:9–11). But Jesus is certainly more than a mere example. His redeeming grace precedes and provides our response of grateful imitation, as Paul and Peter showed in the passages above. In our present text, Paul has already alluded to that glorious union that God's grace forged between believers and the Savior: because we are "in Christ Jesus" (Phil. 2:5), his mind is ours through the Holy Spirit, and that attitude must control how we regard and treat each other, in self-forgetting, others-honoring humility.

Now, some students of the Christ-hymn have questioned whether this final stanza tacitly undermines Paul's appeal to selfless servanthood: When Paul tells the end of the story—Christ's exaltation following humiliation—has he undermined his own pastoral purposes? If the happy ending of Jesus' story implies that our story, too, will end happily, has not Paul invited us, either intentionally or inadvertently, to revert to a self-centered motive?[2] Admittedly, it is a subtler and more patient self-centeredness than the competitive conceit that demands one's own way *this very moment*. But does Paul mean to suggest that as Jesus patiently endured suffering and was then rewarded with glory, so Christians should let others win the little skirmishes of will in the present, for the sake of reaping the big rewards in the future? John Calvin seems to see the subtext of the movement from condescension in suffering to exaltation in glory in this way:

> Now, that all are happy who, along with Christ, voluntarily humble themselves, he shows by His example; for from the most abject condition He was exalted to the sublimest height. Every one therefore that humbles himself

2. I. Howard Marshall, *The Epistle to the Philippians* (London: Epworth, 1991), 55–56.

> will in like manner be exalted. Who will now refuse submission, by which he will ascend into the glory of the heavenly Kingdom?[3]

The connection that Calvin draws between Christ's trajectory from suffering to glory and ours is certainly taught by Paul and Peter elsewhere. In this same epistle Paul links our experience of sharing Christ's sufferings and "becoming like him in his death" with the future hope to "attain the resurrection from the dead" (Phil. 3:10–11). Paul tells the Christians at Rome that those who are coheirs with Christ "suffer with him in order that we may also be glorified with him" (Rom. 8:17). To Timothy the apostle quotes a trustworthy saying: "If we have died with him, we will also live with him; if we endure, we will also reign with him" (2 Tim. 2:11–12). Peter likewise calls Christians to entrust ourselves to God as Jesus did (1 Peter 2:23; 4:19), knowing that as Christ's path led through suffering to glory, ours does as well (1:6–7, 11). So Calvin's explanation fits with the apostles' teaching in other passages: believers in Jesus, like Jesus himself, can endure suffering patiently in the hope of coming glory. The exaltation of Christ portrayed in Philippians 2:9–11 implies our future exaltation with him.

But another explanation of these closing verses of the Christ-hymn might reveal the apostle's pastoral purpose even more clearly.[4] Notice how verses 9, 10, and 11 begin, and how they end. Throughout the downward trajectory of servanthood, suffering, and sacrifice, Christ is the subject of every attribute and the agent of every action: he "was . . . did not count . . . made himself nothing, taking . . . being born . . . being found . . . humbled himself . . . becoming obedient." Now in verse 9, suddenly, as we begin the upswing from Christ's humbling descent to his glorious ascent, *God the Father* takes the initiative and becomes the actor: "God has *highly exalted* him and *bestowed* on him the name that is above every name" (Phil. 2:9). Then, at the finale of the Christ-hymn, as every creature's knee and tongue respond to God's elevation of Christ, the result is nothing less than "the glory of God the Father"

3. John Calvin, *The Epistles of Paul the Apostle to the Galatians, Ephesians, Philippians and Colossians*, ed. David W. Torrance and Thomas F. Torrance, trans. T. H. L. Parker (Grand Rapids: Eerdmans, 1965), 250.

4. Thanks to Micah Renihan, who offered this insightful and, I find, persuasive explanation of Paul's purpose for Philippians 2:9–11 as "a wonderful picture of the intra-Trinitarian selflessness" in an unpublished paper, "The Problem of the Exaltation of Christ: A Defense of the Imitative View of Philippians 2:5–11," submitted in my elective on Philippians in 2009.

(2:11). The Father delights to honor the Son for the Son's accomplishment of his redemptive mission, and the honor bestowed on the Son displays in even greater fullness the glory of the Father.

These two divine persons, who with the Holy Spirit constitute the one triune God, are not in competition with each other. The envy, rivalry, and conceit that threaten our unity (Phil. 1:15; 2:3) have no place at all in the interpersonal relationships within the Trinity. Jesus' supreme glory does not in any way reduce the glory of the Father. When the Father exalts Jesus with the title above every title, to receive worship on bent knees from every creature in heaven above, on earth below, and in even lower regions,[5] we glimpse among the persons of the Trinity that very "mind-set of Christ" that now belongs to us by grace, through our union with him. The Father does not "look . . . only to his own interests" (recall 2:3) but rejoices to bestow supreme honor on the Son. And the Son does "nothing from rivalry or conceit," but instead rejoices that his own exaltation further enriches "the glory of God the Father." The purpose of stanza 3 is to invite us to honor each other above ourselves, reflecting the mutual affection of the Father and the Son (and, by implication, the Holy Spirit) and their delight to enhance each other's glory.

We sometimes speak of close friends as a "mutual admiration society." Each friend sees so much to appreciate in the other that they cannot help but find ways to speak compliments to and about each other, to call others' attention to the friend's kindness, integrity, intelligence, abilities, and achievements. When you meet friends like these, or a married couple who are more in love today than when they exchanged their vows five decades ago, you are glimpsing a creaturely replica, a miniature reflection, of the boundless and endless delight that three persons of the triune God enjoy forever in enhancing the display of one another's beauties and excellencies.

Of course, the mutual love among the Father, the Son, and the Spirit infinitely transcends what we see in the closest of relationships among human beings. The infinite-personal God who is triune is three distinct persons who share one divine substance. The Westminster Shorter Catechism (answer 6) sums up the deep mystery of God's tri-unity as revealed in the Bible: "There

5. Note also John's vision in Revelation 5:13, in which he "heard every creature in heaven and on earth and under the earth and in the sea, and all that is in them, saying, 'To him who sits on the throne and to the Lamb be blessing and honor and glory and might forever and ever!' "

are three persons in the Godhead; the Father, the Son, and the Holy Ghost; and these three are one God, the same in substance, equal in power and glory." We creatures, on the other hand, are distinct persons who are also distinct beings, so our societies, whether of "mutual admiration" or otherwise, are gatherings of discrete individuals. Then when we mix in the complications of our sinfulness—none of us is altogether admirable, none admires as we should, our admiration for another creature can cross over into idolatrous adoration—it is obvious that human "mutual admiration societies" at their best can be no more than faint and flawed shadows of the eternal reality of the perfect and eternal love among the Father, the Son, and the Spirit.

Although our closest human relationships cannot replicate the bond of mutual delight among persons of the Trinity, nevertheless the infinite and incomparable Creator has been pleased to devise a creaturely community in which this triune love is tasted and displayed. This community is Christ's church. The evening before his death, Jesus prayed for the church. Looking ahead through the coming generations of those who would believe in him, he asked his Father on our behalf "that they may all be one, just as you, Father, are in me, and I in you, that they also may be in us, so that the world may believe that you have sent me" (John 17:21). He even repeated the comparison between his oneness with the Father and his followers' oneness with each other: "that they may be one even as we are one, I in them and you in me, that they may become perfectly one" (17:22–23). Though our unity with each other as members of Christ's church cannot be *identical with* the profound unity of the Father, Son, and Spirit, surely our unity can and must be *like* the unity of our triune Creator. In fact, it should resemble the mutual love shared by the three divine persons closely enough and visibly enough that our bond becomes a signpost that points the world toward our Savior and his mission.

The glorious finale of the Song of the Condescending King, therefore, is not a summons to sublimate our self-centeredness for the time being, simply for the sake of satisfying it in the end. We could mistake it as such only by failing to grasp that the incomparable Creator is one God in three persons, whose glory is to enrich and display each other's glory. In *The Pleasures of God*, Dr. John Piper speaks of God's delight in being God: "From all eternity God had beheld the panorama of his own perfections in the face of his

Son. All that he is he sees reflected fully and perfectly in the countenance of his Son. And in this he rejoices with infinite joy."[6] Then Piper asks whether the Father's pleasure in the reflection of his own perfections in the Son means that God is vain, as we would be if we were to spend hours admiring ourselves in front of a mirror. Piper's answer puts our Maker and ourselves into proper perspective. It is vain—empty, hollow (in Philippians 2:3 the KJV captured the sense of the Greek *kenodoxia* in the [now archaic] word "vainglory")—when we mere creatures lavish adulation on ourselves. Such misdirected honor makes us idolaters. Only the incomparable Creator deserves such glory. Precisely because he is uniquely worthy of such glory, it is fitting for us to seek his glory as our ultimate objective: "Man's chief end is to glorify God, and to enjoy Him for ever."[7] And it is right for God himself to seek that same aim: his own glory above all. In fact, Piper observes, if God did not delight in and promote his own glory, if he attributed ultimate value to anyone or anything less than himself, he would be the idolater![8]

Our text, however, celebrates One whom God the Father has appropriately exalted above all, the Son who is eternally the Father's equal and who thus deserves total worship from every creature everywhere. The three persons of the triune God (here, the Father and the Son are specifically mentioned) delight to magnify one another. They are right to do so, for each deserves the highest splendor imaginable—even beyond the bounds of our finite imaginations!

We are created in the image of this God who is One-in-Three. We most closely resemble the God who made us in his likeness when we rejoice to exalt each other, as the Father exalted Christ. We are most like our Maker when we discover that the sweetest dimension of his grace, the mercy that confers on us a share in Jesus' glory, is that both Christ's exaltation and ours find their purpose and goal in "the glory of God the Father" (Phil. 2:11).

Have you discovered the emptiness, the vanity, of setting your sights on yourself—your own agenda, your own reputation, your own satisfaction? The exaltation that God bestowed on his beloved Son, who became the Suffering Servant, sets before you a far grander purpose. He calls you to

6. John Piper, *The Pleasures of God: Meditations on God's Delight in Being God* (Sisters, OR: Multnomah, 2000), 42–43.

7. WSC, answer 1.

8. Piper, *Pleasures*, 43.

break free from the narrow confines of self-interest, to take an active role in the Trinity-reflecting community, Christ's church. Even now, for all of the church's flaws, the risen Lord is working on and in his people to make them a "mutual admiration society" in which the Father's pleasure to honor the Son and the Son's delight to glorify the Father are reflected in gentle words and serving deeds, which reflect a Christ-formed mind-set that counts others more significant than ourselves and attends to others' needs before our own.

The King's Ascent to Wield Universal Authority

Philippians 2:9 marks the dramatic turning point in the hymn, where Christ's downward plunge is reversed by the Father's upward pull. Paul seizes on a striking verb that appears nowhere else in the New Testament. The ESV's "highly exalted" captures his meaning acceptably; but the components of this compound verb—the preposition "above" (*hyper*) prefixed to the root "exalt" (*hypsoō*)—may foreshadow the contrast between Christ's exalted status and all the creatures who are subordinate to him "in heaven and on earth and under the earth." Such a contrast is explicit when this verb appears in the Greek (Septuagint) translation of Psalm 97:9 [LXX 96:9]: "you are exalted [*hyperypsoō*] far above [*hyper*] all gods." Likewise, here Paul explains the verb "exalted above" [*hyperypsoō*] by mentioning "the name that is above [*hyper*] every name," which God conferred on Jesus. Christ's ascent from the depths of despicable death has carried his whole person—including the humanity in which he served and suffered—to a point higher than the highest of all his creatures. As the victorious Redeemer of God's guilty but beloved people, he emerged from the grave the third day, entered heaven forty days later, and soon thereafter celebrated his enthronement by pouring out his Holy Spirit in power on his people. He has carried our humanity, now bursting with new creation life, up from the grave, into the heavens, to take his seat at the Father's right hand.

To the glory that has always been Christ's as the eternal God and Creator of the universe, a new and unprecedented splendor has been added: through his descent he has rescued his enemies and turned us into beloved children of his Father! Christ's request of his Father en route to the cross has been magnificently answered: "I glorified you on earth, having accomplished the work that you gave me to do. And now, Father, glorify me in your own presence with the glory that I had with you before the world existed" (John 17:4–5). And his ascent is not merely a return to the preincarnation status quo. Now, because of his obedient suffering, an enlarged audience adores the glory of God's grace: "Father, I desire that they also, whom you have given me, may be with me where I am, to see my glory that you have given me because you loved me before the foundation of the world" (17:24).

When God highly exalted his obedient Son in reward for his suffering, he "bestowed on him the name that is above every name." Initially we might think that this is the personal name *Jesus*, since Paul goes on to say that every knee will bow "at the name of Jesus." It is better, however, to understand the "*name* that is above every name" not as a personal name but as an official *title*—the title *Lord* that was conferred on Jesus at the time of his resurrection, signifying his supremacy over all as the glorified God-man. After all, the personal name *Jesus* was given to the Son at his birth, in anticipation of his mission to save his people from their sins (Matt. 1:21). It was through his resurrection and ascension that God made Jesus "both Lord and Christ" (Acts 2:36; see Rom. 10:9). After his resurrection, Jesus declared his universal authority as Lord: "All authority in heaven and on earth has been given to me" (Matt. 28:18; see Dan. 7:14). Paul's affirmation that the Son's "name" ranks above all others shows that he is referring to Christ's supremacy over all the powers in the universe. In Ephesians 1:20–21 Paul makes explicit this *titular* supremacy of the "name" bestowed on Jesus: God "raised him from the dead and seated him at his right hand in the heavenly places, far above all rule and authority and power and dominion, and above every *name* that is named, not only in this age but also in the one to come." The superior name granted to Christ in his exaltation is the title *Lord*, as the hymn's climax shows: "every tongue [will] confess that Jesus Christ is *Lord*, to the glory of God the Father."

As Lord, Jesus Christ is the Supreme Emperor of the entire universe, infinitely above the puny Caesars who had the presumption to claim the title *lord*, though they ruled an empire that, in global and historical perspective, proved small and short-lived. As Lord, the exalted Son outranks the superhuman spiritual forces, gods and demons benign and malevolent, that vied for worshipers' fear and allegiance in cosmopolitan colonies such as Philippi, where local Macedonian and Greek polytheism absorbed the influences of Asian East and Roman West.

Rome's imperial dominance long ago succumbed to brutal invaders who showed no deference for the Caesars' glory and the empire's administrative, military, and cultural achievements. On the other hand, the living Lord whom Paul served is still extending his reign to the ends of the earth through the gospel of his grace and the power of his Spirit. Macedonia's homegrown deities, as well as those imported from Achaia (Greece) to the south, from Asia to the east, and from Rome to the west, remain subjects for scholarly research but no longer compete for worshipers' allegiance. Jesus, however, continues to lay claim to the hearts and minds of the peoples that cover the globe, and he does so through a strategy that seems surprisingly fragile. This Lord conquers nations not through force of arms but through the message of his cross, the instrument of his execution and symbol of shameful weakness, carried outward to the nations through his heralds' words and inward into human hearts through his Holy Spirit.

When the Father exalted Jesus his Son, raising him from the dead and installing him as Lord of all at his right hand in heaven, the whole course of history turned a corner, from decay and death toward healing and everlasting life. So how should you respond to Christ's coronation and enthronement? You may have been feverishly slaving to win the "blessing" of other "lords"—financial security, others' approval, romantic love, academic achievement, pleasures of various kinds. If that is true of you, the fact that the Servant who once suffered now wields all authority "in heaven and on earth and under the earth" is your wake-up call. No other lord can deliver on its promises; no other lord deserves your unquestioning allegiance. Only Jesus does. His resurrection from the dead turned human history and cosmic history in a new direction, which is leading to the day when every knee will humbly bow and every tongue express devotion to this living Lord. He already bears the name above every name, the title that transcends all titles. This reality demands that you submit to his dominion today.

The King's Ascent to Receive Universal Worship

The result of the Son's exaltation is "that at the name of Jesus every knee should bow . . . and every tongue confess that Jesus Christ is Lord" (Phil. 2:10–11). This is precisely what Saul was compelled to do when he was confronted by the overpowering glory of Jesus on the Damascus road: he fell on his face and called Jesus "Lord" (Acts 9:4–5).

Saul was dashed to the ground by a blinding light from heaven and could find no other word but *Lord* to address the august Speaker who confronted him. His awestruck behavior was understandable, in view of the intensity of his experience that day. But Paul's epistles show that such expressions of awe and adoration typically characterized the worship services of the early church. The gathered church confessed Jesus as Lord (Rom. 10:9; 1 Cor. 12:3). The conscience-piercing truth of God's Word and the heart-searching presence of God's Spirit in the assembled congregations proved undeniable and irresistible even to unbelieving visitors: "the secrets of his heart are disclosed, and so, falling on his face, he will worship God and declare that God is really among you" (1 Cor. 14:25). A psalmist summoned ancient Israel to worship: "Oh come, let us worship and bow down; let us kneel before the LORD, our Maker!" (Ps. 95:6). Peter and Paul knelt in prayer, and Paul's prayer was in the midst of the church at worship (Acts 9:40; 20:36). Bent knees and confessing tongues expressed a profound sense of the presence of the living God in the homes and halls where Jesus' followers gathered for worship.

Is the worship of our churches today—the worship of your church—focused on Christ's mercy and his majesty so that hearts are bowed in humble adoration and lifted in hope, so that knees bend in humble wonder and tongues joyfully confess that "Jesus Christ is Lord, to the glory of God the Father" (Phil. 2:11)? Of course, physical postures—sitting, standing, kneeling, lying prostrate—may be merely "scripted," either by liturgical tradition or by unspoken expectations about how spontaneous spiritual experience is to be expressed. What we do with our bodies is not an infallible indicator of the state of our hearts. Jesus observed, quoting Isaiah's prophecy: "This people honors me with their lips, but their heart is far from me" (Matt. 15:8). On the other hand, our knees and our tongues are not disconnected from our hearts. Jesus also said, "For out of the abundance of the heart the mouth speaks" (12:34; see 15:18–19). The posture and pronouncement of the earliest Christians challenge you to ask yourself, "As I come to worship, am I alert to the awesome holiness of God among his people? Does his powerful presence, which bent their knees in prayer and set their tongues to praise, grip my heart, too? If so, how can I, how must I, express my wonder with my whole being?"

In our passage, Paul looks forward to a global—rather, universewide—celebration, of which the church's weekly worship is a foretaste. The scene that Paul portrays—the consummation that is sure to come—far outstrips the splendor of the grand finale of any blockbuster cinematic epic. Only the most jaded viewers can keep their pulses from racing and their eyes from moistening at the end of *Star Wars* Episode IV, as Han Solo, Luke Skywalker, and Chewbacca enter the great hall of the Rebel Alliance, to be rewarded by Princess Leia and applauded by crowds for destroying the empire's Death Star. Simpler in ceremony but no less majestic and moving is the climax of Peter Jackson's film trilogy based on J. R. R. Tolkien's *Lord of the Rings* epic, when, at the end of *The Return of the King*, the peaceable peoples of Middle-earth gather under open skies to extol King Aragorn and his hobbit friends for destroying the evil Ring of Power. If you can picture such scenes, then realize that the best that filmmakers can muster with special effects and thousands of "extras" cannot begin to do justice to the splendor of the scene that Paul is portraying. What a jubilant festal assembly that will be, when every creature "in heaven and on earth and under the earth" will bow the knee and acknowledge the utter supremacy of Jesus as Lord of all!

We concluded earlier that the "name" now conferred on the Son is not his personal name *Jesus* but the title *Lord*, signifying his investiture with absolute dominion over the entire universe as he has taken his seat at God's right hand. And yet Paul also wants us to understand that the name *Lord* is not only a title of office. It is also an indication of identity. Paul was well aware that the biblical scholars who had translated the Hebrew Scriptures into Greek (the Septuagint) had used the Greek term *Lord* (*kyrios*) to represent the distinctive name *Yahweh*, by which Israel's covenant God identified himself to his people. Jews and Gentile converts who frequented the Greek-speaking synagogues of the Dispersion were bound to associate the Greek term *kyrios* with the personal name of God, the Creator of the universe and Redeemer of his people Israel. English speakers today do the same when we read "the LORD" in our Bibles (the SMALL CAPS are our translators' signal that the Hebrew original reads *Yahweh*). Even Gentile believers in cities that had no synagogue, such as Philippi, were quickly introduced to the Old Testament Scriptures in Greek translation. (This is why, even when the apostles wrote to congregations with thoroughly pagan pasts, they peppered their epistles with references to the Old Testament.[9]) Even the Gentile Christians at Philippi could be expected to recognize that sometimes *kyrios*, "Lord," was nothing less than a name of God himself. That is how they should understand the name here, as Paul's allusion to an Old Testament passage makes clear.

Paul drew the language about every knee bowing and every tongue confessing from Isaiah 45, where we hear the LORD, the God of Israel, declaring: "Turn to me and be saved, all the ends of the earth! For I am God, and there is no other. By myself I have sworn; from my mouth has gone out in righteousness a word that shall not return: 'To me every knee shall bow, every tongue shall swear allegiance'" (Isa. 45:22–23). In this section of Isaiah, the Lord challenges the pagan idols to do something or say something to back up their claims to be gods. He confidently announces that he alone stands at the beginning and the end of history (41:4; see 48:12–13). He alone can announce the future and bring it about (41:21–24). He is Yahweh, the Lord, the only living God. There is no other (43:10–11)! Therefore, he alone is worthy to receive universal worship (every knee bowed) and a universal confession of absolute allegiance.

Several years earlier, Paul had quoted this very text from Isaiah in his epistle to the Romans, substantiating the sobering truth that everyone will stand before God's judgment seat and give an account to our Creator: "As I live, says the Lord, every knee shall bow to me, and every tongue shall confess to God" (Rom. 14:10–12). Now Paul takes up the words in which the Lord God asserted his supremacy and uniqueness, words that the apostle himself had applied to God's authority as Judge of all. And Paul applies those words directly to Jesus! It is hard to think of a section of Scripture that argues more forcefully and explicitly than Isaiah 40 through 48 that Yahweh, the Lord, is the one and only eternal and living God; that he alone is the source of salvation for his people;[10] and that he alone is

9. Peter, for example, refers to his recipients' past heritage and history in paganism (1 Peter 1:18; 4:3–4), even as the apostle repeatedly quotes Old Testament passages (e.g., 1:24–25; 2:6–8; 3:10–12; etc.) and alludes even more frequently to Old Testament imagery and terminology in order to redefine his once-pagan, still-ethnically-Gentile readers as the new Israel through their faith in Christ ("elect exiles of the dispersion," 1:1; "a chosen race, a royal priesthood, a holy nation," 2:9; etc.).

10. The invitation of Isaiah 45:22, "Turn to me and be saved, all the ends of the earth! For I am God, and there is no other," is preceded by the Lord's declaration of his incomparable uniqueness and

worthy of every creature's complete loyalty and adoration. The fact that the apostle applies such an unmistakably monotheistic text from Isaiah's prophecy to Jesus, who became human and died on the cross, shows who he considers Jesus to be: Creator of the universe and covenant Lord of Israel, equal with the Father.

So when visitors from the Watchtower Society (Jehovah's Witnesses) come to your door and insist that it is blasphemy for you to worship Jesus as though he were Jehovah God, you can calmly—but compassionately!—take them to Isaiah 45, and then to Philippians 2:10–11: The Lord who in Isaiah insists that he alone is God, the God to whom every knee will bow and every tongue confess, is *Jesus*, who stooped to die the cursed death of the cross.[11] This Jesus is the only One to whom the ends of the earth must turn, in order to receive salvation! Peter confessed the same truth to Israel's leaders: "This Jesus . . . has become the cornerstone. And there is salvation in no one else, for there is no other name under heaven given among men by which we must be saved" (Acts 4:11–12). Gently invite your Watchtower visitor to turn with you to Jesus the Lord in humble trust and deep submission, and be saved.

As you invite your visitor (and others as well) to appreciate the divine majesty of Jesus, you yourself need to remember that both submission and trust are included in confessing that Jesus Christ is Lord. You may have noticed that our English version of Isaiah 45:23 reads, "To me . . . every tongue shall *swear allegiance*." That is an accurate reflection of Isaiah's Hebrew. The Hebrew verb *shaba'*, which stands behind the ESV's "swear allegiance," typically refers to the taking of an oath and the resultant demand of loyalty (for example, see Gen. 21:23–24, 31; 1 Sam. 24:21; 2 Chron. 36:13). In the Greek Septuagint, this term was

power to save: "there is no other god besides me, a righteous God and a Savior" (45:21). This motif is immediately echoed after the announcement that every knee will bow and every tongue swear allegiance to him: "Only in the LORD, it shall be said of me, are righteousness and strength . . . . In the LORD all the offspring of Israel shall be justified and shall glory" (45:24–25).

11. The allusion to Isaiah 45 in Philippians 2:10–11 is visible in the Watchtower Society's *New World Translation of the Holy Scripture* (Brooklyn, NY: Watchtower Bible and Tract Society, 1971), in which Isaiah 45:23–24 reads: "By my own self I have sworn . . . that to me every knee will bend down, every tongue will swear, 'Surely in Jehovah there are full righteousness and strength.'" The *New World Translation* renders Philippians 2:10–11 this way: "so that in the name of Jesus every knee will bend of those in heaven and those on earth and those under the ground, and every tongue will openly acknowledge that Jesus Christ is Lord to the glory of God the Father."

rendered *exomologeō*, a term apparently flexible enough to encompass both swearing an oath of loyalty (as in Rom. 14:11) and declaring a solemn conviction (as in Phil. 2:11). In our text, the content of the conviction confessed—that "Jesus Christ is Lord"—implies exclusive and ready commitment to this Lord. Both the bent knees and the confessing tongues of all creatures will one day express their universal allegiance to Jesus the King. You realize, I trust, what this means for you today: If you confess "Jesus Christ is Lord" as his follower and a member of his church, that announcement must be far more than a theological thesis that you affirm and defend. To say that simple but profound sentence is to renounce your independence and submit to the will and word of this Lord, "who is God over all, blessed forever" (Rom. 9:5). It is to bend the knee of your heart, to embrace his agenda for your life and his priorities over your preferences. It is to say and really mean, "To me to live is Christ, and to die is gain" (Phil. 1:21).

Delight in His Glory

The glorious finale of the Song of the Condescending King brings us to the brink of eternity. We look ahead to a future day in which the reality that already defines world history—that God has exalted Jesus the eternal Son and Suffering Servant to the highest place—which remains partially hidden for the present, will be displayed for all to see. Christ's vindication through his resurrection, ascension, and enthronement as Lord at the Father's right hand has reversed the humiliation and suffering that he voluntarily embraced in his incarnation and redemptive mission, his obedience to the Father's purpose, even though it led to the death of the cross. Because Jesus the Son gave all, the Father was pleased to give him the name above every name. Not only has Christ been the Father's eternal equal from the standpoint of his divine nature, but now as the incarnate Son, still sharing our humanity and now abundantly alive from the dead forevermore, he bears the divine name *Lord*. His descent into suffering and ascent to glory have blazed the path for those who trust him, whose spiritual well-being he served in preference to his own interests. To know that he now reigns over all and one day will certainly be acclaimed as Lord by all gives you reason for hope in your current troubles, but it provides even more than the prospect of relief from pain and shame. As you hear of God's pleasure in exalting his Son and of the Son's delight in fulfilling the Father's plan and advancing the Father's glory, you can glimpse your own destiny from a distance. You who trust Jesus can anticipate the day when your love for each other has displaced your inborn self-interest, and the heartfelt unity of affection seen in Christ's church shows the world a reflection of the infinite love among the persons of the triune God. Surely such hope gives you the strongest of reasons to replace competition and conceit with compassionate service to others this very day.

10

Bright Stars in a Dark Sky

Philippians 2:12–18

Therefore, my beloved, as you have always obeyed, so now, not only as in my presence but much more in my absence, work out your own salvation with fear and trembling, for it is God who works in you, both to will and to work for his good pleasure. Do all things without grumbling or questioning, that you may be blameless and innocent, children of God without blemish in the midst of a crooked and twisted generation, among whom you shine as lights in the world. (Phil. 2:12–15)

The human race has shamefully mistreated minorities outside the cultural mainstream. Groups that do not fit into the surrounding society because of race, religion, language, or cultural heritage are marginalized and often excluded. Their rights are ignored, their customs ridiculed, their persons and families harassed—even arrested, assaulted, or murdered. We think of the extermination of Jews in the Holocaust Germany of the 1930s and 1940s. Or the oppression, discrimination, and dehumanization suffered by blacks in the American South both before and after the abolition of slavery. Or the enslavement of Christians in Sudan, or the martyrdom of Christians in Iraq, India, Indonesia, and elsewhere.

Among such marginalized minorities, we could also include Paul's Christian friends at Philippi. When Paul and Silas first brought to them the gospel that had transformed their lives, its opponents retaliated with accusations in which a blend of anti-Semitic prejudice and civic pride can be heard: "These men are Jews, and they are disturbing our city. They advocate customs that are not lawful for us as Romans to accept or practice" (Acts 16:20–21). Whether or not some of the Philippian believers had belonged to the city's tiny Jewish community, the fact that they embraced a faith brought by Jewish messengers, who—to make matters worse—were accused of disrupting law and order and of advocating practices unworthy of Roman citizens, would be enough to put them beyond the borders of acceptable society. So Paul wants to know that they are "standing firm in one spirit, with one mind striving side by side for the faith of the gospel, and not frightened in anything by your opponents" (Phil. 1:27–28). The gospel's enemies were treating believers in ways that could engender fear. Paul compares their trials with his own, noting that they are "engaged in the same conflict that you saw I had and now hear that I still have" (1:30). They had seen Paul and Silas beaten and imprisoned in their own city, and they knew that now Paul was again in Roman custody. Had they been beaten or arrested for their faith? Even if the ferocity or form of "conflict" was different for the Philippians than for Paul, they shared his identity as excluded outsiders. Marginalization is inevitable for those called to live as "children of God without blemish in the midst of a crooked and twisted generation" (2:15).

Christians in North America and Europe also experience mild forms of marginalization by a cultural mainstream in which forces such as secularism, relativism, individualism, and neopaganism are in ascendance. Influential news media have little respect for people who believe that Christ is the only way to God, or that God's revealed moral absolutes bind everybody all the time. Suspicion of or scorn for Christian conviction often characterizes leaders in higher education. Relentless pressure to embrace diversity and toleration as supreme virtues drives an agenda that welcomes any personal choice as acceptable, except one: only those whose consciences cannot affirm religious relativism and ethical indifference are considered intolerable!

These ways that our society marginalizes Christians are not pleasant. Yet in the wider perspective of world history and global Christianity, things have been and are today far worse elsewhere for those who follow Jesus. The forms of scorn that we in the West encounter cannot compare with what Paul and the Philippian Christians were going through when he wrote, or with what fellow believers encounter even today. Yet often it is when the pressure to conform is nonviolent and subtle that it is hardest to recognize and resist. So we, too, need to hear the apostle's encouragement to those marginalized Jesus-followers of the first century, calling them to stand out from the darkness around them.

In Philippians 1:27, when Paul told his friends to conduct themselves as citizens of heaven in courageous steadfastness and compassionate selflessness, he implied the two challenges confronting the church at Philippi: external persecution and internal disunity. He elaborated on the perspective they needed as they faced suffering for Christ (1:28–30), and then on the humility and love that would maintain and deepen their unity (2:1–4). Humility and concern for others are seen supremely in Jesus himself, so Paul led his friends down the descending path of Christ's humiliation, from divine splendor to the nadir of shameful death on a cross; and then Paul pointed up to the Savior's ascent to the pinnacle of glory, to be worshiped by every creature everywhere (2:5–11).

Now in the text before us (Phil. 2:12–18), it feels as though Paul were marching us back down from the "mountaintop" into the nitty-gritty of everyday life. I grew up in the Los Angeles area, and every summer I attended a Bible camp in the San Gabriel Mountains that surround the metropolitan basin. Traveling up to the mile-high campground, I didn't notice the air quality improving until we got out, took a deep breath, and smelled the pines. But at the end of the week, on the way down the mountain, we could see all too well that our "mountaintop experience" was over as we emerged from the forest: there, below, stretched the L.A. basin, blanketed in brown smog. Then I empathized with Simon Peter's longing to erect tents and stay with Jesus on the heights (Mark 9:5–6)! We might prefer that Paul keep us on lofty peaks of Christology, to relish the majesty and mission of Jesus: God become man, become servant, become obedient to death-by-cross, then super-exalted to be acclaimed as Lord. After such a mountaintop experience, Paul's practical coaching about how to respond to others' hostility seems like a disappointing descent into the smog of everyday life: "Keep obeying . . . Don't grumble . . . Your neighbors are a crooked generation . . . I may be executed soon . . . and by the way, through it all I am rejoicing and you should, too!" What an odd assortment of mundane instructions!

Yet this passage is not as mundane or ordinary as it seems. In fact, Paul is building on the mountain peak of his doctrine of Christ, showing us in practical terms what it means to "have this mind among yourselves, which is yours in Christ Jesus" (Phil. 2:5). He is continuing the motif of behavior befitting citizens of heaven, strangers sojourning briefly on the earth as we journey toward the celestial city that is our true home. He focuses on two characteristics of our heavenly citizenship, which make us stand out from the culture around us: (1) sober and hopeful obedience, and (2) patient and selfless contentment.

"Work Out Your Salvation" in Hopeful Obedience

The first implication to be drawn from the condescension and exaltation of Jesus is that Christians' sober and hopeful obedience, fueled by God's grace at work in us, makes us shine brightly against a dark sky. Paul repeats the term *obey*, a form of which he had just used to describe Jesus' faithfulness when he became "*obedient* to the point of death" (Phil. 2:8), to commend the Philippians' past obedience and to call them to current and future obedience. This echo of Paul's song of the King who stooped to conquer shows that the opening word, "Therefore," is full of meaning. The "to-do" list in verses 12–18 is the direct consequence of Jesus' descent to despicable death and ascent to unparalleled supremacy. Everything that Paul is about to urge on the Philippians and us flows out of what Christ did for us in taking human nature, taking the servant's role, becoming obedient to death, and then being exalted as Lord over all.

In light of Jesus' costly obedience for us, we must persist in the path of obeying the Lord, even when obedience seems to bear no fruit and elicit no appreciation, when it demands putting our personal preferences and brilliant ideas on the back burner in preference for others' agendas, when it is painful and inconvenient, when it costs us dearly, as it cost Jesus his very life.

Paul wants his Philippian friends' obedience, though he is absent, to outstrip the faithfulness that he saw in person at the birth of the church and during later visits to Macedonia. That challenge, "not only as in my presence but much more in my absence," reveals Paul's realism. He knows that it is easier to stay fired up and faithful when a dynamic father in the faith is on the scene, cheering young believers on in their newfound faith and eager service. So he anticipates and counteracts that temptation to flagging zeal by expressing his expectation that his friends will obey the Lord not just consistently but increasingly, rather than letting the distance lead them to become weary in well-doing. Although human leaders come and go, the Lord to whom we offer thankful obedience is never absent, and the God who is always present is at work enabling our actions.

Paul describes the dynamic of Christian obedience in the command "work out your own salvation with fear and trembling" and its supportive rationale, "for it is God who works in you" (Phil. 2:12b–13). Together they show the mysterious interplay between divine initiative and enabling, on the one hand, and human participation, on the other. Yet the way that Paul words this command has been troubling for many Christians. They wonder, "Is the apostle of God's free grace suddenly retracting his gospel insistence that God justifies *not* the person who 'works' *but rather* the one who trusts Jesus (Rom. 4:1–5)?" Suddenly Paul seems to be talking about the contribution of *our* working to our salvation. Then he compounds our discomfort when he adds "fear and trembling." Are we not only to *work* for our salvation, but also to do so in an attitude of *nervous insecurity*, terrified that we might lose it all at the last moment?

Indeed, it is implausible to suppose that Paul would so blatantly contradict the gospel of grace that he has proclaimed and defended in his preaching and his letters. In fact, he will express his own complete reliance on Christ alone later in this very epistle, declaring his desire to be found in Christ, "not having a righteousness of my own that comes from the law, but that which comes through faith in Christ, the righteousness from God that depends on faith" (Phil. 3:9). Because God's Spirit of truth speaks God's words through Paul (1 Cor. 2:10–16), we have every reason to anticipate that Paul's exhortation here is consistent with his gospel of grace.

How, then, should this imperative about our work be reconciled with the gospel indicative of Christ's work? One approach would be to say that believers are *justified*—declared right with God—by grace alone through trust in Jesus' obedience and sacrifice alone, but that our *transformation* to become more like Jesus (*sanctification*) largely depends on our own self-discipline and struggle to suppress evil desires and cultivate pure and loving habits of the heart. In short, the idea is that justification is up to Jesus, but sanctification is up to us. Yet this "division of labor" quickly robs Christians of joy and hope. As our first burst of joy at conversion recedes into the distant past as a hazy memory, our daily experience becomes a frustrating struggle to achieve whatever degree of obedience *might* (we hope) win the Father's smile, and we ride a roller coaster between heights of self-satisfaction over moral victories and valleys of self-condemnation over our failures.

Moreover, this "division of labor" in salvation (justification as God's work, sanctification as ours) ignores two features in this text. The first is Paul's selection of the term *salvation.* In biblical theology, genuine salvation is *always* the work of God. Jonah confessed, "Salvation belongs to the LORD!" (Jonah 2:9). Peter and John announced, "This Jesus is the stone that was rejected by . . . the builders, which has become the cornerstone. And there is salvation in no one else, for there is no other name under heaven given among men by which we must be saved" (Acts 4:11–12). Paul teaches that Christians are saved by grace through faith, "and this is not your own doing; it is the gift of *God*" (Eph. 2:8). In fact, in Philippians 1:28, just a few sentences before our present text, Paul had described his friends' suffering and steadfastness as "a sign . . . of your salvation, and that *from God.*" The whole life of the Christian, including perseverance in faith and growth in godliness, is moving toward a comprehensive salvation that is God's gracious gift. Second, notice the rationale that Paul offers in Philippians 2:13: "for it is God who works in you, both to will and to work for his good pleasure." God is not standing back with his arms folded, saying, "I have done my part in giving Jesus to live and die for you. Now, then, work out your own struggle against sin and for holiness!" Paul's point in putting the command and the promise back to back is to profile the process by which the Holy Spirit pursues his lifelong agenda of reshaping our hearts to conform to Jesus' self-sacrificing love.

Another proposal for reconciling Paul's instruction to "work out your own salvation" with his gospel of grace is the view that the command focuses not on pursuing *individual* sanctification but rather on maintaining *corporate unity* in the church.[1] Proponents of this view call attention to the

1. J. H. Michael, "'Work Out Your Own Salvation,'" *Expositor* 9.12 (1924): 439–50, whose view is defended by Ralph P. Martin, *Philippians*, NCBC (Grand Rapids: Eerdmans, 1980), 102–3; by Gerald F. Hawthorne and Ralph P. Martin, *Philippians*, rev. ed., WBC 43 (Nashville: Thomas Nelson, 2004),

emphasis in the context on maintaining unity of conviction and affection in the congregation (Phil. 1:27; 2:1–2), implemented by placing others' needs before our own (2:3–4). In this view, *salvation* refers not to the application of Christ's redemptive work to believers (either with respect to our status—justification—or with respect to our nature—sanctification), but to communal "well-being," to peaceful and respectful relationships with others (who are honored "with fear and trembling"). Paul's assurance of verse 13 refers not to God's work "in" you (as individuals whom the Spirit is sanctifying), but to his work "among" you, as a congregation.

This view rightly notes that Paul's concern here is not merely the individual Christian's quest for holiness, but rather our growth *together* into the pattern of self-sacrificing, others-serving love that Christ has shown us. Throughout the New Testament, growth in godliness is a "group project" (e.g., Eph. 4:11–16). Yet when this interpretation pits corporate unity *over against* personal transformation and contends that Paul's concern is *only* for the community and *not* the individual believer, it divides what Scripture keeps together.[2] Perhaps in some social gatherings a certain kind of unity can be maintained in spite of the personal immaturity of the individuals who constitute the group. Coercive power or peer pressure can sometimes make people conform in external behavior, even when, deep down, they resent the group's or its leaders' expectations. But the health and well-being of Christ's church cannot be severed from its individual members' growth in Christlikeness. That is why Paul has moved from his general summons to oneness of mind (Phil. 2:2) to his specific summons to cultivate the inner quality of humility that will prompt "each of you" to attend not only to each

139–42; and, in a more nuanced form, by Gordon D. Fee, *Paul's Letter to the Philippians*, NICNT (Grand Rapids: Eerdmans, 1995), 231–37, and John Reumann, *Philippians: A New Translation with Introduction and Commentary*, AYB (New Haven, CT: Yale University Press, 2008), 387, 408–9. Peter T. O'Brien, *The Epistle to the Philippians*, NIGTC (Grand Rapids: Eerdmans, 1991), 277–78, aptly summarizes the arguments for the corporate (sociological, ecclesiological) view and against the personal (soteriological) view, although he finds them unpersuasive (as do Moisés Silva, *Philippians*, 2nd ed., BECNT [Grand Rapids: Baker, 2005], 118–23, and G. Walter Hansen, *The Letter to the Philippians*, PNTC [Grand Rapids: Eerdmans, 2009], 173–75).

2. "To state that the passage refers not to individual salvation but to the church's well-being already assumes a conceptual dichotomy that is both false and lethal." Silva, *Philippians*, 119. Fee, *Philippians*, 234–35, comments that the dispute "whether 'salvation' has to do with the individual believer or with the corporate life of the community" is "a false dichotomy. . . . What Paul is referring to, therefore, is the *present* 'outworking' of their *eschatological salvation* within the *believing community* in Philippi" (emphasis in original).

one's own concerns, but also to those of others (2:3–4). The church as a whole receives salvation from disunity only as each of its members experiences the Spirit's salvation from self-centered pride.

Paul's point is that Christ's saving work is comprehensive. Jesus rescues us not only from sin's guilt and punishment, but also from its controlling power; and not only from personal defilement, but also from interpersonal alienation. In rescuing us from sin's guilt and punishment, Christ does it all *apart from* us: he obeys in our place, suffers in our place, rises to victorious life in our place, and even gives us faith by his Spirit (Phil. 1:29). On the other hand, in rescuing us from sin's controlling power, Christ still does it all, but he does it *through* us: his Spirit enlivens, enlists, and enables us as allies. Our salvation from alienation includes not only reconciling us to God through the cross, but also reconciling us to one another through the cross (see Eph. 2:11–18). Members of the family of God will be able to "work out" the family's "salvation" (communal harmony) only as each appropriates the comprehensive "salvation" that Jesus has achieved and is applying to individuals through Spirit-given faith and repentance. We need the Spirit to rescue us from our innate self-centeredness. God is the only One who can change both our desires and resolutions ("to will") and our behavior ("to work"), conforming both motive and action to the Christ-shaped template that evokes his fatherly "good pleasure."

The dynamic of this integral saving work of God—his power to transform our "willing" and our "doing," freeing us not only from sin's guilt but also from its grip—finds expression throughout Paul's epistles. Diagnosing divisive competition in the Galatian churches as spiritual cannibalism ("bite . . . devour . . . consumed by one another," Gal. 5:15; cf. v. 26), Paul prescribed as its cure the fruit of the Spirit—love, peace, patience, gentleness, and other Christ-reflecting qualities that flourish in hearts refreshed by grace. Yet the gospel therapy for the competitive Galatians is not passive inaction. It is to "*walk* by the Spirit," to think and act toward others in light of the fact that, by grace, we now "live by the Spirit" as those redeemed from the Law's curse through Christ's sacrifice (5:16, 25; see 3:13–14). Likewise, writing to the Ephesians, Paul first describes the peace-producing sacrifice of Christ that has united believing Jews and Gentiles and reconciled both to God (Eph. 2:11–18). Then he urges Christians to preserve that unity, which is created and sustained by God's Spirit, by treating each other "with all humility and gentleness, with patience, bearing with one another in love" (4:1–3). God works in us, so that we desire and do the loving deeds that make his saving purposes visible in his church.

The heart-changing work of God's Spirit does not make the struggle against self-centeredness effortless, either in the individual or in the church. Christians can get so snarled in selfishness that we feel helpless to resist its allure, desperately praying with Charles Wesley, "Take away the love of sinning."[3] Or we want to resist impatience or anger, but we feel too tired or are caught off guard when our spiritual resistance is low. So we must ask the Lord to work in us not only the *willingness* to do what pleases him, but also the *doing* itself.

It is the fact that both our willing and our doing lie beyond our own resources and can be found only in God's working that makes this whole project a matter of "fear and trembling." The combination of fear with trembling—perhaps we could say "fear expressed in trembling"—refers to a sober attitude that results from recognizing both our own inadequacy and the life-or-death significance of the situation in which we find ourselves. The author of Psalm 2 counseled the earth's rulers to abandon their rebellion against the Lord and his anointed King and, instead, to "serve the LORD with fear, and rejoice with trembling" (Ps. 2:11). Paul first brought God's Word to Corinth not with the oratorical eloquence of Greek rhetoric but "in weakness and fear and much trembling" (1 Cor. 2:3). Such a sobering fear, however, should not be confused with the abject terror that made the guilty Adam flee from the Lord's presence (Gen. 3:10). Jesus bore our guilt and endured God's wrath for us to set free "those who through fear of death were subject to lifelong slavery" (Heb. 2:14–15). If we are trusting in Christ, we need not fear that God will reject us for our failures. But Christians are aware that the wholehearted obedience that Jesus deserves is so far beyond us that we do not dare let our hearts drift away from him, but rather cling for dear life to his promises in God's Word and his presence in God's Spirit. Conscious of our own frailty and fickleness, we rightly fear to stray from the side of our great Shepherd.

When our commitment to obedience begins to reflect Jesus' costly commitment to obedience for us, that simple shift from self-centeredness to

3. Charles Wesley, "Love Divine, All Loves Excelling" (1747), in *Trinity Hymnal* (Suwanee, GA: Great Commission Publications, 1990), no. 529.

Son-centeredness causes our lives to shine like stars in a midnight sky. In view of the tensions that exist in the Philippian church—rivalry, conceit, preoccupation with one's personal interests—Paul has a very specific form of obedience in mind.

"Work Out Your Salvation" in Patient Contentment

The obedience that the Philippian Christians need to offer in the strength that God gives is *patient and selfless contentment*, which enables them to "do all things without grumbling or questioning." Instead of each brooding over injuries to personal rights or reputation, each must focus on others' needs (Phil. 2:3–4), as Christ Jesus himself has served them (2:5–8), and as they can see exemplified in Paul (2:17), Timothy (2:19–24), and their own Epaphroditus (2:25–30).

As we have seen, marginalized minorities cope in various ways with misunderstanding and mistreatment by the surrounding society. Some scorned subgroups attempt *accommodation*, trying to blend into the surrounding culture by masking their distinctives. Others take the route of *retaliation*. They "fight back" against the mainstream, either with words or with actions, launching an insurgency that pays back wound for wound, insult for insult, as *Time* magazine art critic Robert Hughes observes in *Culture of Complaint: The Fraying of America*.[4] Hughes observes that American society is tattering at the edges, disintegrating into aggrieved victim groups, each resentful that its rights are not shown due respect by others. "Political correctness" and group self-pity are undermining the great American experiment, the attempt to weave a unified fabric of community out of diverse cultural strands. Still other minorities seek refuge in *isolation*, withdrawing into self-contained enclaves with minimal contact with the society outside. Finally, frustration over injustices inflicted by the surrounding society sometimes *turns inward* to be vented on those closest to us, the very reactions that Paul here forbids: "grumbling" against each other and "questioning"[5] the wisdom and care of God. The obedience that results from God's working his rescue deep into

4. New York and Oxford: New York Public Library/Oxford University Press, 1993.

5. Elsewhere in the New Testament, the Greek noun *dialogismōn* (ESV: "questioning") (here plural) refers to spoken disputes (Luke 9:46) or to internal "reasoning" that raises objections or challenges (Luke 5:21–22).

our hearts will be shown in an extraordinary reaction to the pressures of a dominant and unsympathetic cultural establishment. Instead of accommodation, retaliation, isolation, or internal conflict, Jesus' followers will engage one another and their non-Christian neighbors with patient and selfless contentment, neither grumbling in self-pity nor questioning God's purpose.

This surprisingly mellow reaction to opposition will set Paul's friends apart from the spiritual darkness surrounding them in Roman Philippi, and it will pose a striking contrast to ancient Israel's response to adversity in the wilderness, centuries before. In verses 14–15 Paul selects distinctive words and phrases to signal that he is alluding to Moses' description of that generation of Israelites who experienced the exodus from Egypt but then died in the desert. The wilderness generation "*grumbled* against Moses" (Ex. 15:24; 16:2) and against the Lord himself (16:7–8). Moses somberly described those Israelites as "*no longer his children* because they are *blemished*; they are *a crooked and twisted generation*" (Deut. 32:5). Each of the italicized words Paul now embeds in our present text, to call his friends to a radically different response: they must not *grumble*, and their contentment will show them to be "*children* of God *without blemish* in the midst of a *crooked and twisted generation*" (Phil. 2:14–15). The grace of Christ is making former pagans in Philippi into the people of God that Israel should have been. Instead of Israel's complaint and criticism of the way in which God was running their lives, persecuted Christians can exhibit thankful contentment. Instead of Israel's self-centered quibbling with each other, Christians can display humility, honor, and compassion toward one another. Instead of Israel's blemished and twisted unbelief, showing that they were not really God's children, Jesus' people are unblemished "children of God," distinguished by a confidence that holds *fast* and holds *forth* (the verb expresses both ideas[6]) the Word of truth.

As a result, believers in Jesus stand out like sparkling stars in a midnight sky. In painting this vivid portrait of the light-to-darkness contrast between

6. *Epechō* appears only five times in the Greek New Testament. In some contexts it seems to express retention (ESV: "holding fast") (e.g., Acts 19:22: "*stayed* in Asia for a while"), but in others it suggests an outward focus or movement (NIV: "hold out") (e.g., Luke 14:7: "he *noticed* how they chose"; Acts 3:5: "he *fixed his attention* on them"; and 1 Timothy 4:16: "*Keep a close watch* on yourself"). Either connotation fits this context—certainly the Philippians must retain their allegiance to God's Word, resisting cultural pressures to compromise, yet they must do so not by retreating into a fortress, but by boldly extending the good news in grace even to their cultured despisers.

Christians and the society that surrounds and scorns them, Paul invokes the imagery of another Old Testament text. Daniel 12:1–3 predicts the day of resurrection. The Septuagint, the Greek translation of the Old Testament that was widely used in Paul's day, reads: "those who have insight will shine as lights [*phōstēres*] of the sky, and those who are strong by my words as the stars [*astra*] of the sky forever and ever."[7] Daniel's simile paints a vivid skyscape on the canvas of our imaginations. On the day of resurrection, when those who sleep in the dust arise either to everlasting life or to everlasting shame, those who have clung to God's words will shine like bright pinpricks of starlight against the pitch-black backdrop of a spiritually darkened society ("lights" [*phōstēres*] and "stars" [*astra*] are synonymous).[8] Only in Philippians 2:15 does Paul use the term *phōstēres*, telling the Philippians, "you shine as *lights* in the world." We may hear *world* and think of planet earth. Paul's term (Greek *kosmos*) is sometimes used that way, but it can also refer to the entire universe as a display of order and beauty (Rom. 1:20). In combination with Daniel's comparison of God's people to heavenly bodies of light, we should picture *kosmos* as referring to the universe in which sun, moon, and stars are placed. With Daniel and Paul, we should envision a pitch-black night sky, punctuated by a thousand points of light. The NIV vividly captures Paul's imagery: "you shine like stars in the universe."

More and more of us live in cities or suburbs illuminated around the clock by electric lighting, so fewer people have been far enough away from city lights—in a desert, on a mountain, on the open sea—to see that stunning darkness and the countless stars that appear all the more brightly against the backdrop of dark space. Ancient people, who knew real darkness and navigated by stars, could "see" Paul's vivid analogy more easily than we can. To them Paul says, "Those innumerable pinpricks of light, gleaming brightly in the vast ebony 'emptiness' of the night sky, are *you* who belong to Jesus."

This is a new perspective on our "out-of-place" location in this world. Paul puts a surprising, positive spin on the fact that we do not "fit" here. He says, in effect, "The spiritual and moral darkness of the surrounding society may give the impression that *you* are the misfits. But that is only the darkness talking. By radiant grace God has broken through the darkness

7. Author's translation of the Daniel 12:3 LXX.

8. The term "lights" (*phōstēres*) appears also in the account of God's creation of lights in the sky—sun, moon, and stars—in the beginning (Gen. 1:14, 16 LXX).

of your own mind. In his light you have glimpsed yourself as he sees you (a painful, shameful exposure). But he has answered the discomforting truth about you with the comforting sight of his Son, opening your eyes to see the splendor of Jesus. Now he is making you into beacons of light, reflections of Jesus the Light of the World (John 8:12), in a midnight sky. The darker the social setting around you, the brighter your Savior shines through you."

As stars shine brightly in a dark night sky, so Christians must "*[hold] fast* to the word of life." As we persevere, others will see Christ's light in us, as Jesus directed in the Sermon on the Mount: "Let your light shine before others, so that they may see your good works and give glory to your Father who is in heaven" (Matt. 5:16). In many respects, the Roman Empire represented a confluence of admirable human cultural achievement—political and military organization, transportation and communication infrastructure, architectural wonders, and more. But spiritually, the highest levels of Roman society were the heart of darkness. Rome's beauty rested on its brutality, and its prosperity was driven by inhumane cruelty (see Rev. 17:1–6, 18). Only seventy years before Christ was born, Rome not only crushed the slave revolt led by Spartacus but also turned six thousand of the defeated rebels into a grisly example, crucifying them along the Appian Way that connected the capital to southern Italy. Human slaves were an economic asset—listed with other cargo marketed to the great city (Rev. 18:13)—to be exploited, discarded when they outlived their usefulness, and ruthlessly destroyed if they resisted their servile position.

Likewise today, a steady stream of bad news reveals the West's widespread cultural decomposition: greed and implosion in the financial industry, indifference to the suffering of the unborn and the poor, defiant challenges to marriage, addiction to drugs or pornography or entertainment, and the idolatry that worships government as a comprehensive savior. A burgeoning global sex trade kidnaps and exploits vulnerable girls and boys to satiate the lusts of people with power and wealth.

Faced with such alarming trends, we could easily be paralyzed by dismay. There is, however, another way for Christians to view cultural decay. In 1969, at the age of sixty-six, the distinguished social commentator Malcolm Muggeridge abandoned agnosticism and came to faith in Christ. A decade later he described the implosion of Western society in terms eerily prescient of the economic meltdown that has occurred at the end of the twenty-first century's first decade. Yet despite his sobering diagnosis, he sounded a surprising note of hope:

> Let us then as Christians rejoice that we see around us on every hand the decay of the institutions and instruments of power, see intimations of empires falling to pieces, money in total disarray, dictators and parliamentarians alike nonplussed by the confusion and conflicts which encompass them. For it is precisely when every earthly hope has been explored and found wanting, when every possibility of help from earthly sources has been sought and is not forthcoming, when every recourse this world offers, moral as well as material, has been explored to no effect, when in the shivering cold the last [twig] has been thrown on the fire and in the gathering darkness every glimmer of light has finally flickered out, it's then that Christ's hand reaches out sure and firm. Then Christ's words bring inexpressible comfort, then *his light shines brightest*, abolishing the darkness forever.[9]

In Philippians 2:16, Paul says that when his friends meet adversity with patient contentment and winsome witness, he tastes the fruit of the hours and years, the teaching and tears, that he has poured into their lives. It has all been worth it! On the day of Christ, when Jesus returns as the rising Sun and the night is over forever, Paul will present these believers as his gift of love to the Savior, evidence that "I did not run in vain or labor in vain." Again the apostle echoes the Old Testament, specifically the prophecy of Isaiah, who promised that when God's new creation comes, "my chosen shall long enjoy the work of their hands. They shall *not labor in vain* or bear children for calamity" (Isa. 65:22–23). That ancient promise speaks to Paul as a pastor, but it is also a promise to the whole people of God, who live by faith in Jesus. Because God is at work through our service, as we pour ourselves into others' lives, we anticipate that one day they will stand with us before the throne of God and of the Lamb!

"Work Out Your Salvation" in Joyful Sacrifice

In verse 17 Paul transitions smoothly, almost imperceptibly, from "how things are going" with the Philippians back to "how things are going"

9. Malcolm Muggeridge, *The End of Christendom* (Grand Rapids: Eerdmans, 1980), 56 (emphasis added).

with himself. Having urged them to shine in the darkness through their obedience, contentment, and winsome witness, he raises again the prospect of his own imminent execution. His purpose is not to sound a sour note amid his encouragement. Rather, he is leading his dear friends back to his recurring theme of joy by a surprising route. Even Paul's death, whether it occurs sooner or later, will be symptomatic of the suffering entailed in believers' calling to shine like stars in a dark universe, to live for Jesus amid a crooked generation. Yet Paul sees his own suffering—even to the extreme of a violent death—as an opportunity to enhance the service that other believers offer up to their Lord as they hold fast and hold forth his Word: "even if I am to be poured out as a drink offering upon the sacrificial offering of your faith, I am glad and rejoice with you all" (Phil. 2:17).

Paul is evoking another Old Testament scene. In the ancient sanctuary, the daily morning and evening burnt offerings were accompanied by "drink offerings." Before the fire was lit to incinerate the slain lamb, a priest poured wine over the victim to enhance the "pleasing aroma to the LORD" (Num. 28:7–8; see Phil. 4:18). Although both lamb and wine were consumed in consecration to the Lord, the lamb itself was central. This is Paul's point when he now speaks of being "poured out . . . on the sacrificial offering of your faith." In effect, the apostle says to his Philippian friends, "You are the main event. I am 'just icing on the cake.'" Of course, our modern "icing on the cake" metaphor captures only one aspect of Paul's ancient metaphor. Missing from the cake image is the motif of *worship*. Elsewhere Paul says that he is a priest who is privileged to present Gentiles, captured by grace through the gospel, as an offering for God's delight (Rom. 15:15–16). Here, his role is ancillary and the Philippians' consecrated lives are center stage, where their Lord receives their tribute of grateful obedience, enhanced by the apostle's suffering.

Another motif that our cake picture cannot capture is death. The "sacrificial offering of your faith" evokes the slaughter of countless animal victims in Israel's sanctuary, thereby alluding to the Philippians' shared experience of suffering with Paul (Phil. 1:29–30). If their offering of themselves in worship to the Lord—which entails suffering for Jesus' sake—involves further suffering on Paul's part as well, he is pleased to pay the price to enhance their costly praise!

For Paul, "to live is Christ, and to die is gain" (Phil. 1:21), so the prospect of seeing people living their lives for God's glory, *whatever the cost*, turns serious suffering into sheer celebration: "I am glad and rejoice with you all" (2:17).[10] Here again, in the context of suffering and the prospect of impending death, the apostle sounds the recurring note of joy that runs like a golden thread through the tapestry of the epistle. We have heard Paul rejoice in prayer for the Philippians (1:4) and when Christ is proclaimed (1:18). He expects to return to them to enhance his friends' joy (1:25), and he finds his joy fulfilled in their unity (2:2). He will summon them repeatedly to "rejoice in the Lord" (3:1; 4:4). Here he urges them to blend their joy with his in a symphony of sacrificial devotion.

Conclusion

Paul's use of the imagery of Israel's sacrificial system in Philippians 2:17–18 brings his call to obedience, contentment, concern for others, and winsome witness back to the "Therefore" with which he began in verse 12. That "Therefore" linked the Philippians' obedience to the obedience of Jesus in his self-sacrificing, others-serving death. Jesus now shares his resurrection life with those who trust him through the Holy Spirit, who is at work in us both to will and to work for his good pleasure. Paul's willingness to add his sacrifice to the Philippian Christians' offering of their lives, transformed by faith, reflects the Savior who sacrificed all to bring us into the family of God as forgiven and beloved children. Paul invites the Philippians and us to share his Christ-centered joy, and therefore to respond to opposition from without and frustration from within the church not with complaint, but with contentment; not with self-centeredness, but with others-centered sacrifice.

Such hopeful obedience and patient contentment make us bright stars in a dark sky, reflecting the radiance of Christ, the Light of the World.

10. The ESV's "be glad and rejoice with" in verses 17 and 18 reflects two Greek verbs that original hearers would have recognized as even more closely related than "be glad" and "rejoice with," since the second (*synchairō*) is a compound of the first (*chairō*). Options available in English vocabulary and style make it difficult to convey these verbs' family connection: "rejoice and rejoice with" sounds redundant. Settling for less-than-standard English, we might say "rejoice and co-rejoice."

11

Living Replicas of the Servant of the Lord

Philippians 2:19—30

I hope in the Lord Jesus to send Timothy to you soon, so that I too may be cheered by news of you. For I have no one like him, who will be genuinely concerned for your welfare. For they all seek their own interests, not those of Jesus Christ. But you know Timothy's proven worth, how as a son with a father he has served with me in the gospel. (Phil. 2:19–22)

Whom do you want to be like when you grow up? That may seem an odd question, at your stage in life. You may think that you are beyond the point of needing role models to look up to. Of course, children, even teenagers, need role models. As we are growing up, we are trying to figure out what it means to be a mature adult, so we instinctively fix our sights on people we admire. Perhaps we try to imagine what it would be like to be a certain person, with his or her looks or intelligence or charm. We watch the person's actions and facial expressions. We listen to his or her words and tones of voice. We may even make conscious decisions to imitate our "idol" in our hairstyle, clothes, language, or vocation. Or sometimes our role models are so much a part of our lives that we imitate them without even being aware that we are doing it. People have told me for years that on the phone they cannot tell the difference between my dad's voice and my voice. More than once, I've answered the phone in my parents' home, and had the caller immediately launch into: "Hi, Ralph, I was wondering . . ."

Our Need for Role Models

The apostle Paul knows that our need to get our bearings by watching role models operates not only in chronological growth but also in spiritual maturation, and not just for four-year-olds and fourteen-year-olds, but also for forty-year-olds. Whether we are adding inches to our height or to our belt size, whether we are searching for the first whiskers on our chins or mourning departing hair on top, spiritually speaking we are all still "growing up." In Ephesians 4:12–13 Paul implies that none of us is fully a "grown-up" yet. That is why the risen Lord Jesus gives pastors and teachers "for building up the body of Christ, until we all attain to the unity of the faith and of the knowledge of the Son of God, to mature manhood, to the measure of the stature of the fullness of Christ." If the standard of maturity is Jesus himself, each of us individually and all of us together have a long way to go! So the question is not so strange after all: Whom do you want to resemble when you grow up? Paul gives us the right answer in Ephesians 4: "I want to be like Jesus!"

Paul has implied as much in Philippians 2. Rather than acting out of selfish ambition or vain conceit, childishly fixating on your own interests and ignoring others' needs, we need "this mind among yourselves, which is yours in Christ Jesus" (Phil. 2:5)—the Christ who displayed his deity by humbling himself, becoming human, forgoing rights, submitting to the point of death, and then receiving his well-earned exaltation, enhancing the glory of the Father.

But you are saying to yourself, "It is one thing for me to catch a glimpse of Christ's kindness, integrity, and courage from the pages of my Bible and to resolve, 'I want to be like Jesus when I grow up.' It is a very different thing to move toward such maturity in my daily struggles with impatience and worry and preoccupation with my own issues." We face two huge hurdles when God's Word calls us to grow up into the image of our older brother, Jesus. The first challenge is that when Jesus is "the bar," the bar is set impossibly high. Picture a sedentary middle-aged man, viewing his television from his home recliner, watching an Olympic athlete win multiple gold medals, and then saying to his wife, "You know, I'm a bit out of shape. Tomorrow I'll start training so that I can compete in the Games four years from now." This couch potato is so out of touch with reality that it's laughable. Yet the more we know ourselves, the more we realize that his delusional hopes of Olympic glory are more achievable than our pursuit of Christlikeness.

Our text focuses on a second problem that we encounter in our longing to grow up to be like Jesus: Unlike his apostles, we have not been able to observe how Jesus handled stress, or responded to unexpected events, or dealt with an unreasonable supervisor, or coped with whiny children. The Gospels give us glimpses of Jesus in action in similar situations, when his enemies harassed him or his friends' childish attitudes seemed to try even his patience. But still we wish that God would place into our lives today role models whom we could personally watch in action.

Our text teaches, however, that God does embed into our experience living, breathing replicas of Jesus: men and women whose heart instincts are growing by grace so that we can sense the heartbeat of Christ in the way they treat others, react to adversity, and invest their energies. Watching them shows us what growing up to be like Jesus looks like in the nitty-gritty of everyday life. Three such miniatures are profiled in Philippians 2:19–30: Paul himself, his junior colleague Timothy, and the Philippians' own messenger Epaphroditus. None of these miniatures fully matches the Original, of course. Even they are not yet fully "grown up." Paul will frankly admit this about himself in Philippians 3:12–14, where he insists that one mark of maturity is realizing that you are not yet fully mature (3:15). But each of these men reflects Jesus to the Philippians and to us; and as they do, their reflections show us what growing up looks like, and why and how growing up toward the maturity of Jesus is possible.

We have seen that Paul's epistle to his friends in Philippi, like a letter from a loving father and husband to loved ones far away, goes back and forth in addressing his situation and theirs. First, Paul shared encouraging news of the fruit of his custody in Rome (Phil. 1:12–26). Then he turned to their circumstances, urging them to stand together in suffering through humbly caring for each other (1:27–2:18). Now, as he lays out plans for his own, Timothy's, and Epaphroditus's trips to Philippi in the near future, he blends the two "worlds," his in Italy and theirs in Macedonia. He expects that their absence from each other, which has made "hearts grow fonder," will be overcome soon. Even sooner, he will send companions in ministry as foretastes of his own joyful reunion with his Philippian friends.

Paul: Replica of the Submissive Servant

Paul turns the spotlight on Timothy because he plans to send Timothy to Philippi as soon as his own legal case is resolved. But before we give our attention to Timothy, we do well to notice that the apostle couches his hopes and plans for Timothy (Phil. 2:19–23) and for his own expected reunion with the Philippians (2:24; see 1:25–26) in the sovereign will of his divine Master. Paul's forecast of coming events exhibits a deep awareness that Christ the Lord controls our hopes and plans. The apostle's perspective on his own planning process flows so naturally from his lips or pen—"I hope *in the Lord Jesus* . . . I trust *in the Lord*" (2:19, 24)—that we might overlook the phrases that express his deep conviction about who controlled the course of his life. What Paul plans and what he hopes are always "in the Lord." Paul is a replica of the supremely submissive Servant of the Lord who came not to do his own will but the will of the Father who sent him (John 5:30; 6:38).

We know that Christians are *supposed* to attach the caveat "Lord willing" to the plans we form and forecast. After all, biblical wisdom teaches that "the heart of man plans his way, but the Lord establishes his steps" (Prov. 16:9). It is no wonder that the epistle of James (4:13–15) comes down hard on people who imagine that they can map out their own futures. Arrogant fools brashly announce: "Today or tomorrow we will go into such and such a town and spend a year there and trade and make a profit." On the contrary, God's messenger says, "Instead, you ought to say, 'If the Lord wills, we will live and do this or that.'" So it is proper to preface our plans with "Lord willing." But it is one thing to *say* the right words, and another to have our hearts molded by the reality behind the words. What we see in Paul is a deep consciousness of Christ's sovereignty and an eager submission to Christ's control. The apostle is completely aware of the fact that his whole life and every circumstance in his life—things within his control and those outside his control—belong to Jesus. Jesus directs everything in Paul's current and future experience for God's glory and for his people's well-being. So Paul "hope[s] in the Lord Jesus" to send Timothy, and he "trust[s] in the Lord" that he himself will be freed to serve them again, too.

This mind-set of submission to his Sovereign characterized Paul throughout his ministry. Earlier Paul had written to the church at Rome that in his prayers he was "asking that somehow *by God's will* I may now at last succeed in coming to you" (Rom. 1:10). Paul's prayers were answered, but in a surprising way: a riot in the Jerusalem temple, "protective custody" by Roman troops that morphed into arrest, accusations and assassination plots by Jewish authorities, Paul's appeal to the emperor, a long and perilous voyage, shipwreck, and finally Paul's arrival, in chains, in the imperial capital. And every detail was "by God's will."[1]

Plainly, Paul's confidence in Jesus' lordship did not make him passive, lethargic, or fatalistic. Far from it! He was a dynamo of motivation and strategy, brimming with ideas for disseminating the good news about Christ the Lord. At the same time, Paul kept a light grip on his own agenda. In Philippians Paul has already wrestled with the possible outcomes of his hearing before the emperor and concluded that since Caesar's decision was in Christ's control, the true Lord would order his servant's release, to return to his Philippian friends "for your progress and joy in the faith" (Phil. 1:25). Hence he repeats the thought, "I trust in the Lord that shortly I myself will come also" (2:24). But if Jesus orders martyrdom rather than further ministry, Paul will be more than content with that outcome as well: "to depart and be with Christ" will be "far better" for Paul personally (1:23). Christ has the sovereign right and the infinite wisdom to revise his servants' best-laid plans. Whenever and however he does, the outcome will mean greater joy for his servants and greater glory for his Father.

With the simple words that bracket this paragraph, "in the Lord Jesus" and "in the Lord," Paul is a role model for the Philippians and for us. He shows us what spiritually mature, Christlike planning looks like in the routine decisions of daily life: "Shall I keep Timothy here in Rome to help me, or send him away to encourage my spiritual children at Philippi? And when (or if) I am released from Roman custody, shall I pause to see the

1. Even earlier, Paul had written to believers at Corinth that he planned to return to the city and stay with them for an extended time "if the Lord permits" (1 Cor. 16:7).

sights of the city, or set out eastward to bring comfort to the ones I love in Philippi and Ephesus and Galatia?" Because he is resting in the Lord's will and faithful provision, Paul is freed from preoccupation with himself, to live boldly with a priority on God's glory.

How do you approach your planning for the future? Your choice of a college or university, a career path, a husband or wife? Your pursuit of a promotion, a move to a different company, or a new vocation? Your investment strategy, and your spending strategy, and your giving strategy? Your dreams for your retirement years? If you are becoming a spiritual grown-up like Paul, you will formulate your hopes and plans with humility, always aware that Jesus your Sovereign has both the right and the wisdom to overrule your choices and redirect your paths. Along with humility, your planning will express your passionate commitment, not chiefly to your own security and comfort, but rather to Jesus' glory and his mission in the world.

Timothy: Replica of the Selfless Servant

Before Paul's anticipated arrival, the believers in Philippi can expect to receive two other servants of Christ: Paul's younger colleague Timothy and their own servant-leader Epaphroditus. These men will bring the latest news of their beloved apostle's situation in Rome. Paul intends to send Timothy "soon"—that is, "as soon as I see how it will go with me," apparently referring to the verdict in his legal case (Phil. 2:19, 23). He feels even more urgency[2] to send Epaphroditus home immediately, to alleviate the Philippians' alarm over reports of Epaphroditus's life-threatening illness (2:25–28). The imminent timing of Timothy's and Epaphroditus's travels demonstrates Paul's costly commitment to the Philippians' comfort in Christ. He is willing to forgo the support of invaluable companions in order to set his friends' apprehensive hearts at rest.

Although the stated mission of these men is to bring news of Paul's trial and Epaphroditus's recovery from a near-fatal illness, Paul's implicit motive is to place into the midst of the Philippian congregation men who exemplify

2. The comparative adverb *spoudaioterōs*, translated "the more eager" by the ESV (2:28), in this context expresses a sense of urgency as well as eagerness. See Gordon D. Fee, *Paul's Letter to the Philippians*, NICNT (Grand Rapids: Eerdmans, 1995), 280n38, for a brief and balanced discussion of the interpretation of this word group in Pauline literature.

Christlike humility and others-centered concern. In view of the interpersonal frictions to which Paul refers (Phil. 2:1–4; 4:2–3), the church needs to see such reflections of Jesus' servant heart. Timothy and Epaphroditus will be replicas of Jesus the Servant in whom believers can see what selflessness and sacrifice look like in attitude and in action.

Timothy's tender heart will show them how Christ controls our cares (Phil. 2:20–21). In fact, we might even say that Timothy exemplifies how Christ controls our *worries*, for that is the word that Paul uses when he writes that Timothy would "be genuinely *concerned* for your welfare." Our English translators, knowing (I suppose) that Christians are not supposed to worry—at least about food and shelter (Matt. 6:25–34)—have sanitized the Greek word[3] that Paul uses to express Timothy's worries over the Philippians. Later in this letter Paul will use the same verb to direct his friends: "Do not *be anxious* about anything, but in everything by prayer and supplication with thanksgiving let your requests be made known to God" (Phil. 4:6). When this word describes a negative mind-set of faithless fear, our English versions rightly render it "worrying" and "being anxious." When it expresses a *commendable* distress over threats to others, our versions opt for more acceptable alternatives: "care" (KJV) or being "concerned" (NASB, ESV) or, even more calmly, "takes a genuine interest" (NIV). But Paul's Greek-speaking hearers are exposed to the same word in both contexts. They will associate with it the same experience of inner stress over current or potential problems—an experience that "concern" and "taking interest" are too mild to convey.

Think of Jesus' friend Martha, mentally and emotionally pulled in a dozen directions by the demands of serving a fitting meal for over a dozen unexpected guests . . . and then exploding in frustration at Jesus, of all people, for not making her sister pitch in with the preparations (Luke 10:38–42). Martha's emotional investment in getting that meal on the table was more than "genuine interest." She was, as Jesus said, "*anxious* and troubled." Even more to the point, Paul's catalogue of his sufferings as Christ's apostle in 2 Corinthians 11 ranges from social rejection to physical violence before it reaches its climax in "the daily pressure on me of my *anxiety*[4] for all

3. The Greek verb is *merimnaō*, which also appears in Jesus' prohibition of worry in the Sermon on the Mount (Matt. 6:25–34).

4. The ESV rightly—boldly—renders the noun *merimna*, a cognate of the verb in Philippians 2:20, as "anxiety." See also Paul's use of the Greek verb *merimnaō* in 1 Corinthians 12:25–26 to describe a

the churches" (2 Cor. 11:28). Although he knew that those whom God had chosen in love he would justify and eventually glorify (Rom. 8:29–30), Paul was not privy to God's secret saving decree. Therefore, painfully aware of the weakness of those who professed faith in response to his preaching, he worried over the churches. This worry was perhaps the heaviest weight on his pastoral heart.

Timothy shares that inner emotional struggle with Paul, and that selfless concern for others' well-being sets him head and shoulders above some who surround Paul in Rome. When Paul writes, "I have no one *like* him," the word *like* represents a Greek word that appears only here in the New Testament, a word that we could render into English only in an awkward expression such as "of equal soul."[5] Perhaps Paul has chosen this unusual word because he has just spoken of the amazing truth that Christ, who was "equal with God," did not exploit his divine status but instead filled a servant's role. Paul and Timothy also have, on a creaturely level, a special father-son bond of heart and mind. Timothy loves the Philippians so much that his heart aches with Paul's as they hear reports of the Philippian believers' sufferings and rivalries.

Paul goes on to elaborate the selflessness that sets Timothy apart from so many others in Rome (Phil. 2:20–22). When he comments that he has "*no one* like" Timothy, since "they *all* seek their own interests," he is not ignoring his earlier encouraging report that some believers are spreading the gospel "out of love, knowing that I am put here for the defense of the gospel" (1:14–16). Nor is he implying that "the brothers who are with me," whose greetings he will convey in closing, are all self-seeking (4:21). After all, this group may have included faithful colleagues such as Luke, Aristarchus, John Mark, and Epaphras of Colosse, who were with Paul during at least part of his Roman imprisonment (Col. 4:10–14). Rather, Paul has in view those whom he described earlier as operating "from envy and rivalry" (Phil. 1:15).[6] Such people were driven by the same self-centered and competitive

---

commendable mutual "care": "the members [of the body] may *have the same care* for one another. If one member suffers, all suffer together" (ESV).

5. Greek *isopsychos*, which appears in Psalm 54:14 LXX (55:13 English) to describe the psalmist's shock at being betrayed by a close friend, "my equal" (ESV). Compare "equality with God" (Greek *to einai isa theō*) in Philippians 2:5.

6. Often in the New Testament, *all* (Greek *pas*) has a generalizing but not universalizing sense, referring to a great number, or a collective that includes great variety (all sorts or classes). In Mark

motive that Paul diagnosed as a danger to the Philippian Christians, who were likewise inclined to "look . . . only to [their] own interests" and to disregard the needs of others (2:3–4).

Timothy will be an antidote to this spiritual toxemia infecting the Philippians. In his concern for them, they will see a man who seeks not his own interests[7] but those of Jesus Christ. The interests of Jesus Christ are the welfare of his people. In effect, Paul writes, "When I send Timothy to you, you will see in his selfless concern for you the very attitude that you should all extend to each other. He is a miniature of Jesus, the King who did not look out for himself but became a Suffering Servant for you. This mind-set of selfless servanthood is already yours because you are 'in Christ Jesus,' and it is the mind-set that you need to cultivate right now."

Paul implies in verse 19 that Timothy would make a round trip from Rome to Philippi and back again, since Timothy's return would bring news from Philippi to cheer Paul's heart. In today's world of speedy transcontinental air travel, we might not realize the cost in time, comfort, and safety that Timothy's round trip would entail. The most direct route would be overland from Rome on the Via Appia to Brindisi on Italy's southeast coast (over 350 miles). A voyage across the Adriatic Sea (about ninety miles) would bring him to Dyrrachium, the western terminus of the Via Egnatia. Then he would make a 360-mile trek eastward on the Via Egnatia across Macedonia to Philippi. In order to bring news from Paul to Philippi and from Philippi back to Paul, Timothy would invest weeks in order to make this arduous and dangerous trip over land and sea.

The travel itinerary that lies ahead for Timothy illustrates what having the mind of Christ could cost you and me as well, and that may make us uncomfortable. When the Holy Spirit stretches our hearts to embrace God's children far and near, big and small, we begin to pray more often and more urgently for others' needs, finding that their deep crises give us a more balanced perspective on our own worries and discomforts. We work at seeing

1:5, for instance, we are told that "*all* the country of Judea and *all* Jerusalem were going out to [John] and were being baptized by him in the river Jordan, confessing their sins." The wider narrative in the Gospels shows that *not every individual* in the province or the city received John's baptism of repentance. There were notable exceptions, as Luke records: "the Pharisees and the lawyers rejected the purpose of God for themselves, not having been baptized by [John]" (Luke 7:30).

7. Philippians 2:21 ("they all seek their own interests," Greek *ta heautōn zētousin*) echoes 2:4 ("look . . . to his own interests," Greek *ta heautōn skopountes*).

our differences of opinion from the other person's point of view. We put less money into our own investments and more into diaconal offerings to relieve those in greater need than we are. We even find Jesus' love for his people moving us to love them in risky and costly ways, forgoing our precious privacy and giving away our precious hours or days to bring Christ's compassion and correction, packaged in our own persons, to those who need his grace, though they do not deserve it any more than you or I do.

At this point, you may be tempted to think that apostles and their associates and perhaps even pastors can be expected to exhibit the type of selfless service that Timothy displays, but that you—an ordinary Christian, not called to preach the Word—are off the hook. After all, you may assure yourself, you are just "not wired that way." If such a thought is running through your mind, you need to recall two truths.

First, Paul is sending Timothy precisely because his costly compassion for others will model for the whole church the focus on others' needs that the gospel calls *all* believers to have toward each other (Phil. 2:3–4). We cannot evade the challenge of Timothy's example by claiming that only the church's leaders, not its members, are called to demonstrate the mind-set that is ours in Christ (2:5–11).

Second, Timothy himself is not naturally "wired that way," either! To be sure, when Paul recruited Timothy to join his team, the young man was already "well spoken of by the brothers at Lystra and Iconium" (Acts 16:2–3). But Paul's comments in his epistles to and about Timothy show us that despite his home church's and Paul's appreciation for him, Timothy wrestles with fears that work against his readiness to focus more on others' needs than his own. To be blunt, Timothy struggles with insecurity, apprehension, perhaps even vulnerability to a fear of other people and their opinions. In Paul's first letter to the feisty congregation at Corinth, the apostle tried to provide preemptive protection for Timothy: "If Timothy comes, see to it that he has nothing to fear[8] while he is with you, for he is carrying on the work of the Lord, just as I am. No one, then, should refuse to accept him. Send him on his way in peace, so that he may return to me" (1 Cor. 16:10–11 NIV). This caution says much about the eagerness of the

8. The ESV's "put him at ease among you" does not convey as clearly as the NIV Paul's concern that Timothy be "*without cause to be afraid*" (*BAGD*, 127) while in Corinth, as the Greek adverb *aphobōs* signifies.

Corinthian church to beat up on pastors, but it also says something about Timothy. It sends the signal: "Handle Timothy with kid gloves. Don't feed his fears, don't disrespect him, and give him a peaceful send-off." Later Paul wrote to reassure Timothy: "God gave us a spirit not of fear but of power and love and self-control. Therefore do not be ashamed of the testimony about our Lord, nor of me his prisoner" (2 Tim. 1:7–8). After decades of service at Paul's side, Timothy still needs to be reminded that God's Holy Spirit instills power and love, not timidity; and he needs to be encouraged not to be ashamed of the gospel or of Paul himself.

Yet alongside these recurring clues to the fearfulness that seemed to beset Timothy throughout his life, the New Testament also records that he serves the church with consistent faithfulness and outgoing compassion. Though he struggles with fear, he refuses to surrender to it. How can Timothy keep marching forward in the face of fear? How can and should you persevere despite your own misgivings, apprehensions, and fears of failure or of others' disapproval? The answer lies in reminding yourself daily, hourly, even moment by moment, if need be, that you belong to a divine Champion who has dealt the death blow to the worst of our enemies and who stays by our side as our ever-present Protector. God's Son shared our human flesh and blood in order to "destroy the one who has the power of death, that is, the devil, and deliver all those who through fear of death were subject to lifelong slavery" (Heb. 2:14–15). Whatever it is that threatens to paralyze us in fear, this same Champion, Jesus, says to us today, " 'I will never leave you nor forsake you.' So we can confidently say, 'The Lord is my helper, I will not fear; what can man do to me?' " (Heb. 13:5–6).

Fearfulness focuses our concerns inward on ourselves, filling us with anxiety and preoccupation to devise strategies to protect against the loss of our safety, success, or reputations. But something—no, Someone—in Timothy's life is stronger than the innate gravitational pull that would otherwise keep his concerns self-bound, earthbound. By grace Timothy has found power to "push back" against his own timidity and to care more about Jesus, and therefore about Jesus' people, than he cares about himself. So despite Timothy's fears and failings, Paul commended him as a "scale model" of Christ's compassion.

Timothy is serving the gospel with Paul as a son labors alongside his father (Phil. 2:22). In a healthy father-son relationship, a son is eager to please his father, to follow his directions, to earn his confidence, and to grow to resemble his father. The father can trust such a son with challenging assignments. Others recognize that the son reflects the father's character and interests. By invoking the father-son analogy, Paul tells the Philippians, "Timothy's heart beats, as mine does, to see you all grow strong in Jesus, holding fast to and holding forth the gospel to the watching world, and caring for each other in the family of God. When you see him in action, you can be sure that you are experiencing how I care for you, though still separated from you by hundreds of miles."

But there is more than an analogy here: Paul has a special affection for Timothy as his "true son in the faith" (1 Tim. 1:2). By natural birth Timothy was the son of a mixed marriage—his father a Greek and his mother a Jew (Acts 16:1). Yet Paul—once a strict Pharisee of Pharisees, a Hebrew of Hebrews, taught in the Torah by the famous Rabbi Gamaliel—consistently regarded Timothy as his own adopted son and trusted agent. What makes this odd couple now like "father and son"? *The gospel!* On their first missionary journey, Paul and Barnabas came to Lystra, Timothy's hometown, and preached the good news of God's grace (Acts 14:8–20; see 2 Tim. 3:10–11). Through that gospel, the Spirit of God brought Paul and Timothy together from disparate backgrounds to become members of the family of God.

Paul is so keenly aware that he and his "son" Timothy stand together on the ground of God's grace that he chooses his words deliberately and precisely in describing Timothy's role. It is not that Timothy serves *Paul*, as ancient fathers would expect to be served by respectful sons; rather, Timothy serves *with* Paul in the gospel. The verb *serve* (*douleuō*), which appears only here in Philippians, refers to the service of slaves. It is an echo of the opening of the epistle in which Paul describes himself and Timothy together as "servants [*douloi*] of Christ Jesus" (Phil. 1:1). Moreover, it is an echo of his portrait of the King who stooped to take "the form of a servant [*doulou*]" (2:7), through whose suffering and humiliation Paul and Timothy have been rescued by grace and placed into this humble King's service. In Timothy's coming, the Philippians could see Paul's concern for them. Better yet, in Timothy they can glimpse a reflection of how Jesus Christ cares for their welfare, at the greatest of costs to himself.

It is common in human culture for the perks of power and the right to be served by others to gravitate into certain hands. In traditional cultures, younger generations serve their elders. In other societies, power brokers leverage wealth, physical attractiveness, intelligence, personal charisma, or sheer self-confidence to gather a circle of devotees to do their bidding. Jesus told his friends frankly that this is the way that power and prominence work in the world at large (Mark 10:42). But in his church, the church that belongs to the Son of Man, who "came not to be served but to serve," leaders' power is to be placed into humble service to those who are led (10:43–45). The congregation to which you belong is not "Pastor So-and-So's church." It is Christ's church. Every pastor, every elder, and every leader in any sort of ministry must be on guard, lest they forget Paul's indispensable preposition *with*. Those whom you lead do not serve *you*. They serve *with you* in advancing the gospel.

EPAPHRODITUS: REPLICA OF THE SUFFERING SERVANT

Although Paul plans to send Timothy as soon as his appeal to the emperor has been settled, he will not wait for the resolution of his legal status to send another colleague, Epaphroditus, to Philippi. Paul will send Epaphroditus immediately, as soon as the ink dries on the papyrus, which Epaphroditus will hand-deliver to Philippi. Epaphroditus is a member of the Philippian church, "your messenger and minister to my need" (Phil. 2:25). He had delivered the church's contribution toward Paul's expenses in house arrest (4:14–18) and stayed on to help Paul in other ways. But Epaphroditus had fallen gravely ill, and the alarming news that he was at death's door had reached his fellow believers in Philippi. By God's grace and power, the patient had recovered. Now Paul needs to get Epaphroditus himself back to Philippi as soon as possible, in the flesh and in good health, to put to rest the rumors about his deadly disease that are distressing the Philippian church. Epaphroditus is distressed over the report that his friends in Philippi are distressed, so he longs to return to reassure them. And Paul is distressed over Epaphroditus's distress and the Philippians' distress. Paul himself will have less sorrow[9] when the church has welcomed their returning brother with joy (2:28).

9. The Greek *alypoteros*, represented in the ESV by "less anxious," differs from *merimnaō*, which expresses "care" (Phil. 2:20 KJV), concern, worry, and being "anxious" (4:6). Rather, it is derived from *lypē*, which Paul has just used in affirming that Epaphroditus's healing had spared Paul "*sorrow* upon *sorrow*."

So Epaphroditus will set out for Philippi immediately, carrying news of Paul's situation and Paul's thanks for the Philippians' gift, and presenting his own restored health as an occasion for the church to rejoice in God's kindness. But Paul has another, not-so-ulterior motive in sending Epaphroditus home: the Philippians need another human role model to show them, in a man whom they knew well, what it means in the nitty gritty of everyday life to share the mind-set of Christ so thoroughly that one is ready to serve to the point of death, following the Savior's footsteps. As the Philippians will see the *submission* of Jesus the Servant reflected in Paul's readiness to let God direct his plans and the *selflessness* of Jesus in Timothy's concern for their well-being above his own, so they will observe Epaphroditus, one of their own, a man who is prepared to *sacrifice* his very life for the cause of Christ.

Paul draws the Philippians' attention to Epaphroditus by lavishing on him no fewer than five titles of respect and companionship. Three identify Epaphroditus's relationship to Paul, and two focus on Epaphroditus's connection with the rest of the church at Philippi. Epaphroditus is Paul's own "brother and fellow worker and fellow soldier." He had traveled from Philippi to Rome as the church's "messenger and minister," bringing a gift to meet Paul's needs. Together, these titles portray this man as a bridge who connects Christian brothers and sisters together in a partnership of deep love and united mission. When Epaphroditus reached Rome, it was as though the whole Philippian church had arrived to care for the apostle. Now Paul is returning Epaphroditus to the Philippian believers, and in him they will soon see Paul's own love for them, as well as Paul's military mind-set in the spiritual conflict in which both he and they are engaged (Phil. 1:30).

Paul sums up Epaphroditus's bond with himself in the words "my brother and fellow worker and fellow soldier." Paul frequently addresses fellow believers as "brothers" (see Phil. 1:12; 3:1, 13, 17; 4:1, 8)—not because they share biological DNA, but because God has graciously adopted them all as his beloved children (Eph. 5:1–2). Epaphroditus's name implies his pagan past, suggesting that at birth his parents had invoked over him the protection of the goddess Aphrodite. By the grace of Christ, however, Epaphroditus had been born into a new family and invested with a new identity. Now Paul affirms that he and Epaphroditus are brothers. As the Philippians will glimpse in Timothy, Paul's "son," the apostle's deep concern for them, so in his "brother" Epaphroditus they will see Paul's readiness to suffer for them. And of course, in all three men they will encounter replicas of Jesus, the selfless Suffering Servant.

To this family connection Paul adds two more titles that bind him to Epaphroditus: they are "fellow worker[s]" who have labored side by side, and "fellow soldier[s]" who have fought side by side. Having endured beating and imprisonment in Philippi, Paul is well aware of the enemies who harassed the believers in that proud city, which esteemed itself as a "Rome away from Rome." Therefore, earlier in this letter he invoked a military metaphor, marshaling them to steadfast faith, "standing firm in one spirit, with one mind striving side by side for the faith of the gospel, and not frightened in anything by your opponents" (Phil. 1:27–28). Now he is about to send back just such a brave soldier, one who has stood his ground arm in arm with Paul in defense of Christ's gospel and Christ's church.

Probably the most famous lines of William Shakespeare's play *Henry V* are the king's stirring "St. Crispin's Day" speech just before the Battle of Agincourt in 1415. Henry's English troops are outnumbered five to one by French forces, which block their route to the channel. Yet in the speech King Henry exults in the uneven odds, for he declares that the disparity will increase the glory of their courage, whether they prevail or die in defeat. Commoners from the lowest of classes will be ennobled by suffering, fighting, and bleeding alongside their monarch, as the king declares:

> We few, we happy few, we band of brothers;
> For he to-day that sheds his blood with me
> Shall be my brother; be he ne'er so vile,
> This day shall gentle his condition.[10]

Paul likewise honors Epaphroditus as a comrade-in-arms who has soldiered at his side. Then again, Paul would no doubt object to this analogy with the staunch soldiers of every class who won their nobility by striving and suffering beside their monarch on St. Crispin's Day. Paul and Epaphroditus are comrades-in-arms, but what ennobles them both is their identification with the King with whom and for whom they have fought. When Paul reports that

10. William Shakespeare, *Henry V*, Act IV, Scene 3.

a grave disease brought Epaphroditus to the point of death (Phil. 2:30),[11] he is echoing his description of Jesus' sacrificial obedience (2:8). Epaphroditus had been ennobled by God's mercy when his King shed blood to rescue him from wrath. Now Epaphroditus readily risks his life in company with his King of glory and of grace.

Can you appreciate the perspective that Paul's soldiering imagery gives as you respond both to your trials and to fellow believers? Christ the King has honored you by recruiting you into his army. In this army, there are no mere pawns, no foot soldiers whose lives are cheap and easily expendable. The King of kings ennobled us by shedding his blood for us, and by his blood created a "band of brothers" who embrace the privilege of suffering in his cause and serving with and for each other.

Epaphroditus is linked not only to Paul but also to his fellow Philippians, to whom Paul describes him as "your messenger and minister to my need." "Messenger" translates the Greek word that is ordinarily translated "apostle." Jesus' apostles carried out his mission in his authority, so he could say to them, "The one who hears you hears me, and the one who rejects you rejects me, and the one who rejects me rejects him who sent me" (Luke 10:16). So also the church's "apostle," Epaphroditus, had been sent by the church to act on its behalf. Paul wants his distant friends to know that he sees them all as being in Epaphroditus's care. When Paul writes in verse 30 of "what was lacking in your service to me," he is not whining about the Philippians' neglect, but recognizing that a distance of many miles prevents them all from helping him in person.[12] With Epaphroditus's arrival, however, that deficit in their personal presence has been filled.

The other title,[13] which the ESV translates as "*minister* to my need," in the New Testament typically refers to priestly service in a worship setting. Paul is already hinting that he regards the Philippians' donation for his relief

11. The Greek rendered by the ESV "he nearly died" could be translated, more woodenly, "to the point of death he drew near" (*mechri thanatou ēggisen*). Christ became obedient "to the point of death" (*mechri thanatou*) (Phil. 2:8).

12. Peter T. O'Brien, *The Epistle to the Philippians*, NIGTC (Grand Rapids: Eerdmans, 1991), 343–44; Fee, *Philippians*, 283–84.

13. Greek *leitourgos*. See Romans 15:16; Hebrews 8:2 for other uses of this term with priestly overtones. The abstract noun *leitourgia*, "ministry," and the verb *leitourgeō*, "minister," likewise refer to priestly service in worship in Luke 1:23; Acts 13:2; Philippians 2:17, 30; Hebrews 8:6; 9:21; 10:11. A reference to "service" broader than priestly worship might be in view in Romans 15:27; 2 Corinthians 9:12; Hebrews 1:7, 14.

as sacrifices offered to God. At the letter's close, Paul will express appreciation for their gift, calling it "a fragrant offering, a sacrifice acceptable and pleasing to God" (Phil. 4:18). Their contribution to his needs, delivered by Epaphroditus, was an act of worship to the Lord himself.

The measure of Epaphroditus's commitment to his mission of mercy on behalf of believers back home is his readiness to put his life on the line "for the work of Christ" (Phil. 2:30). Whether en route to Rome or after his arrival, Epaphroditus fell gravely ill. Scholars speculate that he could have contracted malaria or even bubonic plague, both of which were frequent and untreatable in the Roman world. Paul does not describe symptoms, except to say—not once but three times—that the disease could well have proved fatal. First Paul comments that Epaphroditus's illness had brought him "near to death" (2:27). Then, as we have seen, Paul portrays his friend's suffering as fitting the pattern of Christ's obedience "to the point of death"[14] (2:30; see 2:8). Finally, Epaphroditus had "risk[ed] his life" (2:30). Of course, Epaphroditus the replica was not identical to his Master, the Original. Whereas Jesus had passed *into* death and *through* death to resurrection life, God's mercy had brought Epaphroditus *back from* the brink of the grave. That divine mercy embraced not only Epaphroditus but also Paul, whose sorrow over his brother's illness would have been compounded by further sorrow if he had died. But God spared Paul "sorrow upon sorrow," and in so doing also gave the Philippians opportunity to rejoice (2:27, 29).

Epaphroditus's illness was the price he paid "for the work of Christ." Some might explain his disease merely medically, as the kind of infection that could strike anyone in the ancient world. His suffering might appear less heroic, less directly related to the cause of the gospel, than Paul's beatings and chains. After all, we might think, Paul got arrested, accused, and mistreated *because* he was preaching the gospel. Paul's was *real* persecution! Epaphroditus, on the other hand, just got sick. Even unbelievers get sick. Yet this clinical perspective on disease, as cogent as it is from a naturalistic perspective, is not the way in which Paul views his colleague's situation. Epaphroditus's sickness invaded his life precisely because he was following Jesus, transporting an expression of Jesus' love from Jesus' people in Macedonia to Jesus' servant Paul in Rome. True, no human persecutor was

14. Greek *mechri thanatou*.

hunting Epaphroditus down. But the price that Epaphroditus paid to bring support to Paul *was* suffering for Jesus' sake, as surely as were the scars on Paul's back from his beating at Philippi.

Epaphroditus has recovered, and he is ready to make the long and risky trip home. Why would he do this? Out of a loving longing for others' comfort and their fellowship, even above his own safety! Epaphroditus's sentiments reflect Paul's own feelings toward the Philippian Christians. As Paul has reported, "I yearn[15] for you all with the affection of Christ Jesus" (Phil. 1:8), so also Epaphroditus "has been longing for you all" (2:26). Paul's own affection for his Philippian friends will be reflected in Epaphroditus's reunion with them. Epaphroditus is being molded into a replica of Jesus' passion to be with his people, to dry our tears and ease the ache of our hearts.

The yearning of Paul and Epaphroditus to be with their brothers and sisters in Christ stands in stark contrast to the individualism and frenetic busyness of many Christians today, who regard involvement with other believers in worship and fellowship as a duty to be discharged as efficiently as possible, lest the family of God unduly interrupt one's private schedule, which is already overfilled with commitments and amusements. The pace of life in urban and suburban communities is driving and demanding, leaving little time to be spent with people for purposes that seem unprofitable for commerce or career advancement. Agrarian cultures often present fewer options and demands, but physical exhaustion and distance pose obstacles to the investment of time with fellow believers. Whether our social context works for or against the church's life together, what will make the difference is not the ease or difficulty of getting together, but the yearning in our hearts. Cultivate that yearning, for it is the appetite that anticipates the very joys of eternity, when the family of God will gather in celebration at our Father's banquet table!

Paul issues two instructions for Epaphroditus's homecoming party. First "receive him in the Lord with all joy," and then "honor such men." These directives strike some commentators as odd. After all, if the Philippians were so worried over Epaphroditus's illness, how could they help but receive him with joy, without Paul's telling them to do so? And if they entrusted their funds to Epaphroditus to carry to Paul in Rome, does not that trust imply that they respect him? Some scholars theorize that the frictions and

15. The same Greek verb, *epipotheō*, lies behind the ESV's "yearn" in Philippians 1:8 and "longing for" in 2:26.

self-centeredness to which Paul referred earlier had hardened into party lines, with Epaphroditus appreciated by one faction and dismissed by the other. On this theory, Paul orders a respectful welcome for Epaphroditus to signal that he has Paul's endorsement in whatever tensions threaten the church's unity.

This theory, though possible, is speculative. A simpler, less hypothetical explanation is that Paul's summons to joy is an invitation to celebrate God's kindness in sparing his children sorrows even in this sorrow-infected age. Paul's summons to "honor such men" is his way of focusing attention on Epaphroditus's willingness to put his life on the line to accomplish his mission from Christ and to relieve others' alarm. Paul is feeding his friends the right answer to the crucial question: "Whom do you want to be like when you grow up?"

Conclusion: Whom Are You Watching?

Look at Paul, and at Timothy, and at Epaphroditus. Let Paul teach you the reality of Christ's control of your life, so that you frame your plans and dreams and hopes in the light of Christ's lordship. Let Timothy show you what it looks like to push back against your own fears, and instead to have your heart opened to the cares and concerns of others. Let Epaphroditus show you the quiet courage that takes risks to health and safety, that puts one's own comforts and convenience in jeopardy, for the work of Christ, in service to the servants of Jesus.

But also look around: do you see fathers and mothers in the faith in your own congregation about whom you would say, "I want to be like him, like her, when I grow up"? Get close to them, watch them, and discover what makes them tick. Or do you see sons and daughters in the faith who are looking to you to be a replica of Jesus, showing them what it looks like to surrender our plans to Jesus' control, to have Jesus turn our cares outward to embrace others, to have the gospel so captivate our hearts that we gladly risk all for Jesus' glory?

As you look at Paul and Timothy and Epaphroditus, look *through* them to the Lord Jesus whose gospel they preached, by whose Spirit they lived and served. In them the Philippians could glimpse—in miniature, but "up close and personal"—the compassion of Christ the Son of God, who loved us and gave himself for us.

12

Trading My Rags for His Robe

Philippians 3:1–11

Indeed, I count everything as loss because of the surpassing worth of knowing Christ Jesus my Lord. For his sake I have suffered the loss of all things and count them as rubbish, in order that I may gain Christ and be found in him, not having a righteousness of my own that comes from the law, but that which comes through faith in Christ, the righteousness from God that depends on faith. (Phil. 3:8–9)

Having served as a seminary dean, I have spent a lot of time reading résumés. We have whole filing cabinets overflowing with pages that list individuals' qualifications and achievements. Some came from applicants who wanted to teach at our school. Others are student applications and transcripts that make the case that the applicants can handle the rigors of Hebrew, Greek, hermeneutics, church history, and theology. These files cry out, "I am qualified; I can do it—choose me!" You know what I'm talking about if you have ever applied for a job. You realize how important it is to make a good impression, to present that irresistible roster of achievements that will put you ahead of the competition, grab the employer's attention, get you the all-important interview, and finally land you the job.

This is the way our world works, so it is not surprising that many people assume that this is the way things work with God as well. Plenty of Paul's Jewish relatives heard that theme as they read their Scriptures (our Old Testament): God has graciously set our people apart from the unclean heathen nations around us. Now it is up to each of us to do our part, to build a solid résumé by keeping his commandments, and thus to obtain God's approval at the last day. For many years Paul, too, had assumed that this is the way God works. In his zeal for the glory of the God of Israel, Saul (his Hebrew name)[1] had striven to establish a record of righteousness through law-keeping, and by proving his zeal for God's honor by hunting down those who claimed that Jesus, a crucified criminal, was the Lord's chosen Messiah. After all, in his public life Jesus had kept company with unsavory characters and challenged rabbinic traditions that Saul held dear. Worse yet, Jesus' death by hanging on a tree confirmed his condemnation by God in the minds of pious Pharisees such as Saul (Deut. 21:23). It is little wonder that Saul, a man "extremely zealous . . . for the traditions of my fathers" (Gal. 1:14), would express his righteous indignation by persecuting the church that heralded Jesus as the Lord's Anointed.

But Saul's whole world had been shaken in a single day, when the risen Jesus appeared to him in blinding light on the road to Damascus. In a flash Saul discovered that the stellar résumé of heritage and personal effort in which he had placed his confidence was useless—even worse than useless. In his misguided zeal for Israel's Lord, he had defied the Lord's Messiah. Like many of his kinsmen, in his quest to establish his own righteousness he "did not submit to God's righteousness" (Rom. 10:2–3). Yet instead of the immediate punishment that he deserved, he was not only pardoned but also enlisted into service as Jesus' witness (Acts 22:14–15). Exposed as "a blasphemer, persecutor, and insolent opponent," Saul received mercy to

1. The book of Acts continues to refer to the apostle by his Hebrew name, Saul, until the venue of his ministry moves to the Greco-Roman world outside Palestine (Acts 13:9). As one who inherited Roman citizenship from his parents (22:28), he would also have borne a Roman cognomen, *Paulus*, now shortened to *Paul* in English. Sergius, proconsul of Cyprus, had the same cognomen (13:7). Although Saul bore both names both before and after his encounter with the risen Christ en route to Damascus, in this chapter I will use *Saul* when referring to the apostle's early life, and *Paul* when speaking of the period of his apostolic ministry of preaching and letter-writing among the Gentiles.

become an example of Christ's "perfect patience," demonstrating to others how amazing God's grace can be (1 Tim. 1:13–16).

That afternoon a radical shift occurred not only in Saul's self-perception but also in his understanding of how right standing with God can be obtained. In the imagery of Isaiah, whom Paul often quoted in his letters, the prophet's prayer that God "rend the heavens and come down . . . to make your name known to your adversaries" had been answered, showing Israel that "all our righteous deeds are like a polluted garment" (Isa. 64:1–2, 6). The people's best efforts at keeping God's commands fell so far short that even their "righteous deeds" only displayed their defilement, like a bloodstained garment or soiled rags. Yet the day of the Lord's descent was also the day when guilt-stained Israel could sing, "I will greatly rejoice in the LORD; my soul shall exult in my God, for he has clothed me with the garments of salvation; he has covered me with the robe of righteousness" (61:10). Saul had experienced that great exchange: his wretched rags of self-striving "righteousness" were stripped away, but his shame was covered by a robe of righteousness, the gracious gift of the glorious Lord whom he had insulted.

In Philippians 3:1–11 Paul rehearses his own experience as an example of God's free grace in Christ. He is safeguarding his friends in Philippi from a plausible spiritual poison spreading among the churches that he had planted among the Gentiles. His letter to the churches of Galatia shows the intensity of his paternal alarm over young believers who were falling under the spell of teachers who insisted that trust in Christ must be supplemented by circumcision and rigorous commandment-keeping, in order for Gentiles to find assurance of their standing among God's people and under God's favor (Gal. 5:3; 6:12). Paul labels this "Judaizing" message—requiring Gentile Christians to adopt Jewish practices specified in the Law—as a "different gospel" (1:6) because it involves a subtle shift of one's trust from Jesus back to oneself. That move would result in being "severed from Christ," and "Christ [would] be of no advantage" to those who submit to circumcision and the obligation to keep the whole Law, imagining that God's justifying verdict depends on a combination of Jesus' obedience and their own (5:2–4).

The spiritual peril posed by the Judaizers' teaching explains the intensity of Paul's tone and language not only in Galatians (see 1:8–9; 5:12) but also in Philippians 3:1–3, where he brands these purveyors of falsehood as "dogs," "evildoers," and "those who mutilate the flesh."[2] From that shocking opening, Paul invokes his own experience to build the case for the gospel of God's gift of righteousness, bestowed through faith alone in Christ alone. First he sums up the distinguished résumé of his own righteousness in which he once rested (Phil. 3:4–6); then he dismisses it all as "loss" and "rubbish"—in Isaiah's imagery, soiled rags—when compared with the robe of real righteousness in which God has enfolded him, the "surpassing worth of knowing Christ Jesus my Lord" (3:7–9). Finally, Paul affirms that his security in God's gift, far from making him complacent, whets his appetite for an even fuller knowledge of Christ, sharing both his sufferings and his resurrection life (3:10–11).

REJOICE—AND LOOK OUT!

Paul's command to "rejoice in the Lord" (Phil. 3:1) sounds again a motif that pervades this letter, forged in the crucible of suffering (1:18, 25; 2:18, 29; 4:1, 4, 10). Although he has just said that he expects them to rejoice over Epaphroditus's rescue from death's door (2:29), Paul makes it clear that the joy that he and the Philippians share is grounded not in their circumstances, but "in the Lord."

From the bright motif of joy in the Lord, Paul transitions to a sobering subject with the comment, "To write the same things to you is no trouble to me and is safe for you." Is Paul referring to what he has just written, remembering that he has called the Philippians to rejoice in 2:18 (and implicitly in verse 29) and is now repeating himself a few sentences later? That is possible, but it does not explain why repeating a command to "rejoice" would make the Philippians "safe." It is more likely that Paul had warned them in person against the spiritual peril of Judaizing legalism and that he is about to repeat his warning in writing (3:2–11).

2. Despite the vehemence of his language in Philippians 3:2, Paul seems less agitated over the believers in Philippi than he had been over the life-or-death peril of the churches in Galatia. Writing to the Galatians, he introduced the topic of the Judaizers' false gospel in the epistle's opening paragraphs, and virtually the whole letter was filled with a passionate and thorough case for faith in Christ and his cross. In Philippians, on the other hand, the apostle addresses the error of those who "put . . . confidence in the flesh" decisively but briefly in ten verses, citing his autobiography as sufficient evidence that self-generated righteousness by law-observance cannot compare with the gift of God's righteousness in Christ.

Paul's sharp shift in tone and topic from "Rejoice in the Lord" to "Look out for the dogs" inclines some New Testament scholars to speculate that Philippians 3:2–11 is a scrap from a different letter, mistakenly inserted here. Yet all the ancient manuscripts of Philippians—going back to the late second century[3]—contain these verses right where we find them. There is no evidence that they were ever absent from any copy of this epistle. Others suggest that, while composing his letter, Paul received news that the Judaizers had reached Philippi. He therefore "shifted gears" abruptly to protect his friends against this plausible but poisonous teaching. An even better explanation is that of Walter Hansen, who observes that throughout this epistle Paul shifts his focus between positive and negative examples: those who preach for love of Paul versus those who preach from selfish motives (1:15–17), citizens of heaven versus their opponents (1:27–28), God's blameless children versus the "crooked and twisted generation" around them (2:15), and those who follow Paul's example versus those who live as the cross's enemies (3:17–21). Here, three imperatives that "call for association with and imitation of good examples"—receive Epaphroditus, honor such men, rejoice in the Lord (2:29–3:1)—are followed by three imperatives that warn against negative examples—"*Look out* for the dogs, *look out* for the evildoers, *look out* for those who mutilate the flesh" (3:2).[4]

The villains of whom the Philippians must beware are described in terms that (in Greek) are tied together through alliteration: *kunas* (dogs), *kakous ergatas* (evildoers), *katatomē* (mutilation). Reinforcing the redundant harshness of their sound is their unsavory sense, for Paul's terms paint people who see themselves as exemplary in loyalty to Israel's Lord in the dark colors of paganism. In the ancient world, most dogs were not pets but street mongrels, scavenging scraps from stinking garbage heaps, including disgusting and defiling carcasses (Ex. 22:31; 2 Kings 9:33–37; Jer. 15:3). In the New Testament, the term *dogs* symbolizes those unworthy to receive holy things (Matt. 7:6; 15:26–27; 2 Peter 2:22; Rev. 22:15). Though the Judaizers no doubt advocated keeping a kosher diet, Paul portrays them as feasting on pollution. Though they claimed to be doers of the Law (Gal. 6:13), they were actually doers of evil, just as Saul had been when he persecuted Christ's church. They wanted to impose circumcision (Greek *peritomē*) on the Gentiles; but by denying its

3. Papyrus 46 is generally dated around A.D. 200.

4. G. Walter Hansen, *The Letter to the Philippians*, PNTC (Grand Rapids: Eerdmans, 2009), 216.

meaning as a seal of the righteousness that is received by faith (Rom. 4:11), they had turned that surgery into the religious equivalent of the "mutilation" (Greek *katatomē*) practiced by pagans, who slashed their bodies to grab the attention of their gods (1 Kings 18:28). God forbade Israel's priests from engaging in such self-mutilation (Lev. 21:5). One commentator says, "Paul uses epithets that 'turn the tables' on them, as to what they think themselves to be about in contrast to what he thinks."[5]

Is Paul overreacting when he repudiates so strongly the religious recipe in which he had been raised, the formula that blended God's initial grace with one's own best efforts at commandment-keeping to yield, so the Judaizers thought, divine approval at the day of judgment? Many people, then and now, have thought Paul was wrong to get so angry over these teachers. Who can fault the goal of trying to build a résumé of righteous living? And is it not plain common sense to motivate ourselves and others to go after that goal by promising a reward for a job well done? After all, it feels so natural to be moved by the fear of failure or by the offer of recognition for strenuous achievement. It may be that the dangerous teachers that Paul had in view would have granted that God graciously provides an initial "nudge" toward holy living. But then, perhaps, they taught, as others after them have done, that God's final approval at the last judgment hinges on how vigorously we capitalize on the start he provided. They may have felt that Paul's gospel of grace, which assures those who trust Jesus *at the outset* that the Savior's blood and righteousness secure our place in the Father's heart, undermines the pursuit of holiness by depriving both nervous uncertainty and ambitious overconfidence of their power to drive us toward that goal.

Yet Paul insists that the only people who can claim the coveted title "the circumcision"—who *really* belong to God's gracious covenant with Abraham—are those who have ceased striving or boasting in their own efforts, and who rather "worship by the Spirit of God and glory in Christ Jesus and put no confidence in the flesh" (Phil. 3:3). Building on the Old Testament's teaching that physical circumcision symbolized deep removal of defilement from the heart, which only God could accomplish (Deut. 10:16; 30:6), Paul had informed the Romans that real circumcision, "a matter of the heart," was performed by the Spirit (Rom. 2:29). When the

5. Gordon D. Fee, *Paul's Letter to the Philippians*, NICNT (Grand Rapids: Eerdmans, 1995), 295.

Spirit of God has performed on you that heart transformation, to which the physical rite of circumcision pointed, your response is to "glory" (ESV) or "boast in Christ Jesus," rather than taking pride in yourself and your accomplishments. Elsewhere Paul unfolds what it looks like to have our basis for boasting redirected from ourselves to Jesus. In the presence of God's infinite glory and grace, we no longer rest our confidence in our intelligence, our influence, our ethnic heritage, or our ethical effort; instead, we gladly heed Scripture's summons: "Let the one who boasts, boast in the Lord" (1 Cor. 1:31, quoting Jer. 9:23–24). Rather than "boasting" that we have persuaded others to follow our lead, we now boast only "in the cross of our Lord Jesus Christ, by which the world has been crucified to [us], and [we] to the world" (Gal. 6:13–14).

Paul's description of those who are truly "circumcised" challenges us to ask ourselves: "In what do I rest my confidence? In whom do I glory and boast?" Though we know and confess that we can receive God's favor through Christ alone, you and I still find it too easy to take pride in personal achievements or qualities that—so we assume—set us a step or two above others. We might hide that pride behind a facade of humble self-deprecation; but amid feats of accomplishment and fearful insecurities, in the secret recesses of our hearts we seek solace in "the flesh," in what our hands have done, to reassure us of peace with God. When you find yourself glorying in anything other than Christ Jesus, you need to return to the gracious gospel that drew you to your Savior at the beginning. You need to see that Christ, and Christ alone, deserves your full confidence, for Christ alone conveys "the righteousness from God" (Phil. 3:9) that silences Satan's charges and your own conscience's guilty discomfort.

The Rags That I Mistook for Robes

Paul now offers his spiritual autobiography as a case study in two ways of trying to get right with God—credentials or Christ. Each route to rightness with God has its own outworking in behavior, in life lived out every day. The routes lead to radically different destinations—and Paul could testify from his own experience that the path commended by the advocates of self-reliant law-keeping leads to a dead end, however promising it may appear at the trailhead. He has also found that resting in Christ and Christ's credentials actually energizes believers to pursue holiness and love with expectant eagerness.

From the standpoint of qualifications that might provide a basis for "confidence in the flesh," Paul could go "toe to toe" with the Judaizers, and he could best their most pious champion. He lists his credentials to show that his embrace of the gospel is not a matter of "sour grapes" or an admission of failure on the part of a sore loser who could not match others' achievement. But he also wants to reassure Gentile believers that Christ and his righteousness are infinitely more valuable than the best that conscientious law-keeping could ever offer. If you are honest with yourself, you will admit that the same spiritual self-reliance that can breed a sense of superiority over others can just as easily feed feelings of insecurity: "Have I done enough? Am I good enough?" Paul shares his own story to commend to us the amazing grace of God, which simultaneously shatters our pride and quiets our qualms by directing our hearts upward and outward, to rest in Jesus the Savior alone.

Paul's résumé opens with four qualifications related to matters of ancestry and parentage, over which he personally had no control: he was circumcised on the eighth day, belonged to the people of Israel and to the tribe of Benjamin, and was a "Hebrew of Hebrews." Modern American individualists might dismiss such advantages of birth as irrelevant to the issue of one's status before the face of God. Often in the West we tend to credit people only with achievements that they have performed themselves. Ancient peoples, including the Jews, were more realistic. They knew that family and heritage mold individuals' personalities and capabilities, for good or ill. Our ancestors, parents, social class, educational opportunities, religious training or lack thereof, and other factors beyond our control mold the persons whom we become. Although Saul did not personally choose his circumcision, his people, his tribe, or his parents' commitment to maintain their Hebrew language and customs, each of these factors contributed to his preeminence within Judaism. First, he was circumcised, as the Law prescribed (Gen. 17:12), eight days after his birth, not belatedly as in a less-observant Jewish family or in adulthood as a Gentile proselyte. (Paul cites his circumcision first, though it came temporally after his heritage from Israel and Benjamin, because the imposition of this rite upon Gentile Christians was the crux of the problem posed by Judaizers, as verses 2 and 3 imply.) Second, Saul was born from the people of Israel, a genetic descendant of the patriarchs Abraham, Isaac, and Israel, not a convert from some pagan nation. Third, his tribe looked proudly back to its ancestor Benjamin, Israel's youngest and beloved son, the only one born in the Promised Land (Gen. 35:16–18). From Benjamin came Israel's first king, the apostle's namesake (1 Sam. 9:1–2). When Jeroboam led most Israelite tribes in revolt against King Rehoboam, only Benjamin and Judah stayed loyal to the Davidic dynasty (1 Kings 12:21; see Ezra 4:1). Finally, Saul's parents, unlike many other Jewish families in the Dispersion, raised him as a "Hebrew" born "of Hebrews." Amid the cultural diversity of Tarsus, they resisted the pressure to neglect the language of Scripture or the rites that set them apart as God's chosen people.

To the advantages of his birth and upbringing—national and tribal identity, circumcision, and conscientious covenantal nurture—Saul added his own strenuous commitment to the Lord and his Law. He maintained his family's association with the party of the Pharisees (Acts 23:6), the sect within Judaism that was distinguished by its precision in adhering both to the written commands in the law of Moses—all 613 of them, as the rabbis counted—and to the oral tradition that was revered as a God-given interpretation and "fence," a secondary barrier to preclude trespassing of the commands themselves.[6] Saul's contemporary, the Jewish historian Josephus, also a Pharisee, said that their party was esteemed by fellow Jews as "most skillful in the exact explication of their laws."[7] To sum up his commitment both to study and to obey the commands delivered through Moses, Saul needed to say no more than "as to the law, a Pharisee."

Not every Pharisee, however, took zeal for God's honor to the length of persecuting the church. Saul's mentor Gamaliel had urged the Sanhedrin to take a "wait and see" approach to the apostles' proclamation of Jesus as Messiah (Acts 5:34–39). Yet Saul himself was enraged by Stephen's announcement that the temple was expendable (Acts 6:13–14; 7:48–50) and that a cursed criminal hanged on a tree (Deut. 21:23; Gal. 3:13; see Acts 5:30; 13:29) was God's Righteous One (Isa. 53:11–12; Acts 7:52). Carrying on the legacy of ancient Phinehas (Num. 25:11–13) and the Maccabees, Saul had launched a

6. Mishnah tractate *Abot* (fathers) 1.1: "Moses received Torah at Sinai and handed it on to Joshua, Joshua to elders, and elders to prophets. And prophets handed it on to the men of the great assembly. They said three things: 'Be prudent in judgment. Raise up many disciples. Make a fence for the Torah.'" Jacob Neusner, *The Mishnah: A New Translation* (New Haven, CT: Yale University Press, 1988), 672.

7. Josephus, *Jewish War* 2.8.14, in *Josephus: Complete Works*, trans. William Whiston (Grand Rapids: Kregel, 1960), 478.

campaign of intimidation, incarceration, and violence to eradicate the Jesus sect (Acts 8:13; 9:1–2). Saul went well beyond the expected compliance with the commandments, expressing his zeal in strenuous opposition to those whom he considered threats to Israel's fidelity.

Paul's heritage and his personal effort converged in a final ground for confidence in the flesh: "as to righteousness, under the law, blameless." That claim may surprise us, when we recall that elsewhere Paul states that God's Law reveals the corruption in every human heart, even the heart of a conscientious rabbi, so that "by works of the law no human being will be justified in his sight, since through the law comes knowledge of sin" (Rom. 3:20; see 7:7–11; Gal. 2:16; 3:10–11). Paul is not claiming that he had achieved sinless perfection during his years as a rigorous law-keeper and zealous avenger of God's honor. Rather, he is speaking, as his colleagues in Pharisaic Judaism would have spoken, of a consistent and conscientious observance of the Law, including its provisions for the removal of guilt and defilement through repentance and sacrifice. In the Psalms, David declares that he is "blameless" (Ps. 18:23, 25, 32) and praises God for dealing with him according to his righteousness and "the cleanness of my hands" (18:20, 24), not because he was the one and only exception to the rule that Adam's natural children share Adam's sin but rather because he sought to express his faith through faithfulness to God's commands. In this sense, Luke's gospel could describe the aged priest Zechariah and his wife Elizabeth as "walking blamelessly in all the commandments and statutes of the Lord" (Luke 1:6), even though Zechariah would dare to doubt God's angel Gabriel. Since animals' blood could not ultimately remove human guilt, the cycle of atoning sacrifices was endless. But to the extent that the Law could offer "blamelessness" to those who observed its rules and its rites, Saul had achieved the apex of such righteousness that lay within the reach of human effort. No one could surpass his record.

In our twenty-first-century ears, for anyone to claim to be "blameless" or even "righteous" sounds arrogant, smug, hypocritical, and perhaps prudish. Yet though we find subtler ways to feel or even express satisfaction with ourselves, we are not strangers to that "confidence in the flesh" that Paul saw so clearly in his own former self. It comes out especially when we compare ourselves with the shortcomings that are so obvious in others. They are lazy and undisciplined; we are diligent and exercise self-denial. They are haughty; we pride ourselves on our humility (though we do not admit it aloud). They boastfully call attention to themselves; we disapprove of their brashness, though we may envy the admiration it attracts. The details may differ between the rigorous Judaism of Paul's day and the casual tolerance of ours, but the drive to reassure our uneasy consciences through our best efforts and sidelong glances at our competitors ties us more closely to Paul's rabbinic colleagues than we care to admit.

The Robe Worth More Than All My Rags

Yet a blinding light from heaven and thundering voice of the risen Lord Jesus had unmasked Saul's complacent self-delusion: "Saul, Saul, why are you persecuting me?" (Acts 9:4). His zealous pursuit of God's approval as recognition of services rendered had made Saul not the Lord's ally, but the Lord's enemy. In that terrifying, paralyzing, pride-shattering encounter on the Damascus road, Saul suddenly made two shocking discoveries.

First, the religious credentials that he had amassed through strenuous effort were worse than worthless. "Whatever gain I had, I counted as loss.... Indeed, I count everything as loss.... I have suffered the loss of all things and count them as rubbish" (Phil. 3:7–8). Paul deliberately employs banking terminology to dramatize how radically the risen Lord had reversed his whole scale of values. Every *deposit* that Saul thought he had been making into his "account" in the presence of God was actually, he now knew, just one more *debit*. All the beautiful fruit that he thought he was producing for God's pleasure was rancid, rotten, repugnant "rubbish"—an off-color term that appears only here in the New Testament. (The ESV's "rubbish" is probably too mild a translation. The KJV's "dung" is more pungent, and more to the point. The Greek term includes both spoiled food and excrement.[8]) As forcefully as he can, Paul pronounces the résumé in which he once rested his confidence to be both personally repellent and objectively defiling before the face of God.

8. For the use of the term to refer to excrement, see Sir. 27:4 (LXX). John Reumann, *Philippians: A New Translation with Introduction and Commentary*, AYB (New Haven, CT: Yale University Press, 2008), 491–92, strongly supports this sense. The term may also entail a wordplay on "dogs" (*kunas*) in 3:2, since Fee, *Philippians*, 319n23, and others believe that *skybala* may have been abbreviated from a phrase describing leftovers "thrown to the dogs" (*eis kysi balomena*).

Saul's second discovery was that his previous credentials had been shown to be worthless—worse than worthless: repellent—by the infinitely more valuable treasure that he had found: "the surpassing worth of knowing Christ Jesus my Lord" (Phil. 3:8). As in Jesus' parable of the man who unearthed a buried treasure and sold all to buy the field in which it lay hidden (Matt. 13:44), so Paul has gladly forfeited all else—his reputation, self-respect, standing in the community, personal safety, and more—in order to "gain Christ," to "be found in him," to "know him" (Phil. 3:8–10).

But should we say that Saul found this treasure? Would he not insist that *the Treasure found him*, when he was not looking for it? The Christ whom he had hated and persecuted had confronted Saul and united Saul with himself by the gift of faith, so that now Paul rejoices to be "found in" Christ (Phil. 3:9). By that amazing union of grace, as Saul was compelled to relinquish his self-reliance and to rely instead on Jesus, Saul received an infinitely better righteousness than he had managed to produce by his own zealous efforts. This is why Paul emphasizes so emphatically the contrast between "a righteousness of my own that comes from the law," on the one hand, and "that which comes through faith in Christ, the righteousness from God that depends on faith," on the other (3:9). By setting *his own* righteousness over against the gift of righteousness that he had received by *faith*, from *God*, and through his new identity in *Christ*, Paul shows how radically the weight of his heart has shifted away from himself and his attempts at obedience. He now rests in the accomplished obedience and sacrifice of Jesus. It is not simply that he has abandoned reliance on maintaining the rites and practices that set Israel apart from the Gentiles (as some recent scholars[9] interpret "works of the law" in such texts as Galatians 2:16; 3:1–5; and elsewhere). It is not that he has resolved to replace adherence to external regulations with a commitment to cultivate deeper character qualities (love, integrity, or justice). Rather, Paul's heart now *looks away from Paul*—away from his heritage and upbringing, his external circumcision and his internal zeal. It has come to rest in "Christ Jesus my Lord" and to receive right standing before God by looking in faith to Christ. The Westminster Shorter Catechism

9. For example, J. D. G. Dunn, *Romans 1–8*, WBC 38a (Dallas: Word, 1988), 188, explains Paul's expression "works of the law" as "characteristically and distinctively Jewish practices ('identity marker')" and "as marking the boundary between Jew and Gentile ('boundary marker')." See Dunn's earlier essay, "The New Perspective on Paul," *Bulletin of the John Rylands Library* 65 (1983): 95–122.

beautifully captures Paul's point when it describes justification as "an act of God's free grace, wherein He pardoneth all our sins, and accepteth us as righteous in His sight, only for the righteousness of Christ imputed to us, and received by faith alone."[10]

In the amazing exchange of grace, Jesus received the curse, condemnation, and execution that Paul deserved. Paul received the divine approval for complete obedience that Jesus deserved. Clothed in the robe of Jesus' righteousness, Paul knew that the shame of his defiance and self-reliance—his "rubbish"—had been carried by the divine Savior down the rugged road of radical obedience, to death on a cross. It is no wonder that Paul describes Christ elsewhere as "the Son of God, who loved me and gave himself for me" (Gal. 2:20).[11] It is no wonder that Paul's supreme desire is that "Christ will be honored in my body, whether by life or by death" (Phil. 1:20).

Notice that the treasure for which Paul has gladly relinquished all his previous claims to fame is not merely forgiveness, nor only a justified status before God in view of the coming judgment. The treasure worth more than everything to Paul is *Christ himself*—to gain *him*, to be found in *him*, to know *him*. In Christ Paul has received the robe of righteousness that his best efforts could never have pieced together; but he had received, was receiving, and would receive so much more. For Paul, Christ is not merely a dispenser of saving benefits: forgiveness, acceptance with God, life-transforming power, future resurrection. Christ is the second person of the triune God for whose fellowship we were designed, and in whose friendship we find our highest joy. This is why the fact that he had received *through faith* the righteousness for which he had long striven *through law-keeping* did not lull him into a spiritual stupor of complacency. The Judaizers no doubt suspected that "pure" grace would have that effect, excusing self-indulgence on the pretext that God's approval is no longer contingent on our performance (Rom. 6:1). But for Paul—and for all others who discover the wonder of the Christ who loved us and gave himself for us—the effect produced by grace is the opposite of self-absorbed passivity. Paul finds his heart set aflame with new ambition, by a passion stronger than his strenuous quest for God's approval.

10. WSC, answer 33.
11. See also Gal. 1:4; Eph. 5:2, 25; 1 Tim. 2:6; Titus 2:14.

FURTHER UP AND FURTHER IN!

Toward the end of *The Last Battle*, the climax of The Chronicles of Narnia, C. S. Lewis subtly challenges the popular conception of eternal life as endless idleness. He shows the redeemed, having passed through the catastrophe that consumed the old Narnia, sprinting with increasing speed and energy and joy ever more deeply into a new Narnia, the real Narnia of which the old was only a reflection. As they run, Aslan's joyful roar rings in their ears, "Come further in! Come further up!" and is taken up by others: "Further up and further in!"[12]

For Paul, the race to explore the wonders of knowing Christ the Redeemer in his fullness has already begun, but it is far from finished. What he has tasted so far of the sweet fruit of Jesus' redemptive mission has whetted his appetite for the complete feast. Having discovered "the surpassing worth of knowing Christ Jesus my Lord," the apostle is eager, more and more, to "know him and the power of his resurrection," to share his sufferings, to be conformed to his death, and finally "by any means possible" to "attain the resurrection from the dead" (Phil. 3:10–11).

In this resounding climax to Paul's autobiography, the apostle shows the connection between his (and our) current experience of knowing Christ, on the one hand, and the future goal toward which we race, on the other. The power of Christ's resurrection, applied by the Holy Spirit, has transformed us who once were "dead in . . . trespasses and sins," and has "made us alive together with Christ" (Eph. 2:1, 5). The Spirit's powerful presence abides in believers as the initial installment and guarantee of our future inheritance (Eph. 1:14)—a foretaste of the final transformation of "our lowly body to be like [Christ's] glorious body" (Phil. 3:21). Paul had written to the Romans: "If the Spirit of him who raised Jesus from the dead dwells in you, he who raised Christ Jesus from the dead will also give life to your mortal bodies through his Spirit who dwells in you" (Rom. 8:11). The Philippians themselves have heard the same promise expressed more concisely: "He who began a good work in you will bring it to completion at the day of Jesus Christ" (Phil. 1:6).

12. C. S. Lewis, *The Last Battle* (London: Macmillan, 1956), 149 (Aslan's shout), alluded to again (158) and picked up in the title of chapter 15, and the shouts of the Unicorn (162) and Farsight the Eagle (163).

In addition to tasting the power of Christ's resurrection in the present, Paul also wants to share Christ's sufferings. Central to Paul's teaching on salvation is the reality that believers are covenantally united to Christ, our representative, both in his death and in his resurrection on our behalf. Because Christ "was delivered up for our trespasses and raised for our justification" (Rom. 4:25), when the Holy Spirit brings us to faith, the death that Jesus died once for all in history is reckoned as our death under the wrath that our sins deserve, and his resurrection on the third day vindicates us as righteous, robed in his righteousness. Moreover, as Paul wrote in Romans 6, our union with Christ is so intimate that his death in history entails our death in current experience to the tyranny of sinful desires, and his resurrection in history ushers us into a new life, into the power of holy living and loving.

In Philippians 3:10, however, Paul has in mind another aspect of his union with Christ when he places "the fellowship of his sufferings"[13] next to "the power of his resurrection." Although Christ's suffering and death accomplished all that would ever be needed to atone once for all for our sin (Heb. 9:26–28), we are still called to endure the wicked world's hostility against our Savior. Perhaps it was as early as the Damascus road that this truth began to dawn on Paul: Christ the Lord counts harm done to his people as harm done to himself: "Saul, Saul, why are you persecuting me?" (Acts 9:4). Paul wrote often of the painful—yet joyful—experience of "always being given over to death for Jesus' sake, so that the life of Jesus also may be manifested in our mortal flesh" (2 Cor. 4:11). Because the Lord has assured him, "My grace is sufficient for you, for my power is made perfect in weakness," Paul is "content with weaknesses, insults, hardships, persecutions, and calamities. For when I am weak, then I am strong" (12:8–10). Paul's sufferings for the sake of communicating the gospel are the way that "in my flesh I am filling up what is lacking in Christ's afflictions" (Col. 1:24).

The full flowering of Paul's current participation in Christ's resurrection power and sufferings is still future, when Paul anticipates "becoming like him in his death" and reaching the final resurrection from the dead. The expression that the ESV renders "becoming like him in his death" contains a verb that means "being conformed" (*symmorphizomenos*). This verb forms a

13. So the KJV and NASB, reflecting more directly the Greek wording (*tēn koinōnian tōn pathēmatōn autou*) than the ESV's "may share his sufferings." See NIV: "the fellowship of sharing in his sufferings."

bridge between the hymn to Christ, who was "in the form of God" (*morphē theou*) and took "the form of a servant" (*morphēn doulou*) (Phil. 2:6–7), and the promise of Christ's future return from heaven to make our lowly bodies "to be like" (*summorphon*) his glorious body (3:21).

The right standing before God that Paul has been granted by grace through faith, wholly apart from his past credentials or current striving and his current experience of resurrection power amid suffering through Christ's Spirit, has whetted Paul's appetite. He longs to experience fully the joy of knowing Christ, which he has tasted in part. Not yet having reached the climax of his race, Paul will press on to take possession of the prize for which Christ has already taken possession of him (3:12).

Paul's resolve to run the race toward resurrection with all his might may seem counterintuitive, if our hearts (like the Judaizers', I suppose) can be motivated only by threat or uncertainty of achieving a reward. Paul's "by any means possible" in verse 11 does not imply uncertainty over the outcome that Christ had already secured when he "made me his own" (Phil. 3:12) and conferred "the righteousness from God that depends on faith" (3:9). Paul simply acknowledges that he cannot foresee the route by which God would bring him to the finish line. But if God in Christ has already erased the threat of judgment and bestowed once and for all the approval that secured Paul's participation in the final resurrection, what could keep Paul running "further up and further in"? Christ's love for Paul has ignited Paul's love for Christ, so Paul—secure in his Savior's grace—longs to know that Savior ever more fully and to love the Savior ever more fervently. The path ahead in Paul's marathon toward the final goal might well entail steep hills and blazing sun. He expects the trail to have its share of suffering, for he forewarned his younger colleague Timothy that "all who desire to live a godly life in Christ Jesus will be persecuted" (2 Tim. 3:12). What could steel Paul—what can steel our resolve—to face and embrace the fellowship of Christ's sufferings, when our natural instinct is to pull away from what brings pain? The once-suffering and now-glorified Savior speaks words of assurance to us today, as surely as he spoke to Paul long ago: "My grace is sufficient for you, for my power is made perfect in weakness" (2 Cor. 12:9). No threat to life, to comfort, or to safety—not "tribulation, or distress, or persecution, or famine, or nakedness, or danger, or sword"—can separate us from the love of Christ (Rom. 8:35). When the pain becomes most intense, the Christ who loves us and gave himself for us draws closest to us through his ever-present Holy Spirit.

The Protestant Reformers saw that the motive stronger than guilt or fear or self-reliant ambition is thankful *love*—the freely offered affection of a child who rests assured in his Father's love. They knew that only the combination of assurance, resting on righteousness not our own, with the Holy Spirit's gift of joy can spark in our hearts a love for God that is not, at root, self-serving. In 1561 Guido de Brès composed the Belgic Confession for the Protestant churches in Belgium and the Netherlands. Six years later he would be martyred for the gospel. In the article on "The Sanctification of Sinners," de Brès wrote:

> We believe that this true faith, produced in man by the hearing of God's Word and by the work of the Holy Spirit, regenerates him and makes him a "new man," causing him to live the "new life," and freeing him from the slavery of sin. Therefore, far from making people cold toward living in a pious and holy way, this justifying faith, quite to the contrary, so works within them that apart from it they will never do a thing out of love for God but only out of love for themselves and fear of being condemned.[14]

Unless I am assured, as Paul was, that my place in God's heart is secured not by "a righteousness of my own" but by the obedience and sacrifice of Jesus the Son, all my best efforts to do and be good—whether through commandment-keeping or character cultivation—will, in the end, be self-serving, not God-loving. But the clearer our grasp of our security in God's love through Christ, the more intense will be our longing to experience in its fullness "the surpassing worth of knowing Christ Jesus my Lord," and the stronger will be our motivation to "press on toward the goal for the prize of the upward call of God in Christ Jesus" (Phil. 3:8, 14).

To Possess by Faith

We had traveled to Africa for the wedding of a very special couple. Our son and his fiancée had come to faith as children. Long before they met,

14. The Belgic Confession, Article 24, in *Ecumenical Creeds and Reformed Confessions* (Grand Rapids: CRC Publications, 1988), 101.

each sensed God's call to convey his gospel to people who had never heard their Creator speak the language of their hearts. Now, as they exchanged their vows, committing themselves to serve Christ together in this cause, we were among a gathering of God's family that included expatriates from many countries and Chadians from several people groups, some highly educated and some not so. The bride and groom had chosen this song, drawn from Philippians 3, for the congregation to sing together:

> All I once held dear, built my life upon,
> All this world reveres, and wars to own,
> All I once thought gain I have counted loss
> Spent and worthless now, compared to this:
>
> Knowing you, Jesus,
> Knowing you, there is no greater thing.
> You're my all, you're the best,
> You're my joy, my righteousness,
> And I love you, Lord.
>
> Now my heart's desire is to know you more,
> To be found in you and known as yours,
> To possess by faith what I could not earn:
> All-surpassing gift of righteousness.
> Oh, to know the power of your risen life
> And to know You in Your sufferings,
> To become like you in your death, my Lord,
> So with you to live and never die.[15]

It was moving to sing those words with those brothers and sisters, as we pondered the price they were paying to be where they were and do what they were doing. It was also humbling, as we realized the implications of those words for everyone who trusts in Jesus: the more we see Jesus as our "all," our "best," our "joy" and "righteousness," to whom nothing can compare, the freer we will be to relinquish all we "once held dear," to know him and to serve him.

15. Graham Kendrick, "Knowing You (All I Once Held Dear)," Copyright: © 1994 Make Way Music (admin. by Music Services in the Western Hemisphere). All Rights Reserved. ASCAP. Used By Permission.

13

The Restless Race of Those Who Rest in Grace

Philippians 3:12–16

Brothers, I do not consider that I have made it my own. But one thing I do: forgetting what lies behind and straining forward to what lies ahead, I press on toward the goal for the prize of the upward call of God in Christ Jesus. (Phil. 3:13–14)

"Christians aren't perfect, just forgiven." If you live in the United States, you have probably seen this motto on the bumper sticker of a car that just cut you off in traffic. What is this slogan really saying? Its intention, I suppose, is to counteract the stereotype (which is sometimes all too true) that Christians imagine that they are morally superior to others, so they can condemn others for lifestyle choices that violate biblical norms. So perhaps it's a way of admitting, "We're not so special. We're flawed, just like everybody else." It expresses humility, while also suggesting that defining Christianity in terms of the impeccable behavior of its adherents is a mistake.

But does the slogan send other signals, too? On the one hand, it appears to minimize the gift of God's forgiveness, since it claims that Christians are "*just* forgiven." On the other, it may imply that forgiveness is all that we wanted from God anyway. Its subtext seems to be: "Since Christians are forgiven, the fact that we are so imperfect doesn't really bother us. It's no big deal." So a motto meant to express humility also suggests complacency.

For the apostle Paul, however, the gifts of forgiveness and right standing with God through faith in Christ made him anything but complacent. To be sure, Paul had abandoned the life of striving and achieving that characterized his life before Christ seized him on the road to Damascus. As a youth, Saul had supplemented the advantages bestowed on him by lineage and upbringing with a vigorous commitment to the law of Moses, a zealous persecution of the church, and a strenuous pursuit of righteousness before God and God's people. From his own perspective and that of his Jewish contemporaries, he had attained the goal of "blamelessness"—not sinlessness, but an acceptable blend of vigorous obedience and rigorous cleansing rituals that (so he thought) would secure his status among "the righteous" who could expect God's approval (Phil. 3:4–6). Yet the blinding flash of Christ's divine glory exposed the futility of Saul's self-serving striving, showing his alleged "gains" to be deficits and his awards to be garbage. Up to that moment, Saul himself suffered from the blindness that he later diagnosed in his Jewish kinfolk: "I bear them witness that they have a zeal for God, but not according to knowledge. For, being ignorant of the righteousness of God, and seeking to establish their own, they did not submit to God's righteousness. For Christ is the end of the law for righteousness to everyone who believes" (Rom. 10:2–4). Suddenly, the Lord of glory stopped Saul in his tracks and put an end to his zealous striving to establish his own righteousness. The gift of grace from the risen Lord Jesus shattered Saul's "confidence in the flesh" (Phil. 3:4) and released him to rest in assurance of God's favor in Jesus the Messiah (3:9).

Nevertheless, as Saul rested in a righteousness not his own, the surprising result was that his assurance became a stimulant, not a sedative, to his passion to follow God's will and further God's glory. Paul is grateful for the gifts of a cleansed conscience and a record expunged of guilt before heaven's Judge. The Spirit of God has begun a good work in Paul, as he has in the believers of Philippi (Phil. 1:6). Yet these first tastes of grace only whet Paul's appetite for the whole feast of fellowship with his Savior that he describes as "gain[ing] Christ," "be[ing] found in Christ," and "knowing Christ Jesus my Lord" (3:8–11).

Paul could sing with Jean Sophia Pigott, "Jesus, I am resting, resting in the joy of what thou art."[1] Yet Paul's resting is, actually, *restless*. The grace that has seized Paul's heart has set him in a lifelong race. This is the imagery he evokes in our text. Paul is overjoyed to be forgiven, but not content to stop short of the *completion* of the Spirit's renovation project in his life "at the day of Jesus Christ" (Phil. 1:6).

A Case Study of the Restless Race

Paul wants more, so he sprints toward the goal and the prize to which God called him in that confrontation on the road to Damascus. Paul's Philippian friends need to see in his own experience the truth that *everyone* who has tasted the firstfruits of God's reconciling and renewing grace will long to feast on the full banquet at their Father's table. So Paul continues his autobiography in Philippians 3:12–16, offering his current restless race as a "case study" in the response that grace evokes when it lays hold of a human heart. Those who are maturing in Christ will share Paul's mind-set, not complacent but rather striving to know Christ more fully and to reflect him more faithfully (3:15). The grace that we have already received invites us to greater grace ahead.

Scholars debate whether Paul's emphasis on his own imperfection was correcting a specific expression of "perfectionism" that was endangering the Philippian church. Some believe that Paul was thinking of observant Jews who were still where Paul once was, confident in their covenant heritage and conscientious law-keeping. Such Judaizers might have boasted, as Paul once did, that they had achieved a "blameless" status through circumcision, commandment-keeping, and sacrificial rituals (Phil. 3:6).[2] Others suspect that Paul was counteracting a different group, who misbehaved sensually, indulging physical appetites and boasting in shameful acts (3:18–19). Such people might base their boasts of perfection not on exemplary behavior but

1. Jean Sophia Pigott, "Jesus, I Am Resting, Resting" (1876), in *Trinity Hymnal* (Suwanee, GA: Great Commission Publications, 1990), no. 188.

2. See Gerald F. Hawthorne and Ralph P. Martin, *Philippians*, rev. ed., WBC 43 (Nashville: Thomas Nelson, 2004), 205.

on claims that their superior knowledge or mystical experience made their treatment of the body and its pleasures spiritually irrelevant. Such views, which came to full flower in Gnosticism over a century later, threatened the church even in the apostles' day. The apostle John warned against people who claimed to be without sin (1 John 1:8–10), on the one hand, while refusing to keep God's commandments, on the other (2:4; see 4:20–21).[3] Ancient dualistic philosophies and religions so segregated spiritual realities from physical activities that such a blend of self-perceived innocence with lawless behavior was not at all far-fetched. Just as Las Vegas, America's legendary "sin city," advertises itself with the absurd motto "What happens in Vegas stays in Vegas," so ancient dualists alleged that "what happens in the body stays with the body," leaving one's spirituality unaffected and undamaged.[4]

Since Paul's time, other versions of perfectionism have arisen. Sincere Christians have taught that a specific crisis of personal resolve or encounter with God's Spirit can usher believers up to a higher plateau of spiritual experience, above all known sin. Wearied as we are by our ongoing struggles and stumbling into sin, we can empathize with believers who thought they had discovered a shortcut to the goal for which we all long.

Yet Paul repudiates every version of "perfection," ancient or modern, that promises relief from the race in exchange for a reward that falls short of the prize that Christ waits to confer at the finish line. Paul refuses to be sidelined short of the finish line and the prize that awaits him there: knowing Christ in all his fullness. Nor will he let his friends at Philippi drop out of the race. He urges them, as he urges us, to keep pace with him all the way to the goal.

In this text Paul invites us to view our Christian life, from its beginning in God's call of grace all the way to its climax at our death or Jesus' glorious return, through the metaphor of an athletic race. Twice he declares, "I press on," repeating a verb that describes pursuit with the aim of laying hold of someone or something (Phil. 3:12, 14).[5] Often in the New Testament, the word expresses pursuing persons for harmful purposes, and in

3. "Perfectionism and antinomianism are not mutually exclusive principles. Very probably, the heresy to which 1 John is addressed consisted of individuals who reckoned themselves to have achieved spiritual perfection, yet perfection of a sort that encouraged wrongful behavior." Moisés Silva, *Philippians*, 2nd ed., BECNT (Grand Rapids: Baker, 2005), 171.

4. See Everett Ferguson, *Backgrounds of Early Christianity* (Grand Rapids: Eerdmans, 1987), 247–49.

5. Greek *diōkō*.

those contexts it is aptly translated "persecute" (Matt. 5:10–12, 44; Acts 9:4–5; 22:4, 7–8; Rom. 12:14; Phil. 3:6; etc.). Here, though, Paul's pursuit is driven by a praiseworthy motive, as in other passages in which he calls us to pursue righteousness, love, and other positive objectives.[6] Our word, *pursue*, captures Paul's purposefulness in running the race. His "pressing on" is not merely stoic stubbornness but a persistent quest for a goal. He surrounds his second declaration, "I press on," with imagery drawn from the footraces of the Olympic athletic competitions for which Achaian cities, seven hundred miles south of Philippi, were renowned: "*straining forward*[7] to what lies ahead," "the goal," and "the prize." Catching the apostle's cue, we will use the components of the race as metaphors to open the meaning of the text: the starting blocks, the running itself, the goal and prize, and the runners' mind-set.

The Starting Blocks: Resting in the Gracious Grip of Christ

Although Paul's eyes are focused forward on the finish line in the future, his confidence that he would reach that goal is grounded in grace already received, in past acts of God's sovereign mercy that set Paul's feet flying from the starting blocks years before. Because God the Son and God the Father, acting through God the Spirit, exerted such invincible power on Paul's heart at the start of his race of faith, the apostle eagerly anticipates that victory will be his at the finale. The grace that set Paul into the race appears twice in our text.

First, in verse 12 Paul uses a wordplay in Greek that is hard to reflect in English (although the ESV comes close). Paul denies that he has already "obtained"[8] the whole treasure of "knowing Christ," and especially the resurrection reunion that he has mentioned in verse 11. Then he uses an intensified verb built on the same root, announcing that the goal of his

6. In other New Testament instances of *diōkō*, people are said to "pursue" righteousness (Rom. 9:31; 1 Tim. 6:11; 2 Tim. 2:22), things that promote peace (Rom. 14:19; cf. 1 Peter 3:11), love (1 Cor. 14:1), and the "good" (1 Thess. 5:15).

7. Although the Greek verb, *ekteinō*, appears with some frequency in the New Testament, especially in the expression "stretch out [one's] hand" (for example, Matt. 8:3; Mark 1:41; John 21:18; Acts 4:30), here Paul uses a compounded form, *epekteinō*, which appears nowhere else in Paul's letters or the rest of the New Testament.

8. Greek *elabon*, an aorist form of *lambanō*.

pursuit is "to *make it my own*, because Christ Jesus has *made me his own*."[9] Paul is striving to take complete and permanent possession of the Olympic medal—in his day, the olive wreath[10]—at the finish line, and his confidence of victory is grounded in the fact that Christ Jesus has already taken complete and permanent possession of Paul.

Commentators and versions differ over the two-word construction in Greek[11] that introduces the last clause of verse 12. Paul's future aspiration to "make it my own" is somehow connected with the fact that "Christ Jesus has made me his own." But how? Some versions (KJV, NASB, NIV, etc.) treat this tiny transition (*eph hō*) as referring to *the object that Paul seeks to seize*: "*that for which*" Christ Jesus laid hold of him. The ESV takes the transition as giving *the reason* that Paul strives so strenuously: "because" he has already been seized by Jesus.[12] Elsewhere in Paul's letters, this transition means "because." In Romans 5:12, for instance, Paul writes that Adam's sin brought death to us all "because [*eph hō*] all sinned." Here in Philippians 3:12 also, Paul's point is that he is striving to seize the prize at the finish line *because* Christ already seized him at the start of his race.

Paul is a prisoner of war, and glad of it. Christ confronted Saul on the Damascus road (Acts 9, 22, 26; Gal. 1:11–16) and laid sovereign claim to his enemy's allegiance. A hymn-writer aptly applied a military metaphor:

> We sing the glorious conquest before Damascus gate,
> When Saul, the church's spoiler, came breathing threats and hate;
> The rav'ning wolf rushed forward full early to the prey;
> But lo! the Shepherd met him, and bound him fast today.[13]

9. The English expression "make [one's] own" represents the Greek verb *katalambanō*. The prefix *kata-* adds intensity to the verb's basic meaning, escalating "obtain" to "*fully* obtain" or "seize." Paul used *katalambanō* also in 1 Corinthians 9:24, employing the same athletic metaphor to urge believers to run in such a way as to *come into full possession* of the prize.

10. Victors in the games held every four years at Olympus received olive wreaths. Other sites of the Panhellenic games in other years awarded laurel, celery, or pine wreaths. *The Olympic Games in Ancient Greece* (Lausanne: Olympic Museum and Study Center, 2002), 11. Accessed at http://www.olympic.ca/education/files/2012/01/The-Olympic-Games-in-Ancient-Greece.pdf.

11. Greek *eph hō*.

12. The ESV's rendering of *eph hō* is supported by Silva, *Philippians*, 176. Also, *BAGD*, 287 (*epi*, II.b.gamma), states: "*eph hō* = *epi toutō hoti* '*for this reason that, because*,'" listing instances in Romans 5:12; 2 Corinthians 5:4. See also Daniel B. Wallace, *Greek Grammar beyond the Basics* (Grand Rapids: Zondervan, 1996), 342.

13. John Ellerton, "We Sing the Glorious Conquest" (1871), in *Trinity Hymnal* (Suwanee, GA: Great Commission Publications, 1990), no. 483.

When the risen Lord sent Ananias of Damascus to the blinded, humbled, fasting, praying Saul, his words resounded in triumph over the enemy whom he had seized by irresistible grace: "he is a chosen instrument of mine to carry my name before the Gentiles and kings and the children of Israel" (Acts 9:15). Because Christ had laid hold of Paul and would never let go, Paul would one day lay hold of the prize awaiting him at the goal—and never let go.

Paul's second reference to the grace that launched him into the race is found in the expression "the upward call of God in Christ Jesus" (Phil. 3:14). Since Paul associates that call with the future goal and the prize that he pursues, some view this "upward call" as referring to the awards ceremony that followed competitions in the ancient games, when the victor was summoned to receive the emblem of his victory.[14] On the other hand, when Paul speaks of "the call of God," he calls Christians to look *back* in gratitude to the sovereign summons by which God brought them to believe the gospel.[15] To the Romans Paul wrote: "For those whom he foreknew he also predestined to be conformed to the image of his Son . . . . And those whom he predestined he also *called*, and those whom he *called* he also justified, and those whom he justified he also glorified" (Rom. 8:29–30). Since we are justified by faith, the call that *preceded* our justification evoked our faith. The Father's invincible call draws his undeserving chosen ones to trust the Son in the life-imparting strength of the Spirit. We have been "called" into fellowship with God's Son (1 Cor. 1:9) and into freedom (Gal. 5:13). Considering our calling is humbling, for it reminds us that God chose foolish, weak, and despised people (1 Cor. 1:26–28). But God's call is also ennobling, for he has summoned us to himself and calls us by name (Rom. 1:1, 7; John 10:3). Christ's conquest of Saul on the road to Damascus marked the moment when God, "who *called* me by his grace, was pleased to reveal his Son to me" (Gal. 1:15–16). So God's call to Saul came long ago.[16]

14. Hawthorne and Martin, *Philippians*, 210.

15. Peter T. O'Brien, *The Epistle to the Philippians*, NIGTC (Grand Rapids: Eerdmans, 1991), 431–33; Silva, *Philippians*, 176–77; G. Walter Hansen, *The Letter to the Philippians*, PNTC (Grand Rapids: Eerdmans, 2009), 255–56.

16. WSC, answer 31, aptly sums up Paul's teaching on God's redemptive call: "Effectual calling is the work of God's Spirit, whereby, convincing us of our sin and misery, enlightening our minds in the knowledge of Christ, and renewing our wills, he doth persuade and enable us to embrace Jesus Christ, freely offered to us in the gospel." Paul attributed the saving call to God (the Father) (see 1 Cor. 1:9; 1 Thess. 2:12; etc.), but he also referred to the Spirit's life-imparting work (Gal. 4:28–29).

The divine call that brought Paul from death to life also secured the complete deliverance that Jesus had won for him: rescue from sin's guilt, sin's penalty, sin's dominion, and eventually sin's sting (death) and sin's presence. God's call was to a future hope (Eph. 1:18; 4:4). God has called us to nothing less than eternal life (1 Tim. 6:12). So Paul strives toward a future prize that is not yet fully in his grasp. He knows Christ, but does not yet know Christ as he longs to know him. He has not "gained Christ" as fully as he desires, and has not "made his own" all that Jesus has won for him.

Nevertheless, the prize was reserved for him by God's call of sovereign grace even before his race "in Christ Jesus" began. Because Christ seized him and God called him in the starting blocks, Paul sprints with all the strength that the Spirit gives him,[17] confident that he will cross the finish line victorious through the grace that will not let him go. The church father Theodoret recast Paul's imagery vividly in hunting terms: "'It was he who first caught me in his net,' Paul says in effect, 'for I was fleeing him and was turned well away. He caught me as I fled. But now I in turn am the pursuer in my desire of catching him, that I may not be a disappointment to his saving work.'"[18] Because Christ has seized me, I long to seize him.

It may seem counterintuitive that confidence of victory at the crack of a starting pistol would stimulate a sprinter's most strenuous effort. Aesop's ancient fable tells how the hare, with a swift start and a long lead, smugly lay down to nap, while the tortoise's perseverance proved the maxim: "Slow but steady wins the race." In other sports, when a dominant team is outscoring its opponent to the point of embarrassment, the second string is sent in to "coast" to the victory. Yet God's sovereign initiative in salvation, far from making our response in faith and love *superfluous*, actually makes our perseverance both *necessary* and *possible*. Paul has already told the Philippians that "he who began a good work in you will bring it to completion[19] at the day of Jesus Christ" (Phil. 1:6). On the basis of that assurance, he has urged them to "work out your own salvation with fear and trembling," grounding that exhortation in the

17. To present everyone mature in Christ, "I toil, struggling with all his energy that he powerfully works within me" (Col. 1:29).

18. Theodoret, *Epistle to the Philippians* 3.12, quoted in ACCS NT 8:273.

19. Paul's verb is *epiteleō*, a cognate of the verb *teleioō* ("am . . . perfect") and *teleios* ("mature") in Philippians 3:12–16. He has had the destination to which God is leading his people in view from the opening lines of the letter.

truth that "it is God who works in you, both to will and to work for his good pleasure" (2:12–13).

An earlier incident in Paul's life illustrates his grasp of the mysterious interplay of God's invincible plan and the indispensable role of human agents in the accomplishment of that divine purpose. Acts 27:23–26 records the encouraging words that the apostle spoke to his terrified fellow passengers as their ship was helplessly driven by a storm on the Mediterranean Sea. He reported that an angel of God had announced not only that Paul would arrive safely in Rome but also that every passenger traveling with him would survive the shipwreck that was sure to come. The ensuing narrative recounts several human decisions that proved to be crucial to the fulfillment of God's promise. As the ship approached land, sailors secretly tried to spare themselves by abandoning the ship. But Paul told the centurion who had custody of him, "Unless these men stay in the ship, you cannot be saved." At that warning, soldiers cut off the sailors' means of escape (27:27–32). Then the soldiers, fearing retribution if prisoners escaped in the chaos of the shipwreck, intended to kill their captives. But they were thwarted by orders of their centurion, who wished to save Paul (27:42–43). Thus the outcome was precisely what God had promised: "all were brought safely to land" (27:44). The means to that outcome were a succession of human decisions and actions.

So also we must live our lives each day in light of the twin truths of God's sovereign control and our own responsibility as the creatures who bear his image. His sovereignty frees our hearts from nervous apprehension over the unseen threats awaiting us in the future. We know that nothing will happen to us—not even traumatic and painful things—that lie outside his plan to do us good and make us more like Jesus. The worst dangers in the world cannot separate us from his love (Rom. 8:31–39). At the same time, the reality that he works out his perfect plan *through* our and others' decisions and actions makes our choices significant and our actions meaningful to the achievement of God's good and invincible goals.

Paul knows that the race of faith, for the prize to which God has already called him and Christ has already seized him, operates on the same mysterious principle. Because the triune God has stopped his self-fueled striving and made him rest in grace, Paul has found in that grace the power to run toward the goal of the fullness of joy to be found in Christ.

THE RUNNING: FORGETTING AND PRESSING FORWARD

In this text Paul repeats himself, making the same point twice in almost identical terms. First, in verse 12, he denies that he has "obtained" his desired goal and expresses his resolve to "press on" to grasp it fully. Then, in verses 13 and 14, he covers the same ground again, in more detail and with more emphasis: "I do not consider that I have made it my own"; therefore, "I press on toward the goal." Through this repetition and the personal address "Brothers," Paul grabs our attention for the point he is drawing from his own Christian experience and applying to his friends at Philippi (and to us): "Let those of us who are mature think this way" (Phil. 3:15). As one commentator observed, by addressing the Philippians as his spiritual siblings, Paul "reaches out, places both hands on their shoulders, looks in their eyes, and says, 'I really mean what I just said, I'm not perfect yet.'"[20] Paul is not feigning humility. He is speaking sober reality, which is as true of every believer on earth as it is of Paul himself.

In verses 13 and 14 Paul paints a vivid picture of his single-minded and strenuous pursuit in the present: "But one thing I do: forgetting what lies behind and straining forward to what lies ahead, I press on toward the goal . . . ." Here is a marathoner who refuses to be distracted by ground already covered or the field of competitors around him. His eye is on the tape across the finish line, and he refuses to waste thought or attention on anything else.

Since Paul does not identify the things that lay behind him, on which he refuses to fix his attention any longer, we might wonder what those things were. No doubt they include the credentials that he has just listed, which make up the impressive résumé in which he boasted until Jesus seized him. Paul has not "forgotten" them in the sense that they were erased from his memory: he can recite them readily. But they no longer occupy his attention as the focus of his concern or confidence, as they did before.[21] True, he had been born into the "right" family, tribe, and nation, had enjoyed the "right" education, belonged to the "right" party, and distinguished himself as an overachiever in the expectations of his community. But the prize ahead is

20. Hansen, *Philippians*, 252.

21. See Gordon D. Fee, *Paul's Letter to the Philippians*, NICNT (Grand Rapids: Eerdmans, 1995), 347n40.

infinitely more desirable, and he dares not look back, lest he stumble in pursuit of the goal.

Yet we blunt Paul's pastoral point if we confine "what lies behind" to the credentials in Judaism that Paul now rejects as rubbish. As a follower of Jesus, Paul does not even dwell nostalgically on past mercies received or ministry offered. When necessary, Paul can invoke a list of his apostolic credentials and accomplishments, as he did for the church at Corinth when his calling and his gospel came under critique (2 Cor. 11:16–29; 12:12). But even when he finds it necessary, for others' spiritual safety, to match the so-called "super-apostles" (11:5) boast for boast, God's grace has so removed Paul's ego from the picture that he can say, "If I must boast, I will boast of the things that show my weakness," for he has the Lord's assurance: "My grace is sufficient for you, for my power is made perfect in weakness" (11:30; 12:9). Grateful for God's past and present graces, Paul fixes his heart and eyes ahead, on the goal of knowing Christ in fullness. As one scholar observes, Paul

> did not keep turning over in his mind the good old days of active service before he was imprisoned; he did not constantly remind himself of all his achievements nor continually recount those special high points of his intimate relationship with Christ. . . . He is not distracted by the trophies of the past . . . . Forgetting is not a passive loss of memory; no, it is an active, continuous discipline of the mind and heart. Although he did not actually forget the past, he emphatically chose to disregard it.[22]

Since Paul's longing focuses on that future grace, he strenuously directs energies toward the goal. "Straining forward" captures perfectly the picture that Paul's Greek portrays: the runner with every muscle engaged, drawing on every energy reserve, in the final all-out sprint to the finish line.[23] Paul is "in hot pursuit" of his prize, as one scholar aptly put it.[24] But what does Paul's "race to the finish" look like in his daily experience? It is not a lifestyle

22. Hansen, *Philippians*, 252.

23. This verb, *epekteinomai*, appears only here in the New Testament or the Greek Old Testament (Septuagint). A related verb, *ekteinō*, describes the action of "stretching out" a hand in the Gospels, in Acts, and often in the LXX. Paul apparently selected the form with the additional prepositional prefix *ep-* because it conveyed more vividly the image of complete and unreserved exertion to reach the goal.

24. Silva, *Philippians*, 175.

of frenetic activity for activity's sake. The point is not so much *in which activities* Paul spends his every waking moment. He can relish both "fruitful labor" (Phil. 1:22) and refreshment in the fellowship of fellow Christians (Rom. 15:32; 1 Cor. 16:18; 2 Tim. 1:16; Philem. 20). The crucial issue is *why* Paul invests his time and energies as he does. He told the Corinthians, "I do not run aimlessly" (1 Cor. 9:26). Paul's eager expectation that "Christ will be honored in my body, whether by life or by death," and his confident confession, "to me to live is Christ, and to die is gain" (Phil. 1:20–21), enables him to assess his current circumstances in relation to the gospel's advance (1:12), to evaluate future outcomes in terms of others' needs (1:25), and to bind his own joy to others' spiritual maturity (2:1–2).

The way that Paul runs his race poses heart-searching questions for you and me. If your thought life, daily routine, spending practices, and interpersonal interactions were translated into Paul's athletic imagery, what would it look like? Are you "running aimlessly"? Is your pace hindered by backward glances toward bygone days, or sideways glances at other runners? If Christ has seized you in his amazing grace, you must not rest on your laurels. Direct your investment of aspiration, imagination, time, energy, and money toward the precious prize that Paul pursues. Rivet your gaze on the goal, on the prize that he finds irresistibly attractive, and run with all the strength and stamina that Christ will give you!

THE GOAL AND PRIZE: TO KNOW CHRIST IN HIS FULLNESS

What is the finish line toward which Paul runs with such single-minded determination? What prize awaits him there? He began verse 12 with the disclaimer that he had not already "obtained," having just spoken of the resurrection of the dead (Phil. 3:11). Resurrection—the rescue of his body from death, "the last enemy" (1 Cor. 15:26), and reunion of his person in the life of the age to come—was integral to Paul's hope for the future. In a few sentences he will remind the Philippians that "we await a Savior, the Lord Jesus Christ, who will transform our lowly body to be like his glorious body" (Phil. 3:20–21). To that end Paul is running his earthly race, sharing Christ's sufferings and sustained by the Spirit with a foretaste of Christ's resurrection power, until his union with Christ on earth climaxes in conformity to Christ's death (3:10).

In a sense, however, the deliverance of Paul's body from death's clutches is only the decorative wrapping around the true treasure that Paul prizes above all. When God called him decades earlier, it was a summons that would carry Paul "upward" into the presence of the Son of God, who (as he says elsewhere) "loved me and gave himself for me" (Gal. 2:20). Paul's new treasure is to "gain Christ" and to know Christ in the fullness of his glory (Phil. 3:10). Paul's definition of "perfection" is to be with Christ and to be like Christ forever and ever. This, for Paul, is to be "perfect," to be "complete."[25] Until Jesus' return, Paul will not be whole, not complete. He still has gaps and lapses. He still finds his own heart divided between wanting to delight the Lord and wanting, even now, to satisfy himself. But Paul cannot be content with his divided heart. He wants the *whole* rescue that Jesus promised: he wants to see God—and, in the seeing, to become like his sovereign Savior.

In the materially comfortable West, we might find it easy to dismiss earlier ages' hopes for resurrection and heaven as old-fashioned and escapist. We might grant that such otherworldliness can help people to cope when life expectancies are short and the standard of living is wretched, with meager food, polluted water, unsafe housing, or violence on the streets (conditions that still exist in much of the developing world today). But images of a world purged of evil may be less interesting to people who enjoy comfortable homes, ample and tasty food, reliable healthcare, and trustworthy emergency services. Is that your unspoken reaction to talk about heaven or resurrection or the life to come? On the other hand, though, has the accumulation of "creature comforts" in affluent societies really raised the level of our contentment? Hasn't the stuff that we hold only whetted our appetites for stuff that lies beyond our grasp but that, we imagine, would finally satisfy our hungry hearts? Some years ago, social commentators assigned the label *affluenza* to a disorder that they described as "an array of psychological maladies such as isolation, boredom, passivity and lack of motivation engendered in adults, teenagers and children by the possession of great wealth" and "an unhappy condition of overload, debt, anxiety, and waste resulting from the dogged pursuit of more."[26] Fixating on earthly things not only blinds us to future

25. The Hebrew *tam* and the Greek *teleios* express "perfection" as "completeness." Paul uses the related verb *epiteleō* in Philippians 1:6: "will bring it to completion."

26. Michael Good, "Affluenza's Gonna Get You," *KPBS On Air* 29:11 (September 1997): 14–15, citing John de Graaf, producer of the public-broadcasting documentary *Affluenza*.

joy and glory, but also dulls our taste to the flavors of the present. Even in the first century, when lives were more fragile and conditions more brutal, Paul knew of people who had their "minds set on earthly things" (Phil. 3:19), the horizons of their hopes constricted to the here and now that they saw and touched. But Christ's apostle insists that we take a deeper look at God's goal for the people he loves. The prize for which we run is so much more than merely escaping this world's miseries! It is bigger and better than never going to bed hungry again, or having shelter from the rain and cold, or being free from the pain of cancer or arthritis, or having a reunion with lost loved ones. The best thing about the prize that awaits us at the finish line is not the taste of the food at the Lamb's wedding supper, as delightful as that will be. It is not having tears of sorrow wiped from our eyes, never to return. It is not streets of gold or "mansions over the hilltop" that never need repairs or alarm systems to discourage thieves. The most intense pleasure of heaven is found in John's vision of the coming New Jerusalem, in which God's "servants will worship him. *They will see his face*" (Rev. 22:3–4). As a result, as John writes elsewhere, "we shall be like him, because *we shall see him as he is*" (1 John 3:2). This is the prize that Paul sees as he stretches his stride toward the goal: that he, his Philippian friends, and all others who trust in Jesus will know Christ as they have been known by Christ (1 Cor. 13:12), and therefore "be pure and blameless for the day of Christ, filled with the fruit of righteousness that comes through Jesus Christ, to the glory and praise of God" (Phil. 1:10–11).

Does that goal of beholding God's face and basking in his favor captivate you? It must, for no lesser goal will satisfy you everlastingly—since no lesser goal fits the purpose for which your Maker designed you: "to glorify God, and to enjoy Him for ever."[27] You may assume that your affections and desires are beyond your control and impervious to change. That assumption is simply false. We rightly observe that enjoying certain foods or musical styles is "an acquired taste." The same is true of spiritual affections and appetites, as the imagery of the Bible's authors illustrates. It was self-destructive folly when the people of Israel abandoned the Lord, "the fountain of living waters," and hewed out for themselves cisterns that held no water (Jer. 2:13). But such dietary disorders can be reversed: "Oh,

27. WSC, answer 1.

taste and see that the LORD is good!" (Ps. 34:8). Busy schedules, relentless distractions, and entertaining diversions glut our mental palates with spiritual junk food, but this unhealthy diet does not have to continue. You may and you must cultivate a "taste" for the soul-satisfying friendship of God through time spent, even now, in his presence through the Word, prayer, and worship.

No wonder Paul wants to protect his friends from settling for second best, from stopping short of real completeness! The alternatives—"perfection" as good deeds, or as insider knowledge, or as mystical experience—all fall short of the best gift that God promises to those who trust Jesus. In "The Weight of Glory," C. S. Lewis wrote:

> Indeed, if we consider the unblushing promises of reward and the staggering nature of the rewards promised in the Gospels, it would seem that our Lord finds our desires, not too strong, but too weak. We are half-hearted creatures, fooling about with drink and sex and ambition when infinite joy is offered us, like an ignorant child who wants to go on making mud pies in a slum because he cannot imagine what is meant by an offer of a holiday at the sea. We are far too easily pleased.[28]

Lewis goes on to survey the astonishing destiny of joy that utterly overshadows the physical pleasures and passions that often distract us:

> The promises of Scripture may very roughly be reduced to five heads. It is promised, firstly, that we shall be with Christ; secondly, that we shall be like Him; thirdly, with an enormous wealth of imagery, that we shall have "glory"; fourthly, that we shall, in some sense, be fed or feasted or entertained; and, finally, that we shall have some sort of official position in the universe . . . .[29]

God's glorious goal for us, the prize of being with Christ and all that will flow from it, dwarfs not only our current physical pleasures, but also each of the "second-best" images of spiritual perfection that we surveyed earlier. To tell ourselves that we have arrived because we've kept the rules better than some, or because we are above the rules, or because we had an emotionally

28. C. S. Lewis, "The Weight of Glory," in *The Weight of Glory, and Other Addresses* (Grand Rapids: Eerdmans, 1965), 1–2.
29. Ibid., 7.

overwhelming experience—they are all just too small. Paul will not settle for second best, and neither can we!

THE RUNNERS' MIND-SET: HUMBLE, PATIENT, AND HOPEFUL

Competitive running is as much mental as it is muscular. Physical speed and stamina must be fortified by internal confidence and commitment. Because the Christian race of faith is a team sport, more like cross-country or relays than individual sprints, Paul deftly transitions in verses 15 and 16 from his own resolve to run "all out" his not-yet-completed race to the mind-set that must characterize all believers. Paul has been showing the mental resolve that will lead us all to the prize.[30] That mind-set blends realistic humility, patience, and disciplined persistence.

The *realistic humility* that distinguishes those who are mature in Christ comes from sharing Paul's appreciation of the magnificence of the prize that awaits us at the end. When we share Paul's hunger and thirst for all that is entailed in "gaining Christ," we realize that we are far from having "arrived." Paul's Greek contains a little pun that is hard to put into English. In verse 12 Paul had said that he was not "already perfect" (*teleioō*). Now in verse 15 he uses the related adjective, including himself among "the perfect ones" (*teleioi*). Trying to avoid the appearance of self-contradiction, the ESV translates the adjective as "mature." The word can mean "mature"—no longer childish, but short of flawless (1 Cor. 14:20). In that sense Paul invites the *teleioi*, "mature," to join his admission that he has not arrived.[31] But Paul wants his choice of language to jar us a bit. His point is that the only ones who are "perfect" are those who see themselves as Paul does: *imperfect*. And he is not necessarily picking out a special class of Christian believers—the "perfect ones" or even the "mature ones," in contrast to spiritual babies. No, he is saying to *all* Christians: "The closest to perfection that *any of us* will get in this life is to see that you have not reached the finish line but must keep running toward the goal." Mature Christians must think as Paul does, painfully aware of our pockets of immaturity but eagerly anticipating the

30. He reinforces his call to take a cue from his own mind-set in Philippians 3:17: "Brothers, join in imitating me, and keep your eyes on those who walk according to the example you have in us."

31. The NASB has "become perfect" in Philippians 3:12 and "us . . . , as many as are perfect," in verse 15, making the link in Greek transparent in its English wording.

moment when we cross the finish line and receive the prize: seeing Jesus clearly and being changed by that sight.

Taking our own incompleteness seriously gives us *patience* to address our own and others' failings without frustration, and yet without complacency. Paul displays this delicate balance in his intriguing comment, "If in anything you think otherwise, God will reveal that also to you" (Phil. 3:15). Paul's language is so terse that his meaning is hard to nail down. Is he saying that some Philippian Christians might not "buy" Paul's perspective that no believer in this life has reached the goal of "perfection"? Or that the Philippians might still have disagreements among themselves (as they apparently do, 2:1–2; 4:2)? Whatever the issues on which believers might not see eye to eye, Paul is so confident in the Spirit's quiet but invincible persuasion that he can wait for God to complete the "good work" that he began in them long ago (1:6). Paul is no postmodern relativist, timidly offering his perspective and then inviting his friends to express theirs, as though neither has a take on truth that trumps other viewpoints. He does not say, "If in anything you think otherwise, God will reveal that also to *us*." Rather, "God will reveal that to *you*." In other words, "God will show you that your perspective is out of sync with the norm that God has revealed through me, his apostle."

But Paul realizes that he and his brothers and sisters are still "in process," so we don't all grow at the same pace or in the same places in our convictions and affections and attitudes and actions. Paul can be patient with others' pockets of immaturity because God holds Paul fast, despite Paul's own pockets of immaturity. And Paul knows that God will bring every member of the body of Christ to the goal, "the unity of the faith and of the knowledge of the Son of God, to mature ['perfect'! *teleios*] manhood, to the measure of the stature of the fullness of Christ" (Eph. 4:13). Not every disagreement is a "hill to die on." On the central truths of the gospel, of course, Paul is uncompromising. He has described those who preach a rival gospel as "dogs," "evildoers," and "those who mutilate the flesh" (Phil. 3:2), and he will soon write in tears about enemies of the cross who are destined for destruction (3:17–18). But not every issue among us is a life-or-death gospel issue. On peripheral matters Paul can relax and wait for the Spirit to give growth, knowing that God can change minds in ways that we cannot.

This is a hard balance to maintain. Some wings of the church embrace a mushy tolerance that seems unwilling to stand for gospel truth and to say

"NO" to any divergent opinion at all. Others, rightly alarmed by an open-mindedness that refuses to draw the line anywhere, swing to the opposite extreme and start drawing lines on every issue, large and small. How can we tell the difference? It is not always easy, but a good starting point for discerning the nonnegotiable core of the faith revealed in God's Word is with the creeds and confessions that the church has formulated over the last two millennia. The Apostles' and Nicene Creeds and other early ecumenical confessions affirm that God is Trinity, that Christ the Redeemer is God and man, and other vital truths. Confessions flowing from the Reformation[32] explain the gospel truths that salvation is by grace alone, through faith alone, in Christ alone, so that God alone receives all glory. These truths that believers "together-speak" (which is what "con-fess" means) might not address every issue on which some have strong convictions. On those topics Paul's approach does not preclude discussing our differences frankly; but in such areas we can imitate the apostle's patience in the way in which we speak and listen to each other, knowing that one day we will agree even on these issues, as we do now on the gospel itself.

Finally, the mind-set of mature racers is marked by *disciplined persistence*. Though each is at a different point in the marathon, Paul's concluding exhortation applies to all: "Only let us hold true to what we have attained" (Phil. 3:16). Paul's Greek paints an even more vivid, more athletic image: the ESV's "hold true" reflects Paul's original "keep pace" or "run in line."[33] Paul blends encouragement with exhortation to urge his friends not to be disheartened by the daunting distance to the finish line, not to let their energy flag or their pace lag. The hare in the fable showed that overconfidence and complacency can sideline those who should be striving to win. Hopelessness can be just as disabling. To combat the despair that could paralyze us when we see how far we are from the finish line, Paul reminds us that by God's grace we have already "attained" a degree of progress toward the goal. To be sure, the Philippian Christians need courage in the face of suffering and

32. *Reformed*: Heidelberg Catechism, Belgic Confession, Canons of Dort; *Presbyterian*: Westminster Confession and Catechisms; *Lutheran*: Augsburg Confession, Formula of Concord; *Baptist*: London Baptist Confession.

33. The metaphor implied in *stoicheō* is most explicit in Romans 4:12, where Paul speaks of those who "walk [*stoicheō*] in the footsteps [*tois ichnesin*] of the faith" of Abraham. In Galatians 5:25 and 6:16, the ESV more overtly reflects the ambulatory imagery in this verb by rendering it "walk." (Compare the NIV's "keep in step with" at Galatians 5:25.)

humble compassion toward one another. Yet Paul can and does give God thanks for these believers' growing and overflowing love. They are not still in the "starting blocks." They can praise God for how far he has brought them, as they keep running. The progress that we, too, have experienced so far must steel our relentless commitment not to lose ground, but instead to live in ways that fit the gospel and its implications, as we have grasped it so far.

Conclusion

As you watch Paul run for the goal, savoring the hope of gaining Christ, knowing Christ, being with Christ (which is better by far), and becoming like Christ, do you realize that you have been setting your sights too low? Have you settled for a modicum of respectability, courtesy, and self-restraint in a safe and pleasant lifestyle? You were made for far more than this!

If you are not a Christian, Paul has confirmed what you suspected all along. The bumper sticker is half true: "Christians are not perfect." If we have acted as though we were, blame us and not our Master. But you need to know that there was one perfect person in the history of the human race—Jesus of Nazareth. He invaded Paul's life when Paul was plotting and striving to erase Jesus' name from the face of the earth. Paul once thought he had mastered God's expectations: "Perfection? Been there, done that." But Jesus grabbed Paul, "caught him in his net," as Theodoret said, and whipped his heart around 180 degrees. In the process, Paul discovered that he was not as blameless as he thought he was. He learned that Jesus, who knew him through and through, had loved him and had lived the life of *real* perfection on Paul's behalf. Then Christ died on a cross, under God's curse, to rescue Paul and you and me from the wrath we deserve for our self-worship. Having raised Jesus from the dead, God had credited his Son's perfect record to Paul the imperfect, as God's Spirit launched a lifelong process to make Paul perfect, inside and out.

You see, the bumper sticker is only half true: Christians are not "*just* forgiven." Forgiveness—deep forgiveness that sets your guilty conscience free from regret and fear—is a priceless treasure. And contrary to what you may have been told, this forgiveness is not a prize to be won for a lifelong race well run. It is a free gift bestowed at the starting line, received by trusting Jesus. By placing that gift in our hands at the start, God fires our desire for every other gift of grace that he plans to lavish on us, until he has brought us to the goal, the prize of glorifying and enjoying him forever.

All of this might sound too good to be true. I suggest that it is too good to be fiction. Who of us could have dreamed up such a bizarre arrangement? It's not the way in which our duty-bound minds work. Who but the God who is so different from you and me could have come up with a plan to set us *running* with all our might . . . by teaching us first of all to *rest* in the race that his Son has run and won on our behalf?

If you are a Christian, can you recall that first taste of God's love, when Christ, through the call of the gospel, made you his own? Then you must share Paul's longing to become a person who responds eagerly, instinctively to God's love by loving your Redeemer "all out" and loving other people sacrificially. Resting in the mercy that he has already lavished on you should whet your appetite for more, spurring you to sprint toward the goal of knowing him as he knows you. Again, 1 John 3:2 says that "we shall be like him, because we shall see him as he is." In the very next sentence, John draws the conclusion: "Everyone who thus hopes in him purifies himself as he is pure" (3:3). If you have been seized by Jesus, it is because he intends to give you one gift of grace after another—his righteous record, his resurrection life in "foretaste" form through his Holy Spirit, and finally his smiling face when he appears to transform our lowly bodies to resemble his glorious body and to make our divided hearts whole and complete, delighting to be and to do what delights him.

14

A Tale of Two Citizenships

Philippians 3:17–21

But our citizenship is in heaven, and from it we await a Savior, the Lord Jesus Christ, who will transform our lowly body to be like his glorious body, by the power that enables him even to subject all things to himself. (Phil. 3:20–21)

Imitation is the sincerest form of flattery." So the saying goes. Like any other effective proverb, it has that little twist of paradox that hooks us, makes us look again, and prods us to think. After all, isn't flattery, by its very definition, *insincere*? Isn't it flattery when we lavish compliments that we don't really mean on people who don't deserve them, but who can give us something we want? But the adage says that imitation is a *sincere* form of flattery, and it is right. The people whom we embrace as role models are not usually those whom we secretly despise but manipulate through false praises. They are people whom we sincerely admire. For that very reason we imitate them, modeling ourselves after their pattern. The principle that imitation expresses real admiration holds true, whether we think of:

- A little girl disciplining her "naughty" doll with the words that Mommy uses to correct her;

- A teenager whose clothing, hair color, and posture mimic her favorite rock star;
- A young husband who treats his wife as his father treated his mother; or
- A budding artisan watching her mentor mold a lump of clay into a beautiful vase, and then reproducing the master's movements at her own potter's wheel.

Of course, we often imitate people without even being aware that we are doing it. Perhaps their character, their attitudes, and their treatment of others are so attractive that we instinctively gravitate toward their way of doing things, almost whispering to ourselves, "I want to be like him [or her] when I grow up." On the other hand, sometimes we find ourselves following others' example even when we dislike them, either because of the force of their personalities or because we cannot imagine functioning any other way. How else can we explain why those abused in childhood so often grow into abusers themselves?

In this text, Paul teaches that the people whom we imitate are indicators of our true *citizenship*, of the political[1] entity that defines our identity and our status, our privileges and our duties. The term *citizenship*, which appears in verse 20, is related to the verb that Paul used earlier, in Philippians 1:27, to summon his Christian friends to behavior befitting their privileged status as citizens.[2] Paul now makes explicit that believers are citizens not merely of Caesar's city, Rome, but of heaven itself, where Jesus the true Savior and Lord now reigns. Throughout our present text, therefore, the theme of citizenship provides the backdrop for the contrast that Paul draws between his own positive example and the polluting pattern seen in the lifestyle of others (Phil. 3:17–19). The apostle's point is that the models who mold your desires and behavior are telltale signs of the town you call "home." When

1. The Greek term rendered "citizenship" is *politeuma*. It is derived from the term *polis*, "city." Today we use *political* to refer to governments at various levels: local (city), regional, state or provincial, and national. In Paul's day, the primary political unit was the *polis*, the city-state, of which there were many not only in Macedonia and Achaia but throughout the Roman Empire. Citizens' "political" standing and rights were derived not from residence within the boundaries of a nation (or empire) but from their legal status in the governance network of a specific *polis*.

2. As we saw in chapter 6, in Philippians 1:27 the ESV's "let your manner of life be" represents the politically charged Greek verb *politeuomai*, "behave as citizens."

we hear a Southern drawl or a Scottish brogue, we know in an instant where the speaker was raised. Her accent and inflection echo the voices that surrounded her as she grew up. So also, Paul implies, your values and choices will reflect the accents of your spiritual "hometown."

Now, on first hearing, that analogy and other examples of the ways that our past has placed its stamp on us may seem bleakly deterministic. We want to protest: Must Southerners forever drawl, or Scots forever roll their *R*s? (Not that either would trade their charming pronunciation for less distinctive accents, of course!) Are we so programmed by our past that we *cannot help but* reproduce the patterns that we have experienced, even those that we find repellent? In Paul's inspired hands, however, the connection between our citizenship and our exemplars is actually full of hope. This is because Paul is speaking of a "hometown" legacy that, surprisingly, flows *back from the future* into the present. The city that defines your identity is neither the one in which you were born nor the one in which you were raised. It is the city *toward which* you are moving, drawn forward by the glory of the Savior and Lord, Jesus, who already rules there in resurrection life.

Writing to people raised in the pagan milieu of a Macedonian metropolis, Paul daringly declares that their *real* identity is now defined by a city that they have never seen or visited, a city that lies ahead as their joyful destiny: "our citizenship is in heaven" (Phil. 3:20). Residents of ancient Philippi would be familiar with the concept that one's identity could be determined by a city far distant from the place of one's birth and residence. Philippi stood out among the cities in its region because Philippi was a "colony" of Rome, the imperial capital in Italy (Acts 16:12). In 42 B.C. Octavian, who would become Caesar Augustus, defeated his rivals in a decisive battle just outside the city. He settled many of his loyal troops in Philippi and conferred on their city the status of "colony" and on its citizens the privileges of citizenship in Rome (the city) itself—including exemption from imperial taxation and the protections of due process under Roman law (see Acts 16:37–39; 22:25–29). Of course, not all residents of Philippi were citizens of Philippi. Slaves, freedmen, and peasants typically did not have status as citizens, so they lacked the privileges and responsibilities that citizenship entailed. But Philippi's "citizens" were Rome's "citizens," and Paul capitalizes on this connection to illustrate a life-changing reality: though his friends have not yet set foot in the heavenly city where their

Lord now rules, that city already defines their dignity and determines their destiny. It must therefore control their character and conduct in the present, even before they reach "home."

People who trust Jesus, whatever their origins in terms of earthly cities, have had another, higher citizenship conferred on them. Christ, who radiates God's glory as God's equal, stooped as Savior to suffer the depths of indignity and death on earth, rescuing his people from the divine wrath that we deserve. Then, in his resurrection and exaltation, he granted to those he rescued the privileged status of citizenship in heaven, where he reigns as Lord over all. Although we have not yet seen that city, our status as its citizens is already making a difference, transforming our values and drawing us to examples worth imitating. Paul's call to follow his footsteps implies the promise that God's grace is strong enough to snap us out of the patterns that the past has imprinted on our minds and hearts. For those who trust in Jesus Christ, *who we are* is no longer determined by *where we have come from*, but instead by *where we are going*.

In these verses Paul speaks of two competing *destinations*, of contrasting *paths* that people "walk" to reach them, and of two rival groups of *guides* that show the way. On the one hand, Paul and those who shared his Christ-centered commitment are role models to be followed by heaven's citizens as they walk their earthly pilgrimage (Phil. 3:17) toward the glorious coming of our Savior from heaven at the end of history (3:20–21). On the other, Paul describes people whose "walk" exhibits enmity toward Christ's cross and preoccupation with earthbound interests, and whose "end is destruction" (3:18–19). Paul directs our eyes first in one direction, toward mentors worth mimicking (Phil. 3:17), and then in the opposite direction, toward the dangerous examples whose trajectory leads downward into the pit of hell (3:18–19). Finally, he points our gaze upward and forward to the heavenly metropolis that is our true home, and to the Savior-Lord who will come from there, to impart his own glory to all who belong to his celestial city (3:20–21).

We will examine Paul's powerful message of exhortation, warning, and hope by exploring first the contrast that he draws between the two *ways of walking*, two patterns of conduct rooted in contrasting systems of value, exhibited in the lives of contrasting companies of mentors and models. Then we will consider the contrast between the *two cities*, the two destinies, to which those paths lead.

Two Ways of "Walking," Two Groups of "Guides"

Paul's opening instruction, "Join in imitating me," and his reference to "the example you have in us" call our attention to the strategy of teaching by example that we have seen him use throughout this letter. Admittedly, much can be learned from books and lectures, but Paul knows that the mind-set of Christ can be received only by God's gift of grace (Phil. 1:6, 29; 2:12–13), and that it can be cultivated only through observation and practice. We learn wisdom—that rare blend of insight and skill—first by watching as others exhibit it, and then by trying to replicate what we have seen. So far in this brief epistle Paul has modeled how a Christ-focused heart responds to suffering (1:12–18, 27–30) and values others' needs more highly than its own (1:19–26; 2:1–4). He has lifted his friends' sights to the height of Christ's lowly servanthood, exhibited in the depth of his death on the cross (2:5–11). He has commended Timothy and Epaphroditus, men who would soon come to Philippi, as miniature replicas in which Jesus' concerns for his people and his readiness to risk all could be seen in person (2:19–30). Paul's own life story has shown not only the futility of striving toward one's own righteousness, but also the priceless treasure of gaining Christ and resting in his righteousness (3:1–11). Finally, Paul has just summoned the mature to mimic his maturity by admitting that they are far from fully mature and by joining him in the race for the prize that lies ahead (3:12–16).

Now, in case anyone has missed his earlier hints, Paul spells it out point-blank: "Imitate me—think as I think, do as I do, pursue what I pursue." Paul called other churches to imitate his life of faith, as he was imitating Christ (1 Cor. 4:16; 11:1; 1 Thess. 1:6), but only here does he stress so strongly how imperative it is for his friends to follow his lead *together*. Paul's Greek word order underscores "togetherness" by opening the sentence on that note. Although the ESV's "Brothers, join in imitating me" conveys Paul's point clearly in natural English, the apostle's *emphasis* could be rendered more dramatically (though more woodenly) as: "*Co-mimics* of me become, brothers."[3] Not just "copy me," but "*Together* copy me." Indispensable to

3. Greek *symmimētai mou ginesthe, adelphoi*. Our English *mimic* is not an accurate replacement for the Greek *mimētēs*, from which it was derived, since *mimic*, unlike *mimētēs*, typically connotes mocking disrespect. Since the compounded form, *sym* + *mimētēs*, does not appear in extant Greek literature before this text, Paul may have coined the compound to emphasize the importance of the

imitating Paul is "standing firm in *one spirit*, with *one mind* striving *side by side* for the faith of the gospel" (Phil. 1:27). Such unity demands the humility to "count others more significant than yourselves" by attending first to others' interests and concerns (2:3–4).

The imitation of Christ (through imitating Paul) is not an individual quest but a family project. This point is reinforced by the way that Paul fixes the Philippians' gaze not merely on himself but also on others who "walk according to the example you have in us" (Phil. 3:17). Timothy and Epaphroditus are such role models. No doubt the circle is larger, including members of the Philippian church and believers passing through from elsewhere.

Paul's forceful directive "keep your eyes on" is related to the noun "goal," which Paul has just applied to the focal point of his own single-minded pursuit (Phil. 3:14).[4] Earlier Paul urged the Philippians to focus their attention on others' concerns (2:4).[5] Now he says, "Watch carefully: we'll show you how to run this eager, restless race to grasp more completely the Christ who has grasped us, in whose grace we rest." It is possible, as we have seen, to "absorb" the influences of others' example unwittingly and passively. But God is calling you here to be intentional and deliberate: find folks in whom Jesus' love and purity shine, and fix your gaze on how they live out their gratitude for his grace in the way they treat others, respond to setbacks and sorrows, and aim for God's glory in every situation.

You will need both discernment and courage to identify and imitate the role models whose lives reflect the character of Christ. The entertainment industry invites you to emulate individuals who personify edgy rebellion, flashy self-promotion, and immediate gratification of any and every desire. The advertising industry subtly signals that image and appearance are the measures of personal worth and significance. The workplace often rewards those who are clever and assertive. You will have to look deeper than surface appearances to notice people who genuinely reflect the deeper beauty

Philippians' *partnership* in their imitation of Paul. G. Walter Hansen, *The Letter to the Philippians*, PNTC (Grand Rapids: Eerdmans, 2009), 261.

4. "Keep your eyes on" reflects the Greek *skopeō*, and "goal" represents *skopos*, which appears in the New Testament only in Philippians 3:14. In the Greek text of Job 16:12 and Lamentations 3:12, *skopos* refers to the "target" on which an archer sets his sights.

5. In 2 Corinthians 4:18 Paul used *skopeō* to affirm that believers fix their gaze on eternal realities that are invisible to the physical eye. Paul also used this verb to draw attention to dangers to be noticed and avoided (Rom. 16:17; Gal. 6:1; see Luke 11:35).

of the great King who walked among us as a despised servant, "gentle and lowly in heart" (Matt. 11:29). Picture in your mind the older Christians with whom you worship weekly as you read Paul's profile of the fruit of the Spirit (Gal. 5:22–23) or his lists of the marks of maturity to be sought in the church's elders and godly women (1 Tim. 3:1–11; Titus 1:6–9; 2:2–3). Then, when you identify such mentors, find ways to get close to them and ask the Lord for boldness to follow their lead, swimming upstream against the current of the crowd.

At this point Paul's concern is not so much with beliefs to be embraced as it is with behavior to be emulated, although the latter always reflects the former. He evokes the biblical imagery of "walking," contrasting those who "walk" in line with Paul's example to those who "walk as enemies of the cross." Old Testament people of faith "walked with God" (Gen. 5:22; 6:9; Mic. 6:8). They were instructed to walk in the ways of the Lord (Deut. 8:6; 10:12; 30:16; 1 Kings 2:3; etc.), and not to walk "in the way of sinners" (Ps. 1:1). In other letters, Paul's shift from expounding God's saving work to enjoining our fitting response was marked by his summons to a new way of "walking": "As you received Christ Jesus the Lord, so *walk* in him" (Col. 2:6); "*Walk* in a manner worthy of the calling to which you have been called" (Eph. 4:1).[6] Those saved by grace through faith are God's workmanship, created in Christ Jesus to *walk* in good works (Eph. 2:8–10). We are no longer to "*walk* as the Gentiles do" in mental darkness, calloused consciences, and insatiable sensual appetite (4:17–19). The way a person "walks" is the way in which he or she approaches everyday life. That pattern of conduct displays the habits of one's heart, one's core commitments and deep desires.

If we ask what aspects of his own and his colleagues' examples Paul longs to see replicated in the Philippians' "walk," the closest clues are found in the autobiography that he has just offered and in the negative features of the alternative way of "walking" that he is about to describe with tears. Positively, we must imitate Paul by continuing to rest in Christ's righteousness, not succumbing to the Judaizing error or other seductions that offer self-grounded confidence in exchange for self-reliant striving. Instead of straining to make himself look good by his religious credentials and achievements, Paul now seeks to be found "in Christ," robed in a righteousness woven of the Savior's

6. See also Rom. 6:4; 8:4; 13:13; Gal. 5:16; Eph. 5:2, 8; 1 Thess. 4:1, 12; etc.

seamless obedience—the gift of Jesus' perfect obedience from manger to tomb. Resting in Christ's achievement, rather than making Paul listless or complacent, has whetted his appetite for more, for all that Jesus has earned for him: "that . . . I may attain the resurrection from the dead . . . [becoming] perfect" (Phil. 3:11–12). So Paul wants us to "catch" from his example that same blend of assurance and anticipation, humbly admitting that we have not yet arrived and striving eagerly for more (3:12–14). That blend will make us patient with others who, like us, have not yet arrived (3:15). When you are assured that your identity is defined by Christ's cross and resurrection in the past—removing your guilt and conferring his righteousness—and by the heavenly city from which he will return in the future, you can pursue your own growth in godliness without nervous frenzy, and you can wait for growth in others without impatient frustration. You can follow the lead of Timothy, who cared more for others than for his own convenience. You can become like Epaphroditus, who put his life on the line for the gospel. You can imitate Paul, who gauged the conditions of his captivity not by how it hindered mobility, but by how it opened opportunities to tell soldiers about Jesus.

On the other hand, Paul has to issue a sharp and sorrowful warning against a different set of guides, whose "walk" leads down a very different path, to a very different destination: "For many, of whom I have often told you and now tell you even with tears, walk as enemies of the cross of Christ" (Phil. 3:18). Paul is painfully aware that the Philippians are exposed to a different company of mentors and models, whose motives and behavior contradict everything that his friends have seen at work in Paul and his colleagues.

Students of Scripture over the centuries have reached different conclusions about the identity of the dangerous role models described in verses 18–19. Some church fathers and later commentators have believed that Paul was still speaking of Judaizers who tried to impose the law of Moses on Gentile Christians, as he did earlier in the chapter (Phil. 3:2–11). Ambrosiaster, for example, commented, "Those who bring him to tears are the very ones who had already overthrown the Galatians . . . ."[7] Paul had warned the Gentile believers of Galatia that the Judaizers wanted to avoid being persecuted for the cross and that they hoped to "boast in your flesh" (Gal. 6:12–13). So it

7. Ambrosiaster, *Epistle to the Philippians* 3.19, quoted in ACCS NT 8:276.

is conceivable to apply the features that Paul lists in Philippians 3:18–19 to Judaizers. They were enemies of Christ's cross, certainly, so their end would be destruction. Perhaps they found "glory in their shame" by boasting over circumcision, a procedure that should have been treated privately.[8] In Philippians 3:2, as we saw in chapter 12, Paul had implied that when circumcision is viewed as a work necessary for salvation, it is equivalent to pagan mutilation rites such as those practiced by the prophets of Baal in Elijah's day (1 Kings 18:28). It is also conceivable that "their god is their belly" could refer to the imposition of Israel's kosher dietary laws on Gentile converts to Christ, as part of their commitment to comply with the whole Law after their circumcision.

On the other hand, many commentators believe that the danger posed by these enemies of the cross is not legalism's fixation on commandment-keeping, but the opposite: antinominianism's abandonment to lawless indulgence in physical pleasure. From this perspective, "their god is their belly" captures their devotion to bodily appetites—not only for food but also for sex—as though satisfying such urges were human beings' highest aim. The Greek term[9] rendered "belly" here refers elsewhere not only to the stomach in which food is digested (1 Cor. 6:13) but also to the womb in which infants are conceived (Luke 1:41, 44; John 3:4). Hence the term may represent both dietary and sexual appetites. Paul uses the same term in Romans 16:18 to characterize divisive people who "do not serve our Lord Christ, but their own *appetites*" (ESV). "Glorying in their shame" would be boasting in indecent acts that would embarrass anyone with a functioning conscience. Lawless pleasure-seekers are as opposed to Christ's cross as legalistic commandment-keepers, since the cross exposes the ugliness of sin and the gravity of its consequences. Those who indulge bodily desires, no less than those who strive to rein in such urges to establish their own righteousness, will meet their end in destruction.

On balance, it seems that at this point in his letter, Paul has in view people who are identified with the Christian community at large but whose pattern of *behavior*, whose "walk," contradicts their confession of Christ and his cross in practical ways. They may be travelers passing through Philippi

8. Chrysostom was aware of this interpretation but rejected it. Chrysostom, *Homily on Philippians* 14.3.18–21, cited in ACCS NT 8:277.

9. Greek *koilia*.

rather than members of the congregation, since Paul calls them "many" rather than "some of you" (cf. 1 Cor. 15:12).[10] Yet Paul's weeping over them suggests that he grieves over their destiny with an empathy that he does not feel for the Judaizers. Judaizers provoke Paul's righteous indignation because their grace-denying doctrine puts Christians' very lives at risk. For the Judaizers' doctrine of "salvation-by-Jesus-plus-law-keeping" Paul feels outrage, not sorrow. He expresses it in the strongest terms: dogs, evildoers, mutilation (Phil. 3:2), "accursed" (Gal. 1:8–9), and even "I wish those who unsettle you would emasculate themselves!" (Gal. 5:12). Here, on the other hand, what alarms Paul is not this group's divergence from the gospel in *doctrine*, but their departure from the cross in their *lives*.[11] And here Paul *weeps* as he writes about them. He feels a tenderness toward them that he never expresses for Judaizers. True, Paul also wept for his Jewish kinfolk who rejected their Messiah Jesus (Rom. 9:1–3). But here he is moved to tears because people claim to belong to Jesus, yet their "walk" contradicts Christ's cross, whatever words they may speak.

Although the group in question might have been "traveling" under a "passport" that claimed citizenship in Christ's kingdom, their conduct fit more naturally into a "city" that was anything but celestial. Ancient Rome did not issue passport documents as governments do today, but the Roman world was aware of cultural differences that set one ethnic group or region apart from others. Racial stereotyping is not new. Long ago certain nationalities were "typecast," much as we find today. An ancient cultural critic on Crete captured his countrymen concisely in the line: "Cretans are always liars, evil beasts, lazy gluttons"—and their neighbors concurred (Titus 1:12–13). Athens had a reputation for idle intellectual curiosity. The orator Demosthenes chided his fellow Athenians for constantly asking the question, "Is anything new being said?" and the historian Thucydides described the citizens of Athens as "slaves of each new paradox."[12] Things had not changed much by Paul's day, for we read in Acts: "Now all the Athenians and the foreigners who lived there would spend their time in nothing except

10. Gordon D. Fee, *Paul's Letter to the Philippians*, NICNT (Grand Rapids: Eerdmans, 1995), 367–68.

11. Peter T. O'Brien, *The Epistle to the Philippians*, NIGTC (Grand Rapids: Eerdmans, 1991), 453.

12. Demosthenes, *First Philippic* 10; Thucydides, *History of the Pelopponesion War* 3.38, quoted in Dennis E. Johnson, *The Message of Acts in the History of Redemption* (Phillipsburg, NJ: P&R Publishing, 1997), 208n18.

telling or hearing something new" (Acts 17:21). Of course, refreshing exceptions were recognized then, as they are today. Nonetheless, underneath the stereotypes, whether ancient or modern, lies a substratum of accurate observation: certain groups do tend to display shared characteristics even as they have other cultural features in common.

It is distressing when the tendency to paint whole people groups with a broad brush adversely affects individuals who are exceptions to the common caricature, as accurate as that caricature may be in terms of the "big picture." You are probably familiar with the stereotype of the "Ugly American": rude, brash, demanding, too loud, and laden down with too much stuff. Not all Americans are like this, but enough of us fit this stereotype that people from other cultures, especially societies that value patience and respect and restraint, often see us this way. If we feel the sting of injustice when others view us as Ugly Americans just because our passports identify us with our rude fellow citizens, we can appreciate why Paul is grieved that some people carry a "Jesus passport" but display priorities and passions that contradict the citizenship they claim. That is why he focuses his friends' attention on his own pattern of life and that of those who share his love for Jesus.

In a sense, the fact that Paul's description of behavior that conflicts with the cross can fit either self-disciplined law-keepers or sensual lawbreakers reinforces Paul's deeper point: *fixating on anything other than Christ* is behaving like an enemy of his cross and displaying a mind-set captivated by merely "earthly things." Think, for instance, about the description "their god is their belly." Again, "belly" stands for physical appetites—not only food and drink, but also sexual desire, leisure, and all the perks of affluent living. Now, there are two ways to make your "belly" into your god. On the one hand, you can link your *happiness* to what you consume. This is the form of belly-idolatry that Paul probably has in view here. On the other hand, you can link your *holiness* to what you refuse to consume. Or people might "glory in their shame" either by boasting openly in their sexual exploits or by basing their standing before God on the circumcision of an organ that should be discreetly covered—a surgery that, when abstracted from faith in God's promise, is tantamount to pagan "mutilation" (Phil. 3:2).

Linking happiness to consumption is often "the American way"—greed, gluttony, lust, laziness, and material overload. In the 1970s it found expression in the tagline of Schlitz beer commercials: "You only go around once in life, so you've got to *grab* for all *the gusto* you can . . . in a great light beer." In 2009 Dos Equis beer advertising delivered the same sentiment from "the world's most interesting man" (so we are assured): "You only live once. Make sure it's enough." It is hard to imagine a clearer expression of an earthbound mind-set (even if it is presented with a tinge of self-mocking irony)!

Or you can treat your appetites as divine by linking your *holiness* to what you refuse to consume. This is the way of ascetic self-denial commended by many religions, East and West—pursuing superior virtue by rejecting the pleasures of everyday life. This certainly seems more virtuous than sensual indulgence does, but it is just another form of idol worship, another alternative to trusting Jesus, another expression of "minds set on earthly things." Paul wrote to Timothy, "The Spirit expressly says that in later times some will depart from the faith by devoting themselves to deceitful spirits and teachings of demons, through the insincerity of liars whose consciences are seared, who forbid marriage and require abstinence from foods that God created to be received with thanksgiving by those who believe and know the truth" (1 Tim. 4:1–3). Who would have guessed that demons *disapprove* of good marital sex and tasty food? Who would have dreamed that self-denial can be the sign of a numbed conscience? In fact, though, whether the devil and his demons tempt us to believe that we cannot be happy without physical pleasure or persuade us that we can become holy by rejecting God's good gifts, either way their aim is the same: to pull our hearts away from Christ's cross.

If the grave is the farthest horizon you can see, if your citizenship is limited to this earth, this life, and what its few decades have to offer, a certain set of priorities flows from that: "Grab what you can now—don't miss out on anything that might bring your restless heart some satisfaction, or at least distraction." But citizens of heaven walk a different path, following the footsteps of Paul and his companions, to a different destination. We enjoy God's good gifts in this life without idolizing them. Therefore, we can freely forgo them when he directs, not because we trust our own abstinence to save us but because our hearts belong to the Savior whose appearance we eagerly anticipate that day. We echo a Puritan's prayer:

> Let no strong affection wantonly dally with the world.
> May I live high above the love of things temporal,
> sanctified, cleansed, unblemished, hallowed by grace,
> thy love my fullness,
> thy glory my joy,
> thy precepts my pathway,
> thy cross my resting place.[13]

Paul and his colleagues have shown us how our heavenly citizenship works out in earthly situations and relationships, turning chains into springboards for testimony (Phil. 1:12–13), enabling us to receive from God's grace not only the gift of faith but also the gift of suffering for Jesus (1:29–30), transforming others' needs into opportunities to reflect the humility of the Suffering Servant (2:1–11), and even making death itself into "gain" (1:21).

Two Cities, Two Destinies

In verse 20 Paul names the astonishing reality that has been his subtext throughout the paragraph: "our citizenship is in heaven." To be a citizen of Philippi was to be a citizen of distant Rome, where Caesar ruled his far-flung empire, with all the attendant privileges and responsibilities. So also Jesus-followers in Philippi, whether their status in society was slave or citizen or something in between, were citizens of a distant cosmic capital, of heaven itself, where their Savior and Lord, Jesus Christ, infinitely mightier than Roman emperors, was ruling the universe.

Paul draws two contrasts between those whose conduct contradicts Christ's cross and the citizens of heaven who cling to the cross. First, as we have seen, he sets the *earthbound* mind-set of "many" over against believers' *heavenly* identity and citizenship. Writing to the Colossians, he drew the distinction in similar earth-vs.-heaven terms:

> If then you have been raised with Christ, seek the things that are above, where Christ is, seated at the right hand of God. Set your minds on things that are above, not on things that are on earth. For you have died, and your life is hidden with Christ in God. When Christ who is your life appears, then

13. "Vocation," in *The Valley of Vision: A Collection of Puritan Prayers and Devotions*, ed. Arthur Bennett (Edinburgh: Banner of Truth, 1975), 307.

> you also will appear with him in glory. Put to death therefore what is earthly in you: sexual immorality, impurity, passion, evil desire, and covetousness, which is idolatry. (Col. 3:1–5)

He went on to show that setting our minds on things above is not escapist daydreaming, but having the beauties of heaven and heaven's King permeate and transform our values and relationships in the present, on the earth:

> Put on then, as God's chosen ones, holy and beloved, compassionate hearts, kindness, humility, meekness, and patience, bearing with one another and, if one has a complaint against another, forgiving each other; as the Lord has forgiven you, so you also must forgive. And above all these put on love, which binds everything together in perfect harmony. (Col. 3:12–14)

Because God's sheer grace has conferred on heaven's citizens unimaginable privileges, we are humbled, not haughty; patient, not proud; eager to serve, not demanding service; outgoing toward others, not turned in on ourselves. Heavenly-mindedness is humility in action on earth, serving, forgiving, loving others in everyday life.

The second contrast concerns the different *destinies* that await those whose thoughts are earthbound and those whose hopes are heavenbound. About the former, Paul states tersely and bluntly, "Their end is destruction." Schlitz and Dos Equis know how to exploit the desperate longing of shortsighted self-indulgence. If your citizenship is merely earthly, and the grave marks the boundary of your hopes and dreams, you had better grab the gusto and hope it's enough. But it will not be. For "citizens of earth," whose perspective never rises beyond the horizon of the here and now, the future is bleak: "Their end is destruction."

The world wept for Michael Jackson, a pop-music star who died too young, but it could not keep him alive or even turn the brief life he did live into a fairy tale that others would want to live. Long ago a man died at an even younger age, without sedatives or painkillers, his wrists and ankles pierced by spikes driven into a Roman cross. No one had to speculate about the cause of his death: blood loss and asphyxiation did their gruesome task in crucifixions. Yet that young man had already died voluntarily—had chosen to die—by the time his executioners thought they would finish their job. He had predicted, "No one will take my life from me. I lay it down of my own volition." And he did. But he also took it up again, as he said he would (John 10:17–18). Jesus is the risen and living Lord who makes people like us into citizens of heaven.

In telling the story of Jesus through this epistle, Paul has been rotating our line of vision from the past, when Jesus became "obedient to the point of death, even death on a cross" and God then "highly exalted him" (Phil. 2:8–9); through the present, as we strain forward to seize the prize for which Christ Jesus seized us (3:12–14); and now to the future, when the Savior and Lord who loved us, died and rose for us, and is now working his good pleasure in us (2:12) will return from heaven, our home, to "transform our lowly body to be like his glorious body" (3:21).

The portrait of believers' destiny that closes Paul's "tale of two cities" focuses not on a place, but on a person. Other Scriptures give us tantalizing glimpses of heaven and previews of the new heavens and earth, purged of sin and sorrow, that Christ's glorious return will inaugurate. Here, however, Paul wants us to see that what makes heaven heavenly, what makes our citizenship a source of boundless joy, is the presence of "a Savior, the Lord Jesus Christ."

Paul applies to Jesus two titles that Philippians, as citizens of Rome, would have heard applied often to the emperor: *Savior* and *Lord*. Not surprisingly, Roman coins bore the image of the reigning emperor (Matt. 22:19–21). Those minted in Greek-speaking regions often bore the Greek titles, *sotēr*, "Savior," and *kyrios*, "Lord." In times of local disasters—earthquakes, famines, military attacks—cities throughout the empire looked to Rome and its Caesar for "salvation," just as Americans today seek protection, relief, and reconstruction from the resources of the federal government.

But Christians have an infinitely more generous Savior than Caesar, and our Savior has brought, is bringing, and will bring an infinitely fuller salvation than Rome or Washington could supply—more generous relief and more total rescue from all that is wrong with our lives.

We don't have to grab all the gusto, drink all the beer, experience all the pleasures, or visit all the exotic vacation spots. Those diversions are ever so brief, and never as satisfying as we hoped they would be. Sooner or later our "lowly bodies," already in a slow but steady process of decay, won't be able to enjoy them anyway. We don't even have to escape the frustration, pain, grief, suffering, and sorrow that this life brings in order to live lives of meaning and hope and even joy—as Paul did, locked in Roman chains—in the midst of this world's mess and misery.

Our Savior has already rescued us from God's wrath on the cross—that is wonderful, but that is not all that he has in store for his heavenly citizenry. He will also rescue us from the toxic by-products of our own and other people's rebellion. When he comes back from heaven, he is going to raise our sin-stained, suffering-scarred bodies from the dead. As he emerged triumphant from the grave on the third day, with a real, material body that could never, ever again be harmed by pain or disease or death, so our Savior will give us glorious bodies like his!

Do you ever pause to think about that? No more muscle aches, broken bones, arthritis, memory loss, indigestion, heart disease, or cancer. Best of all, no more appetites bent in on themselves in guilty self-indulgence: a body wired to desire what our Creator designed us for, what he is pleased to give us, and what he is pleased to see us enjoy through and through, forever and ever. *Hallelujah, what a Savior!*

Then alongside the title *Savior* Paul puts another: *Lord*. The residents of the empire, whether citizens or slaves or others in between, had been taught to call Caesar "Lord." Along with their dependence on the emperor's generosity as savior came the demand to submit to his authority as sovereign. Christians, too, must submit to governmental authorities ordained by God (Rom. 13:1–7), but we have an infinitely mightier, more majestic Lord. In verse 20 Paul echoes the confession that concluded the glorious hymn to Christ, the humbled and exalted Servant of the Lord: "that at the name of Jesus every knee should bow . . . and every tongue confess that Jesus Christ is Lord, to the glory of God the Father" (Phil. 2:10–11).[14] Moreover, Paul underscores the implications of Jesus' title *Lord* by connecting the transformation of our lowly bodies to resemble his glorious resurrection body with "the power that enables him even to subject *all things* to himself" (3:21). Our new, glorious bodies, replicas of his resurrection splendor, are only one part of the panorama for the future that Paul brings into view. We will inhabit a new world, a "new heavens and a new earth in which righteousness dwells," as Peter describes it (2 Peter 3:13), in which everything—absolutely everything—is completely subject to Jesus, the Last Adam. The psalmist

14. In Greek the word order makes the parallel between 2:11 and 3:20 even more striking: *kyrios Iēsous Christos* (2:11) and *kyrion Iēsoun Christon* (3:20).

spoke in wonder at the honor that God lavished on humanity: "you have put all things under his feet" (Ps. 8:6). The author to the Hebrews shows that this statement is not a nostalgic glance back at paradise lost, but rather a preview of "the world to come," to which Jesus is leading his brothers and sisters (Heb. 2:5–8, 10). When our Savior and Lord appears from heaven, we will receive bodies impervious to the hurts of this earth and this age. Better yet, the current sources of harm will themselves be gone: no more tears or mourning or crying or pain, no more death and no more curse (Rev. 21:4; 22:3–5).

Do you trust Jesus? Then this is your destiny, Citizen! In the very next verse (Phil. 4:1), Paul will charge you again, as he did in his earlier summons, to stay united in the battle (1:27): "Stand firm" in this hope! Let your values and behavior befit your status as citizens of heaven! Standing fast as citizens of heaven means living daily life now in humble patience, eager hope, and selfless service to others—as King Jesus humbly served us.

Or perhaps your life is not now in Jesus' hands. But as you have been exposed to the truth of what it means to be a citizen of heaven, you found yourself thirsting for a hope that reaches beyond the horizon of the beer advertisements, beyond the few brief decades that this life offers. Are you tired of trying to chew taste out of pleasures that don't last, or trying to silence an uneasy conscience by your own self-discipline? Christ's cross unveils an ugly truth about what you, and we all, deserve from the King who rules in the city that your heart longs for: we deserve the death that Jesus died (the death that he alone, in all of human history, did not deserve!). But that same cross shows you the lovely truth that God so loved the world that he gave his only Son, so that whoever believes in him (that "whoever" can include you right now) should not perish—that miserable end of those whose horizon is earthbound—but, instead of perishing, should have everlasting life (John 3:16). Trust him today, and he will make you a citizen of heaven, with a future as bright as eternity is long.

15

Stand Together

Philippians 4:1–3

Therefore, my brothers, whom I love and long for, my joy and crown, stand firm thus in the Lord, my beloved. I entreat Euodia and I entreat Syntyche to agree in the Lord. Yes, I ask you also, true companion, help these women, who have labored side by side with me in the gospel together with Clement and the rest of my fellow workers, whose names are in the book of life. (Phil. 4:1–3)

We speak words by the thousands every day. Who would have imagined that they could be so powerful? Sometimes words stab to the heart and leave scars that last for years. Sometimes words convey balm to ease the pain and start the healing process. Words spoken rashly divide friends and lovers. Words spoken gently and wisely dispel estrangement and draw people together again. When difficult issues must be raised and discussed, we need to choose both our words and the tone of voice in which we speak them with special care.

I suppose every parent, now and then, mentally "replays" painful scenes that unfolded as our children were growing up, wishing that we could go back and rewrite lines that we spoke in those dramatic contests of wills. I have mercifully forgotten the details of many of these "showdowns," but one left a lasting impression and a valuable lesson about the power of well-chosen words. I don't even recall which of the "house rules" I was sure that one of our adolescent daughters had broken, but I can still see the brief "trial" in the living room in which I, the judge, was so sure of the facts of the case that I gave the defendant no opportunity to present exonerating evidence. When I pronounced my verdict and banished the convict to her room, she accepted her punishment submissively. Case closed . . . or so I thought. Sometime later, still with deference, she emerged from her "cell" to hand me a letter, in which she had calmly explained the *actual facts* of the situation, about which I had made a snap judgment earlier, and she registered her gentle protest against the injustice she had received. As I read her words, I was humbled by my daughter's meek stand for truth. The truth of two biblical proverbs hit home:

> If one gives an answer before he hears, it is his folly and shame. (Prov. 18:13)
>
> A soft answer turns away wrath, but a harsh word stirs up anger. (Prov. 15:1)

She spoke truth softly in that letter, and her words melted my heart. I asked her forgiveness. So much depends on the words that we speak or write, and on the tone in which we convey them!

In the text before us, Paul is preparing to close his letter of friendship to the family of God at Philippi. The "Therefore" with which he opens Philippians 4:1 is his way of marshaling the great gospel truths that he has been expounding in what we now call the first three chapters to serve as pillars and prods, as foundation and motivation, for a series of specific directions that will bring his epistle to its conclusion. The commands in chapter 4 spell out more particularly what it means for Christians to follow the example of Paul and his like-minded, cross-focused colleagues (3:17), resting in God's grace already received (3:2–11) and racing together toward God's grace to be consummated when our Savior appears from heaven, the site of our true citizenship (3:12–21). He will offer counsel to help them meet their sufferings with joy, gentleness, prayer, thankfulness, and meditation on the beauties of Jesus (4:4–10).

But he must also address a more delicate issue, fraught with the dangers of misunderstanding and offense. Therefore, Paul must choose his words with the utmost care. Spirit-filled apostle and pastor that he is, Paul does so with consummate tenderness and skill. His wording and tone not only reach our hearts on the issues he addresses, but also teach us much about how we ourselves can speak "the truth in love" (Eph. 4:15) when others may not wish to hear it.

The distressing difficulty that demands such pastoral diplomacy is this: two women whom Paul treasures for their courageous partnership in the cause of the gospel are now at odds with each other. Their friction threatens the congregation's unity at the very time when they all need to stand "firm in one spirit, with one mind striving side by side for the faith of the gospel" (Phil. 1:27). Paul must correct these dear sisters in the Lord, and he has to do so not in person but in ink on papyrus. Paul sometimes chose to write rather than coming immediately to confront a wayward church, praying that God's Spirit would use his written words to soften hearts before his arrival (2 Cor. 1:23–2:4). At other times, correspondence was the only mode of communication open to him. Generally Paul preferred to address hard issues in person, as he said to the Galatians: "I wish I could be present with you now and change my tone, for I am perplexed about you" (Gal. 4:20). For the Philippians, Paul has laid the foundation for his loving entreaty to Euodia and Syntyche in his earlier appeal to the whole congregation, "Complete my joy by being of the same mind, having the same love, being in full accord and of one mind. Do nothing from rivalry or conceit, but in humility count others more significant than yourselves" (Phil. 2:2–3). Now he approaches the problem head-on, naming names and appealing for reconciliation. What will he say, and how will he say it?

This brief text contains three commands: "stand" (Phil. 4:1), "agree" (4:2), and "help" (4:3). We could state these more fully as: "stand firm . . . in the Lord," "agree in the Lord," and "help these women, who have labored . . . in the gospel . . . , whose names are in the book of life." Like these sisters in first-century Philippi, you and I need to hear Christ's summons to steadfastness, unity, and mutual aid: "stand, agree, and help." And we need to hear those directives in the context of God's grace in Christ: "in the Lord, in the gospel, in the book of life."

Stand Firm in the Lord

As we already observed, the "Therefore" that opens Philippians 4:1 marks this as an ethical inference from Paul's previous discussion.[1] The summons to stand firm flows from the realities of Christ's gracious gift of righteousness, the supreme treasure of knowing him, the racecourse still set before us with its finish line ahead, and the competing exemplars who vied for believers' admiration and imitation—those like Paul who walked the way of the cross toward heaven, or those whose behavior contradicted the gospel, whose goals were earthbound, and whose end would be destruction.

As we hear Paul's outpouring of affection for his Philippian brothers and sisters in Jesus, we can feel that his transition is more than logical or rhetorical. His mood has swung from the depths of sorrow to the heights of delight. Just moments before, his heart was heavy and his eyes moist with tears as he spoke somberly of "many" who were walking "as enemies of the cross of Christ" (Phil. 3:18–19). Then the clouds lifted and the sun broke through as he revealed our heavenly citizenship and spoke of the coming Savior and Lord whose return we eagerly await (3:20–21). Now the dawning light of the destiny that he shares with his Philippian friends floods his heart with intense affection, and words of tender love flow freely from his lips: "my brothers, whom I love and long for, my joy and crown," and then again, "my beloved."

As he does often in his letters to churches,[2] Paul has previously addressed the Philippians as "brothers" in this letter (Phil. 1:12; 3:1, 13, 17) and will do so once more (4:8). Although he once placed preeminent value on his ancestral lineage from Israel (3:5), Paul now treasures a higher and stronger set of family ties, which bind him together with everyone, of every race, who through faith is united to Jesus the Son of God. From what we know

1. Compare the use of the same Greek conjunction, *hōste*, in Philippians 2:12 to draw ethical application from the great gospel indicatives of Christ's incarnation, humiliation, death, and exaltation (2:5–11).

2. With the exception of his epistle to the Ephesians—where "brothers" appears only in the closing benediction, "Peace be to the brothers" (6:23)—Paul addresses every church, whatever the ethnic composition of its members, as "brothers."

of the ethnic composition of the city of Philippi and what Acts tells us about the first converts who became the core of this church, it is unlikely that many of them were Paul's Israelite kinfolk. But Jesus' blood, shed for us on the cross, binds believers together more tightly than DNA could ever do. We are now brothers and sisters (Paul's Greek term *adelphoi* implies inclusion of his female siblings) in Christ, and Paul rejoices in that bond, reinforcing it as he prepares to speak correction into the lives of two treasured sisters.

Although natural siblings sometimes spar and experience estrangement, Paul twice uses the word *beloved* to underscore the depth of his affection for his Philippian brothers and sisters. (The ESV translates the first occurrence with "whom I love.") Paul called the Philippians "my beloved" in Philippians 2:12, as he urged them to respond to Jesus' sacrificial obedience (2:8) with their own obedience. This pastor loves his flock intensely, and that love is the wellspring from which he is about to speak words that may cause discomfort to two dear colaborers. In fact, Paul's love for the Philippians runs so deep that he feels the pain of their separation: "my brothers, whom I . . . *long for*." At the beginning of this letter Paul had mentioned the longing he felt to be with his Philippian friends again: "For God is my witness, how I *yearn for*[3] you all with the affection of Christ Jesus" (1:8). Between these expressions of Paul's yearning or longing for his friends, the apostle spoke in the same terms of Epaphroditus's "longing"[4] for his fellow believers in Philippi and his eagerness to return and allay their fears for his health (2:26). Christ's love for us ignites in us an affection for our brothers and sisters so intense that our absence from each other *hurts*. Our hearts ache for our reunion, as the old hymn rightly says:

When we asunder part,
It gives us inward pain;
But we shall still be joined in heart,
And hope to meet again.[5]

3. In Philippians 4:1 Paul employs an adjective, *epipothētoi*, "longed for"; so his Greek would read, woodenly, "my brothers beloved and longed for." In 1:8 he had used the related verb *epipotheō*, "yearn for."

4. Again, Paul uses the verb *epipotheō*.

5. John Fawcett, "Blest Be the Tie That Binds" (1782), in *Trinity Hymnal* (Suwanee, GA: Great Commission Publications, 1990), no. 359.

Paul has not yet exhausted the terms of endearment that come to his mind as he contemplates God's family at Philippi. Besides being brothers beloved and longed for, they are "my joy and crown." Paul felt the same way about and used the same terms for the Christians in the neighboring Macedonian city of Thessalonica (1 Thess. 2:19–20). The Jesus-followers in Macedonia were both poor and persecuted (2 Cor. 8:1–2; Phil. 1:27–30; 1 Thess. 1:6–7; 2:14). Perhaps for that very reason, it seems, Christ's strength shone more brightly in their weakness, so that these impoverished and beleaguered believers had a special place in the apostle's heart. Joy, of course, has run like a golden thread through the tapestry of this epistle. Since Paul is returning to the motif of standing firm in unity, we recall his earlier appeal to "complete my joy by being of the same mind, having the same love, being in full accord and of one mind" (Phil. 2:2). These brothers and sisters are Paul's joy already, and they can fill up his joy even fuller by pursuing and preserving heartfelt oneness with each other.

In calling the Philippians his "crown," Paul brings the Philippians and us back to the Panhellenic games initiated among the Achaian cities to the south. He speaks not of a king's diadem but of the wreath[6] woven of various plants—wild olive, laurel, celery, or pine—that would be placed on a victorious athlete's head.[7] Earlier in the epistle he invoked the athletic metaphor, encouraging his friends that their perseverance in faith would prove that Paul "did not run in vain" as he taught them God's Word (Phil. 2:16). Then he brought us back to the racetrack, as it were, setting the pace for every believer through his single-minded resolve to "[strain] forward to what lies ahead" and "press on toward the goal for the prize of the upward call of God in Christ Jesus" (3:13–14). By evoking Olympic imagery here, Paul is saying to his beloved brothers and sisters, "*You* are the prize for which I am running with all my might!" Of course, as Paul's "victory wreath," they are not the *whole* prize. Paul sprints to "gain Christ and be found in him." As he runs, Paul has his eyes turned toward heaven, from which he awaits his Savior. But when he calls the Philippians "my joy and crown," Paul shows

6. Greek *stephanos*, as in 1 Corinthians 9:25: "Every athlete exercises self-control in all things. They do it to receive a perishable wreath [*stephanos*], but we an imperishable." See 2 Tim. 2:5.

7. The olive wreath (*kotinos*) was awarded to victors in each event of the games held at Olympia once every four years. In other years the Pythian games were held at Delphi (laurel wreath), the Nemean games at Nemea (celery), and the Isthmian games at Corinth (pine). See Everett Ferguson, *Backgrounds of Early Christianity* (Grand Rapids: Eerdmans, 1987), 75–77, and sources cited there.

that the prize for which he runs is not an individualistic achievement but the redemption of Christ's whole church. As Paul envisions that day, his warm embrace of his fellow believers as his crowning joy shows how out of place it would be for them (or us) to succumb to self-centered rivalry, placing our personal interests above others' concerns. As you dream—I hope you dream—about the day of Christ's return, are your brothers and sisters in Christ, with whom you live and sometimes struggle week in and week out in the church, central to that celebration scene on your imagination's secret screen? Has Paul's insight captivated your heart so that you see your spiritual siblings, for all their flaws, as your joy and crown, loved and longed for even in their imperfections?

At last, in an atmosphere aglow with deep love, Paul reaches the command: "stand firm thus in the Lord." This summons to "stand firm" repeats Paul's summons in Philippians 1:27–28, where he augmented his call to stand with the charge to unity, "striving side by side for the faith of the gospel, . . . not frightened . . . by your opponents." So again the apostle evokes the imagery of military combat against enemies intent on our defeat and destruction. Likewise here Paul will follow up his comprehensive call to "stand firm . . . in the Lord" with a summons to two women who have "striven side by side"[8] with him in the gospel to exhibit unity of mind (4:3).

How are we to "stand firm" in the face of hostile enemies? Paul gives two answers, in the tiny adverb "thus" and in the phrase "in the Lord." First, "thus" (or "in this way") invites us to look both backward and forward in the flow of the letter.[9] Looking back, we see the example of Paul and those who share his mind-set, ready to share Christ's sufferings in anticipation of sharing his resurrection glory. But "stand firm thus" primarily points us forward[10] to the concrete instructions that follow: first, "agree" and "help" in

8. Paul uses the verb "strive side by side" (Greek *synathleō*) in Philippians 1:27 and 4:3—and it appears nowhere else in the New Testament. In view of the military overtones of the term (*BAGD*, 783, proposes the gloss "*contend* or *struggle along with*"), I cannot explain the ESV's rendering it as "*labored* side by side," obscuring the term's combat connotation.

9. Commentators debate whether the adverb *houtōs* refers readers back to the preceding discussion (Phil. 3:17–21 or 3:12–21 or even 3:2–21?) or forward to specific imperatives to follow in 4:2–9. Moisés Silva, *Philippians*, 2nd ed., BECNT (Grand Rapids: Baker, 2005), 186, concludes, "Whatever the precise syntactical force of *houtōs*, however, one should not assume, with regard to 4:1 as a whole, that a backward and a forward reference are mutually exclusive."

10. Gerald F. Hawthorne and Ralph P. Martin, *Philippians*, rev. ed., WBC 43 (Nashville: Thomas Nelson, 2004), 237, 240, suggest persuasively that the adverb primarily points readers forward to the series of imperatives by which Paul "puts meaning into the word *houtōs*, 'thus.'"

Philippians 4:2–3, and then "rejoice," "do not be anxious" but "pray," "think about" Christlike virtues, and "practice" the pattern seen in Paul (4:4–9).

In other words, Paul, Timothy, Epaphroditus, and others have shown how Christian soldiers stand their ground in the battle. Now Paul puts their example and our duty into words: reconciliation, unity, and mutual aid; joy and gentleness; replacing worry with thankful prayer; saturating our thoughts with heaven's beauties and our practices with heaven's values. To those whose minds are set on earthly things, this battle plan—so lacking in aggression, deception, and self-advancing strategy—may seem counterintuitive and destined for defeat. Heaven's citizens know better, having been rescued by "weakness" of the Suffering Servant, who is Lord of all.

Therefore, second, the source of our strength to stand is "in the Lord." Jesus is the Lord who can and will subject all things to himself, when he transforms our lowly bodies to resemble his glorious resurrection body (Phil. 3:21). He has begun a good work in us by his Spirit, and he will bring it to successful completion (1:6). Because God is at work in us, we can both want and do what brings him pleasure (2:12–13). Whatever our circumstances, whether abounding or abased, we can find our joy in the Lord (3:1; 4:4, 10). In him who strengthens us we can respond with thankful contentment in every situation (4:13). Christ does not give his followers commands that he has not kept in our place, nor does he issue directives without accompanying them into our lives through his life-giving, heart-transforming Spirit. We *must* stand firm. The Lord commands it. We *can* stand firm because the Lord who issues the command has bound us to himself by faith, so we now stand "in the Lord."

Agree in the Lord

In Philippians 1:27–2:4 Paul first introduced the civic and military metaphors to which he has now returned. In that earlier passage he called his friends to fulfill their responsibilities as citizens in a way worthy of the gospel, to stand firm, striving side by side in defense of the gospel as they faced the onslaught of opponents, not caving in to cowardice. Throughout that section Paul repeatedly made the point that the key to victory is unity: "in one spirit, with one mind . . . being of the same mind, having the same love, being in full accord and of one mind" (1:27; 2:2). He went on to describe the mind-set that fosters such unity: "in humility count others more significant than yourselves. . . . Look not only to [your] own interests, but also to the interests of others" (2:3–4). Presumably he made these points so forcefully so early because the pressures of suffering had exposed stress points of self-centeredness in the congregation at large.

One of those stress points was a fissure that had developed between two women who occupied well-deserved places of prominence in the congregation: Euodia and Syntyche. Paul tells us nothing about the occasion or the gravity of their disagreement and little about the women or their role in the church, although the apostle's silence has not forestalled later speculations on such questions. Presumably the Philippian church generally, as well as each of the protagonists, was aware of the incidents or issues that had disrupted these women's relationship. Did the problem begin with a personal affront, or perhaps a disagreement over priorities or strategies in ministry? We do not know. The apostle devotes no space to resolving the issue itself, as he had when Corinthian Christians disagreed over meat offered to idols (1 Cor. 8–10). The church at Corinth also seemed to be on the verge of shattering because of the intense rivalries of competing parties at Corinth, each swearing allegiance to a different leader (1 Cor. 1–3). But it is hard to imagine that the frictions at Philippi had reached that ignition point, in view of the gentleness of Paul's appeal to these sisters.

What were the roles of Euodia and Syntyche in the congregation? Again, we cannot speak with confidence. In the book of Acts, women receive special mention in connection with Paul's and Silas's planting of the churches in Macedonia. Lydia hosted Paul's team and the church's meeting in her home in Philippi (Acts 16:14–15, 40). Other Philippian women, with whom Lydia would gather for prayer each Sabbath, may have joined her as the core of the church (16:13). Prominent women responded to the gospel in Thessalonica (17:4) and Berea (17:12) as well.

Egalitarian scholars interpret Paul's commendation of Euodia and Syntyche as those who "have labored side by side with me in the gospel" to signify that Euodia and Syntyche held office as ministers of the Word,[11] "full members of [Paul's] mission team."[12] But construing their description in this way would prove too much, for then Paul's earlier application of the same

11. Gordon D. Fee, *Paul's Letter to the Philippians*, NICNT (Grand Rapids: Eerdmans, 1995), 398.
12. G. Walter Hansen, *The Letter to the Philippians*, PNTC (Grand Rapids: Eerdmans, 2009), 285.

wording to the whole congregation (Phil. 1:27) would mean that he expects every believer to fulfill the special pastoral office of teaching and oversight.

Paul maintained that spiritually qualified men should be called to exercise distinctive teaching and oversight authority as elders (1 Tim. 2:12; 3:1–7). Yet Paul also encouraged all believers to teach and admonish each other from the Word (Col. 3:16). All sorts of believers also spoke the gospel to those outside the church (Acts 8:1–4). The New Testament shows that Christ, the head of the church and source of his Spirit's gifts, has designed his community in such a way that the distinctive authority and responsibility of leaders (Eph. 4:11) are complemented by the participation of all members as we serve each other (Eph. 4:12–16; 1 Cor. 12). Leaders who hold "special" office are not in competition with all believers who hold a "general" office by virtue of their union with Christ. One Reformed catechism teaches all Christians to affirm that "I share in [Christ's] anointing . . . to confess his name, to present myself to him as a living sacrifice of thanks, to strive with a good conscience against sin and the devil in this life, and afterward to reign with Christ over all creation for all eternity."[13] Since Paul honors every member of Christ's body as providing service that all others need, Paul values Euodia and Syntyche as fellow combatants in defense of the gospel, as he wants every other follower of Christ to be. And apparently these women had enough "visibility" in the Philippian congregation that their falling out, whatever its cause or its severity, was public enough to warrant the apostle's direct mention in a letter to the whole church. Paul's robust understanding of the diverse gifts of the Spirit and the contribution of every member to the church, the body of Christ, enabled him to recognize these women as valued partners in the advance of the gospel, struggling and laboring at his side. His perspective cautions Christians today against excessively restricting the role of unordained believers, women or men, in promoting the good news of Christ.

As we overhear Paul appealing for reconciliation to these women whom he admires, we wonder whether he reflected back on the breach that he himself had experienced with a close colleague years before. When Paul first evangelized Philippi, his ministry partner was Silas and their young assistant was Timothy; but the team might have been quite different if Paul and his

13. Heidelberg Catechism, Lord's Day 12, Q & A 32.

previous colleague, Barnabas, had found a way to agree on one important decision. Having proclaimed the gospel to Gentiles in Asia Minor, Paul and Barnabas had returned to Antioch in Syria. Both there and in a council of apostles and elders at Jerusalem, Paul and Barnabas stood shoulder to shoulder and arm in arm, staunchly defending Gentile believers' freedom in Christ. Their oneness of mind and soul seemed unbreakable, until they started to lay plans to revisit the churches planted previously and they encountered a simple personnel decision.

On the first trip, Barnabas's cousin John Mark had served as the missionaries' aide until, for reasons left unexplained in Acts, he abandoned them in Pamphylia (Acts 13:5, 13). Barnabas wanted to give John Mark a second chance to prove himself, but Paul did not think he could be relied on. The disagreement became so sharp that their team split down the middle. Barnabas took Mark to Cyprus, while Paul recruited Silas and headed overland into the interior of Asia Minor, where Timothy joined their entourage (Acts 15:32–16:3).

Luke narrates the dispute matter-of-factly, neither taking sides nor excusing this breach of the pattern set earlier in Acts, when "the full number of those who believed were of one heart and soul" (Acts 4:32). In view of Barnabas's generous character (4:36–37; 9:26–27), it is no surprise that he was willing to give Mark the benefit of the doubt and mentor him further. His influence apparently bore fruit. At last Paul himself, writing from Rome, would endorse Mark and his ministry (Col. 4:10).[14] Is Mark still with Paul as he writes to the Philippians? Perhaps his presence is one factor prompting Paul to offer his tender appeal to Euodia and Syntyche.

Having set the mood through his expressions of affection for the whole congregation, Paul now broaches the subject of reconciliation and unity. First, he calls his sisters by name. In fact, in his Greek word order their names come first: "Euodia I entreat and Syntyche I entreat." In the twenty-first century, it may strike us as rude to name individuals in a public document and then to correct them, compounding their embarrassment. Leading a home Bible study on this text many years ago, I commented, "Can you imagine the looks on Euodia's and Syntyche's faces when this sentence was

14. In a later Roman imprisonment, Paul wrote to Timothy, "Get Mark and bring him with you, for he is very useful to me for ministry" (2 Tim. 4:11). Not only Mark but also Paul's estimation of him had taken a 180-degree turn!

read aloud in the congregation at Philippi?" Our children would sit on the family-room floor with pencils and paper to keep them occupied, as we adults discussed the Scripture. At the end of this particular session, my son handed me his work of art for the evening, and I could only smile. He could indeed imagine their faces! His sketch showed a small congregation seated in Mediterranean dress, and front and center were two women with mouths agape and eyes staring wide in shock!

I love that sketch, but I now know that I misled my son. Euodia and Syntyche would probably have felt some discomfort when their names and Paul's appeal were read aloud. But for ancient writers and for Paul in particular, to *avoid* naming these women would have been far more harsh. It would have meant placing them at arm's length. As one scholar says, "That he names them at all is evidence of friendship, since one of the marks of 'enmity' in polemical letters is that the enemies are left unnamed, thus denigrated by anonymity."[15]

Not only does Paul show them the respect of calling them by name, but he also honors them by conveying his request using a Greek verb that the ESV rightly translates "I entreat." As Christ's apostle, Paul had authority to issue commands (2 Thess. 3:6, 12). But to close friends he expressed his wishes (which would please Jesus, too) in less "coercive" terms. As he interceded for the runaway slave Onesimus, now a newborn believer, with Onesimus's master Philemon, Paul's friend and frequent host: "though I am bold enough in Christ to *command* you to do what is required, yet for love's sake I prefer to *appeal* to you—I, Paul, an old man and now a prisoner also for Christ Jesus—I *appeal* to you for my child, Onesimus, whose father I became in my imprisonment" (Philem. 8–10). Here "appeal" represents the same Greek word[16] that is reflected by "entreat" in our text. Paul is not pulling rank and throwing his apostolic weight around. He is humbly asking respected colaborers to behave in keeping with the gospel that they have all defended. And he entreats each of them, repeating his appeal with each woman's name. "It is as if, suggests [eighteenth-century scholar J. A.] Ben-

15. Fee, *Philippians*, 389–90.

16. Greek *parakaleō*, which has a broad semantic range. *BAGD*, 617, suggests as English glosses for *parakaleō* such words as "*appeal to, urge, exhort, encourage*" (listing Philippians 4:2 in this category), but also "*request, implore, appeal to, entreat*" and "*comfort, encourage, cheer up.*" Peter T. O'Brien, *The Epistle to the Philippians*, NIGTC (Grand Rapids: Eerdmans, 1991), 477, says that in Philippians 4:2 *parakaleō* "signifies 'to beg, beseech, appeal to.' "

gel, he is exhorting each separately face to face."[17] Who is obligated to take the first step in pursuing reconciliation? Jesus placed that burden squarely on the shoulders of the one who had initiated the offense (Matt. 5:23–24). But then Jesus turned things around, calling victims to seek out those who had sinned against them like a shepherd seeking a lost sheep (18:15; see vv. 10–14). Both the offender and the offended must take the first step! So without taking sides or distributing blame, Paul issues equal appeals to Euodia and to Syntyche: please let the friction cease, and let the one mind that is yours in Christ prevail!

Paul's appeal to Euodia and Syntyche to "agree in the Lord"—his Greek says "the same thing *think* in the Lord"—echoes his earlier exhortation to the entire church: "Complete my joy, that the same thing you may *think* . . . the one thing *thinking*" (Phil. 2:2). We have noticed that Paul uses the Greek verb here rendered "think" with striking frequency in this brief letter.[18] As Paul uses the term, it is not coldly cognitive. Though it certainly includes one's mind-set or perspective, as Paul uses it, it also includes one's emotional response to others. He uses it to express how he feels toward the Philippians (1:7) and how they think with concern about him (4:10). Here Paul invites and urges Euodia and Syntyche not merely to decide who is "right" or even to come up with a compromise acceptable to both. His concern is that their disagreement, over whatever issue and for however long it has persisted, has disrupted their ability to exhibit in their relationship to each other the unity that is theirs "in the Lord."

Earlier, when Paul asked the Philippians to be like-minded (Phil. 2:1–4), he pointed his friends to Christ, who not only sets the standard for our self-less servanthood but also transforms our minds and motives by his Spirit: "Have this mind among yourselves, which is yours in Christ Jesus" (2:5). Now again, Paul reminds these sisters, who have contended for the gospel, shoulder to shoulder with himself and each other, that they are—together—"in the Lord."

When your relationship with a Christian brother or sister hits an impasse, when you cannot resolve a disagreement, when it is a strain even to be in the same room with him or her, at that moment you both need to pause and

17. O'Brien, *Philippians*, 478.

18. Greek *phroneō* appears ten times in Philippians (1:7; 2:2 [2×], 5; 3:15 [2×], 19; 4:2, 10 [2×]—once in every 10.4 verses. The verb appears twelve times in the rest of Paul's epistles.

take to heart Paul's gentle reminder to Euodia and Syntyche that there is a third person involved. The tense situation includes not only believers who disagree with each other and hurt each other, but also *the Lord*, in whom you both now live as citizens of heaven.

Help These Women . . . in the Gospel . . . in the Book of Life

Paul's final request is to another leader in the church, whom he calls "true companion" or "genuine yokefellow." The petition is simply that this leader "help" Euodia and Syntyche work toward reconciliation, in view of the way they have distinguished themselves in contending alongside Paul for the cause of the gospel (Phil. 4:3). Paul trusted this colleague to have the wisdom to discern the sort of help their sisters need in order to restore their oneness of mind. Perhaps the disagreement had become so sharp that they had stopped hearing each other, each defensively marshaling her own arguments rather than listening to the other. Perhaps the perspective of a third party could bring clarity and balance to the one-sided perception of one or both of them. Perhaps the involvement of Paul's "genuine yokefellow" would alert them to the sobering truth that their interpersonal dispute, whatever its origin, was adversely affecting the wider congregation.

Paul now works out in a very practical way the implications of his imagery of the church as the body of Christ, in which every member grieves when any member suffers (1 Cor. 12:26). The disaffection of these two prominent Christian women has so adversely affected others that it must be named publicly in Paul's tender letter to the congregation. So now others must pitch in to help these sisters make peace with each other. This is not Euodia's and Syntyche's private problem, to which others can turn a blind eye. Their need calls for the aid of the family of God.

Paul quietly belabors the "with-ness" of the Christian life—that we live and grow in Christ *with* one another—by peppering verse 3 with a series of Greek words that begin with the prefix *with* (*syn-*). The leader who must help is in "yoke with" (*sy[n]zygos*) Paul, pulling alongside the apostle like oxen in tandem. The "help" that he is to give is to "hold with" (*sy[n]llambanō*) them. They have "contended with" (*synathleō*) Paul for the gospel, as soldiers at his side. And they are part of a larger circle of "fellow workers," who "work with" (*synergoi*) Paul. *With, with, with, with* . . . Paul drives home the point that we are in this together.

The identity of the "genuine yokefellow" whom Paul enlists to assist with the women's reconciliation is unknown to us, though it was no doubt evident to the Philippians. Timothy and Epaphroditus have been nominated (though, since they are still with Paul in Rome as he writes, it seems strange that he would address either of them in the epistle itself, and in such a cryptic way). Silas and Luke, each of whom spent time in Philippi, have been suggested, and they cannot be excluded. It has been speculated that "Yokefellow" (*Syzygos*) is a proper name (though it does not appear as such elsewhere), and that Paul adds "genuine" (*gnēsios*) to confirm that his character fits his name: he is a real partner, who pulls his weight. Some suggest that "true companion" is Paul's way of addressing the whole congregation, or of inviting any member to prove himself or herself a "true companion" by pitching in and helping along the peacemaking process.[19] Since the Philippian church had several "overseers" (Phil. 1:1), maybe one of these stood out as having worked closely with Paul. In the first-century churches, among the circle of elders who ruled well, some were called especially to "labor in preaching and teaching" (1 Tim. 5:17).[20] Perhaps Paul's "true companion" was one of the overseers whose primary role was to teach the Word. His partnership with Paul in plowing and planting the Word for a gospel harvest (like oxen in yoke) was all that was needed to identify him as the minister best suited to aid the reconciliation process for these treasured sisters in Christ.

More important than our guessing the identity of the "true companion" is our getting the point of Paul's emphasis on the *togetherness* that should characterize believers. Euodia and Syntyche have stood together with Paul for the gospel. The brother who must help them has "pulled the plow" alongside

19. Epaphroditus is supported by J. B. Lightfoot, *Saint Paul's Epistle to the Philippians* (1913; repr., Grand Rapids: Zondervan, 1953), 158, and John Reumann, *Philippians: A New Translation with Introduction and Commentary*, AYB (New Haven, CT: Yale University Press, 2008), 628–30; Luke by Fee, *Philippians*, 393–95; a leader named Syzygos (very tentatively) by O'Brien, *Philippians*, 480–81; the congregation as a whole by Hawthorne and Martin, *Philippians*, 242; various members by Silva, *Philippians*, 193. Hansen, *Philippians*, 284–85, mentions these options and others before concluding that we do not have enough information to speculate persuasively about the identity of Paul's "true companion" at Philippi.

20. See Edmund P. Clowney, "A Brief for Church Governors," in *Order in the Offices: Essays Defining the Roles of Church Officers*, ed. Mark R. Brown (Duncansville, PA: Classic Presbyterian Government Resources, 1993), 43–66.

the apostle. They belong to a greater company of fellow workers, including Clement and others—too many to name.[21]

This company of coworkers has been gathered not by the courage or character of its members but by God's grace. The adhesive of their unity, which makes reconciliation between Euodia and Syntyche so imperative, is the mercy that God has shown them in the gift of his Son. The gospel has taken root and borne fruit in their lives because the living God, in his sovereign grace, wrote their names "in the book of life" before he created the universe. This vivid image of the book of life is the note on which Paul ends his appeal for his sisters' oneness of mind and heart, through the assistance of others in the family of God.

The theme of God's book, in which he has written the names of those whom he chose to enjoy eternal life, is an ancient one. At Mount Sinai Moses mentioned God's book as he pleaded with the Lord to forgive idolatrous Israel (Ex. 32:32). Isaiah described Israel's survivors as those who had "been recorded for life in Jerusalem" (Isa. 4:2–3). A psalmist declared that when the Lord "registers the peoples," he will inscribe even the names of Gentiles from the surrounding nations—Egypt and Babylon, Philistia and Tyre and Cush—as having been born in Zion, as citizens of God's city (Ps. 87:4–6). Daniel foresaw the rescue and resurrection of "everyone whose name shall be found written in the book" at the end of time (Dan. 12:1–4). When seventy-two disciples returned to Jesus, rejoicing over the power of his name to dispel the forces of darkness, their Lord directed them to a deeper wellspring of joy: "rejoice that your names are written in heaven" (Luke 10:20). John's visions on Patmos revealed that those whose names are written in the Lamb's book of life will be vindicated at the last judgment (Rev. 3:5; 20:15; cf. 13:8; 17:8).

Paul has just reminded us that "our citizenship is in heaven" and that we await a returning Savior and Lord from that celestial city that defines our identity (Phil. 3:20). Now, he impresses on Euodia and Syntyche, on Paul's true companion (whoever he may be), on Clement and other fellow workers, and on the whole family of God at Philippi the amazing truth that God himself has enrolled each of us who trusts Jesus—personally, by name—in

21. Elsewhere Paul applies the title "fellow worker" (*synergos*) to Epaphroditus (Phil. 2:25); Prisca and Aquila (Rom. 16:3); Urbanus and Timothy (16:9, 21; see 1 Thess. 3:2); Titus (2 Cor. 8:23); Aristarchus and Mark and Jesus Justus (Col. 4:11); Philemon, Epaphras, Demas, and Luke (Philem. 1, 23–24).

his heavenly Zion's registry of citizens. The source of our enrollment, our election, is the unfathomable love of the Father. Its basis is the blood of the Lamb, shed for our forgiveness and cleansing. The life-giving presence of God's Spirit powerfully implements God's electing grace in our lives. In one sense, as Paul told Timothy, only "the Lord knows those who are his" infallibly (2 Tim. 2:19). Yet the fruit that the Spirit bore in believers' behavior gave Paul a glimpse of God's hidden choice, as he wrote to Christians in nearby Thessalonica:

> For we know, brothers loved by God, that he has chosen you, because our gospel came to you not only in word, but also in power and in the Holy Spirit and with full conviction. . . . And you became imitators of us and of the Lord, for you received the word in much affliction, with the joy of the Holy Spirit, so that you became an example to all the believers in Macedonia and in Achaia. (1 Thess. 1:4–7)

So in our text Paul names names—Euodia, Syntyche, Clement—affirming confidently that their names are written in the Lamb's book of life. The divine grace that wrote our names on the electing heart of God before the creation of the universe infinitely overshadows any and every interpersonal friction that threatens to divide us. Because we are one "in the Lord" and his mercy, we can and we must stand together!

16

Antidotes to Anxiety

Philippians 4:4–9

Do not be anxious about anything, but in everything by prayer and supplication with thanksgiving let your requests be made known to God. And the peace of God, which surpasses all understanding, will guard your hearts and your minds in Christ Jesus. (Phil. 4:6–7)

Paul's Christian friends at Philippi had things to worry about. They faced both external and internal threats to their peace and progress as heaven's citizens, trekking as pilgrims through earth's unfriendly terrain. From the outside, they were confronted by opponents whose intimidating aggression was daunting, putting at risk their courage to stand together. Paul, their spiritual father, was chained in Rome, awaiting the emperor's life-or-death verdict. Back in Philippi, they, too, were engaged in the conflict that they had witnessed in Paul and Silas's experience when their congregation was first planted. Within the church, individuals' preoccupation with their own agendas jeopardized their unity of mind and affection toward each other. Earlier in his epistle, Paul had addressed suffering and the threat it poses to our joy and peace

(Phil. 1:27–30). He also spoke to the problem of self-centeredness and its insidious effect on the unity of the church (2:1–4). For both problems, the apostle's prescription was for believers to refocus on Christ himself, and Paul showed them how from his own example.

As Paul draws his note of friendship and thanks to a close, we have heard him reassert his call to courage and to unity, summoning every follower of Jesus to "stand firm" in the Lord and urging two prominent members of the congregation to pursue reconciliation and oneness of mind in the Lord (Phil. 4:1–3). Now in Philippians 4:4–9 Paul continues the "wrapping-up" process in what sounds like a last-minute bullet-point "to-do" list: rejoice, be gentle, don't worry, pray, think good thoughts, do good deeds. As he does in other letters, Paul fires off every piece of parting advice he can think of before his closing benediction (see 1 Cor. 16:13–18; Col. 4:2–6; 1 Thess. 5:12–22). He sounds like a parent giving last-minute instructions to a freshman son before he sets off for college: "Start your papers early. Write us. Watch your checkbook balance. Phone us. Take showers. E-mail us. Wash your underwear. Text us. Go to church. Keep us posted. Get enough sleep. Stay in touch. Don't do drugs. Write or phone or e-mail or text us." Ideally, rather than tuning out or rolling his eyes, in that list the young man hears their loving hearts speaking!

Paul's list certainly springs from love. He has just described his friends in Philippi as those whom "I love and long for, my joy and crown . . . , my beloved" (Phil. 4:1). Moreover, although at first glance the list seems as random as a last-minute parental "stream-of-consciousness" farewell, these bullet-point directives are actually linked by more than the apostle's affection.[1] We see the list's deeper unity when we recognize these instructions as God's antidote to the anxiety that so often disrupts our joy and deprives us of peace. Paul prescribes the remedy to the stresses that we experience both from the pain of suffering from outside and from the strain of friction within the church.

There are two ways to handle the stresses of life. One approach comes "preloaded" at birth on the "hard drive" of our hearts. The other can come

1. Peter T. O'Brien, *The Epistle to the Philippians*, NIGTC (Grand Rapids: Eerdmans, 1991), 484, calls attention to the lack of conjunctions (asyndeton) connecting these commands. He therefore rejects the idea that a logical or causal relationship links Paul's exhortations in Philippians 4:4–9, contending rather that "through the use of asyndeton, the apostle's commands take on an individual importance; each is isolated and so made emphatic" (484–85). Our exposition will seek to show that there are more thematic interconnections than O'Brien seems to recognize.

only from a radical change of heart and perspective, produced by the gracious intervention of God. The first approach is rooted in the desire to control the variables of our own lives through diligence, ingenuity, and hard work. This is the "Invictus" approach immortalized in William Ernest Henley's famous nineteenth-century poem, celebrating Victorian England's stiff-upper-lip self-reliance. The poem's brave motto, "I am the master of my fate, the captain of my soul," exudes nobility, dignity, and resolute courage. But when its "I'm in control, I can do it" optimism bumps up against the harsh reality that so many factors lie beyond our control, it has nothing to offer but stress and frustration. Ask the employee and breadwinner who gave decades of competent and loyal service to a company, only to be "downsized" in a time of recession. Ask the cancer patient whose oncologist sadly changes the subject from treatment to pain management. Ask the homeless victims of tornadoes and hurricanes, of floods and droughts and wildfires.

Henley's braggadocio may sound quaint and out of touch as news media constantly bombard us with images of chaos in our complex and troubled world. Yet politicians know that voters' hearts still long for the stirring strains of can-do, it's-all-under-control confidence. No one wins elections by sounding uncertain over the prospect of solving the nation's problems upon entering office. Maybe that's why Bobby McFerrin's calypso ditty, "Don't Worry, Be Happy," won the Grammy for best song in 1988. Its verses actually described the miseries of homelessness, poverty, and loneliness, but few people recall the verses. What everyone remembers is the lilting, four-word refrain, "Don't worry, be happy" (which may have been more ironic than sincere). It lifts the heart to hum this little hymn to hollow hope, doesn't it? And if we need more to calm our uneasy minds, media sages and celebrity psychologists offer stress-management strategies to help anxiety-prone people cope with the uncontrollable forces in our present and the uncertainties in our future.

Jesus' servant Paul, however, writing God's truth, commends to us a radically different approach to the troubles that tempt us to worry. Paul presents a far stronger antidote to anxiety than politicians' promises, cheery self-coaching, or calming meditation. He directs his Philippian friends and us to a life-anchor that goes deeper than the surface storms of circumstances, even deeper than whatever emotional equilibrium we could muster through happy talk or mellow mantras or any other stress-management technique.

Paul offers us an anchor that secures our well-being eternally in the life and love of the ever-living God. He commends to us the joy that he has found through having his life defined by Christ, his cross, and his resurrection power. From that joy flow calm gentleness, thankful prayer, and the pondering and practice of the character of Christ. The result is protection from worry through the peace of God, conveyed to our troubled hearts through the living presence of the God of peace.

As we listen to "Dr. Paul's" prescription to remedy the anxiety that threatens our joy and peace, we must remember that Paul is not offering an ivory-tower theory from the armchair comfort of a tranquil university campus. He is writing from imprisonment, with the possibility of brutal execution on the horizon, and he writes to people who face real-world threats. So a brief review of the troubles facing Paul and his Philippian friends will set the context for us as we hear Paul's counsel. We will consider first, therefore, the Philippians' reasons to worry. Then we will explore the remedies to worry that this Word of God presents.

Reasons to Worry: Rejection, Resistance, Recession, Relationships

As we comb through this brief letter, we turn up clues that Paul and his friends encountered sources of distress that still tempt us to worry today: rejection, resistance, recession, and relationships.

Paul and his friends faced the pain of *rejection*. In Philippians 3:5–6 Paul rehearsed the stellar résumé that had once brought him honor in the community of his birth: he had belonged to the right nation, clan, family, and religious party. But when Jesus invaded Paul's life, Paul lost all that he had once considered "gain," forfeiting his standing in the Jewish community. Now he was accused of leading people away from the law of Moses and violating the sanctity of the temple (Acts 18:13; 21:21, 28). So Paul encountered rejection from the very people whose high opinions he had once treasured. The believers in Philippi faced similar social rejection. Their first church-planters had been "smeared" as disturbing civic order and advocating anti-Roman practices—grave accusations in a city that boasted in its status as a colony of Rome (Acts 16:20–21). When Paul wrote this letter, the Philippians themselves were facing social ostracism, still "engaged in the same conflict that you saw I had" (Phil. 1:30). Christians in Asia Minor, to whom Peter wrote his first epistle, were likewise suffering social stigma from the pagans among whom they resided: "they are surprised when you do not join them in the same flood of debauchery, and they malign you" (1 Peter 4:3–4). If the pain or fear of rejection—by family members, coworkers, former friends, or civic leaders—has put you on the defensive, you need Paul's prescription as much as the first-century believers of Philippi did.

For Paul (certainly) and the Philippians (probably), things had gone beyond interpersonal rejection and had escalated to overt, even violent, *resistance*. When Paul and Silas first brought Christ's message to Philippi, they were not only falsely accused, but also beaten and jailed. Now Paul writes from Rome in chains (Phil. 1:13). Paul's Philippian friends also face opponents whose threats tempt them to cowardice and retreat (1:27–30). Whatever the form of this opposition, it goes beyond social rejection to overt resistance. You may not have faced prison or bodily harm for your faith, but people who dislike your allegiance to Jesus can express their resistance in subtler ways. Since even mild opposition can easily intimidate and silence us, when we meet resistance we need Paul's antidotes to anxiety.

Paul will conclude his epistle thanking the Philippians for their financial contribution (Phil. 4:10–20). In his thanks, Paul alludes to a third factor that can generate anxiety: financial hardship or *recession*. Before the church's donation arrived, Paul's finances had been tight. He calls it "my trouble" (4:14). The Philippians, too, knew tight budgets. Elsewhere Paul described the "extreme poverty" of the churches in Macedonia (Philippi, Thessalonica, Berea) (2 Cor. 8:2). Money is an "equal-opportunity" anxiety-producer. Christians and non-Christians alike, poor and rich alike, fret about where funds will come from to cover life's necessities. Paul will share the secret of contentment that he has found in "plenty and hunger, abundance and need" (Phil. 4:12), to help his friends at Philippi resist the temptation to worry over economic hardship. Do you worry over finances? Your resources and budget may be at subsistence level, with no margin to absorb unforeseen emergencies. Or you may anticipate adequate income in your employment years but fret over fragile investments that may fail to provide the secure retirement they once promised. When you wonder where next month's rent check or your children's college tuition or the doctor's payment will come from, you need God's antidotes to anxiety.

Friction in interpersonal *relationships* has also reared its ugly head in the Philippian congregation. Their conflict is not as sharp as at Corinth, where Christians split up into competing parties (1 Cor. 1:10–14). But Paul has needed to urge them to replace rivalry with humility (Phil. 2:1–4), and he has just appealed to two beloved coworkers to resolve their differences "in the Lord" (4:2). Are relationships the source of your sleepless nights? Perhaps your husband or wife seems distant or every conversation degenerates into criticism and arguments. You fear that your marriage itself may be crumbling. Or your child, who was so open and loving and trusting just months ago, is becoming increasingly defiant, or sullen and secretive. What is going on behind that dour, expressionless face, inside her mind and heart? Or perhaps the sweet unity of your congregation is jeopardized by misunderstandings or competing agendas among members, as the oneness of the Philippian church seems to have been at risk when Paul wrote. If strained relationships cause you stress, you, too, need God's antidotes to anxiety.

Rejection, resistance, recession, relationships—the Christians of Philippi had plenty to worry about, as we do! What antidotes to anxiety does Jesus prescribe for our troubled hearts through his servant Paul?

ANTIDOTE TO ANXIETY #1: REFOCUS ON YOUR FAITHFUL LORD

Paul's parting directives—his parental "to-do" list for his spiritual children, now far from his fatherly eye—touch on a variety of themes: joy and gentleness, prayer, pondering, and practicing. Yet through them all, underlying each instruction, runs a motif that binds them all together: *refocus on your faithful Lord.*

If you have attended weddings in the last decade or two, you have probably heard Johann Pachelbel's famous "Canon in D Major." You can probably hum from memory the stately ground-bass motif that moves from D to A to B to F sharp to G to D to G to A, even if you couldn't reproduce all the variations built on that recurring theme. Likewise, the motif that pervades each "movement" of Paul's parting instructions and binds them all together is the presence of the true, triune God in the lives of those who trust in Jesus. Notice how persistently in these few verses Paul mentions the God who has loved and rescued us through Christ: "Rejoice in *the Lord* always" (Phil. 4:4). "*The Lord* is at hand" (4:5). "Let your requests be made known to *God*" (4:6). "The peace of *God* . . . will guard your hearts and your minds in *Christ Jesus*" (4:7). Even "what you have learned and received and heard and seen in me" (4:9) is the gospel, which is "preaching *Christ*" (1:15–18). And finally, "*The God of peace* will be with you" (4:9).

Through different variations Paul keeps playing one tune: the antidote to anxiety is to have the living God deeply involved in your life. Whether he is addressing how to find emotional equilibrium in trouble (joy, Phil. 4:4), how to respond to those who reject or resist us (gentleness, 4:5), how to petition the Father (pray, 4:6–7), or how to cultivate Christ-centered "habits of the heart" (ponder, 4:8) and patterns of behavior (practice, 4:9), at every turn Paul shows us another facet of the anxiety-banishing constancy and compassion of our Creator and Redeemer.

Paul meets us at every turn with a reminder of the God of grace because he knows that our anxiety is not merely the product of poor coping strategies. It is symptomatic of *misplaced trust*. Anxiety shows that our hearts are so set on something that we are terrified of losing it, desperate to hold onto it for dear life. That "something" that we cannot bear to lose is our heart's foundation, its "center of gravity." Even good things—love, family, knowledge, success—cannot last through thick and thin because they are creaturely and finite. In his 2009 book *Counterfeit Gods*, Pastor Timothy Keller told recession-terrified New Yorkers (and the rest of us) that what we worry about is symptomatic of the counterfeit gods that we instinctively count on, though we sense uneasily that those idols cannot bear the weight of our hopes. Keller says:

> Anything that becomes more important and nonnegotiable to us than God becomes an enslaving idol. In this paradigm, we can locate idols by looking at our most unyielding emotions. What makes us uncontrollably angry, anxious, or despondent? . . . Idols control us, since we feel we must have them or life is meaningless.[2]

On the other hand, those who trust in Christ can face every threat and wound that this twisted world can inflict—"tribulation, or distress, or persecution, or famine, or nakedness, or danger, or sword," death and life,

2. Timothy Keller, *Counterfeit Gods: The Empty Promises of Money, Sex, and Power, and the Only Hope That Matters* (New York: Dutton, 2009), xxii.

things present and things to come, or "anything else in all creation"—because we are assured that nothing can "separate us from the love of God in Christ Jesus our Lord" (Rom. 8:35–39). Refocusing on your faithful Lord, treasuring Jesus and his grace as your life's foundation imparts joy and gentleness, enabling us to combat worry by praying with gratitude, by pondering Christ's character, and by practicing the pattern of gospel-shaped conduct.

ANTIDOTE #2: JOY IN THE LORD

Joy is interwoven like a golden thread throughout this letter from prison. Paul prays with joy over his Philippian friends (Phil. 1:4). He is filled with joy when others preach about Jesus, even from unworthy motives (1:18). His friends' unity of heart will fill up his joy (2:2). Even if death for Jesus' sake is imminent, Paul rejoices and wants his friends to join his rejoicing (2:17–18). Though he was content when funds were few, he rejoiced when the Philippians' contribution arrived (4:10). Paul has commanded his beloved brothers in Philippi to rejoice "in the Lord" (3:1), identifying the deep well from which joy springs, whatever the vicissitudes of life's surface circumstances. Now in Philippians 4:4 he explicitly states that Christians can and must rejoice "in the Lord" and do so "always." Because our joy is rooted "in the Lord" who will never leave us, we are to rejoice at all times and in all circumstances.

Paul may have in mind the song that closes Habakkuk's prophecy. That prophet was upset that the wicked in Judah seemed to escape justice. God promised to punish unfaithful Judah through an even more evil empire, Babylon, but that only compounded Habakkuk's distress. Yet Habakkuk also received good news, "The righteous shall live by his faith" (Hab. 2:4); so his oracle closed with a song of joy, even in adverse circumstances:

> Though the fig tree should not blossom,
> nor fruit be on the vines,
> the produce of the olive fail
> and the fields yield no food,
> the flock be cut off from the fold
> and there be no herd in the stalls,
> yet I will rejoice in the LORD;
> I will take joy in the God of my salvation. (Hab. 3:17–18)

It is natural to link our happiness and hopes to juicy figs and ripe olives on trees, sweet grapes on vines, wheat in fields, sheep in folds, cattle in corrals, a robust stock portfolio, or a healthy retirement account. But a moment's thought shows how fleeting all such resources are. Habakkuk knew that what lasts through boom and recession, success and bankruptcy, is the commitment of God to his people. So does Paul.

Paul's "rejoice in the Lord always" should not be confused with Bobby McFerrin's "Don't worry, be happy." Rejoicing in the Lord does not mean that we never experience sadness or grief over loss. Paul himself felt sorrow over Epaphroditus's almost-fatal illness (Phil. 2:27) and wept over those who behaved as enemies of the cross (3:18). Paul was no Stoic, coolly shielding his composure from the ebbs and flows of emotion, keeping people and their problems at arm's length. The Stoics, a prominent school of Greek philosophy in Paul's day, commended and embraced the virtue of *apatheia*, "lack of feeling." Such a cool intellectual aloofness, the Stoics believed, insulates individuals from the wide range of emotions from pleasure to pain. What Paul is commanding, however, is completely different from the Stoics' anesthetized emotional life. Biblical joy, as God commands it, is compatible with the whole spectrum of emotions that fit the range of situations that confront us in this sin-stained world. Pastor Keller is right:

> "Rejoicing" in the Bible is much deeper than simply being happy about something. Paul directed that we should "rejoice in the Lord always" (Philippians 4:4), but this cannot mean "always feel happy," since no one can command someone to always have a particular emotion. To rejoice is to treasure a thing, to assess its value to you, to reflect on its beauty and importance until your heart rests in it and tastes the sweetness of it. "Rejoicing" is a way of praising God until the heart is sweetened and rested, and until it relaxes its grip on anything else it thinks it needs.[3]

To "rejoice in the Lord" is to resist the instinct to focus on visible pleasures and problems. It is to concentrate our minds deliberately on treasuring the Lord Jesus Christ, to focus thought on his majesty and mercy, his purity and power, to "see and savor the glory of God in the face of

3. Ibid., 173.

Christ"[4] until our hearts are profoundly persuaded that he really is all we need in every situation.

ANTIDOTE #3: GENTLENESS IN HOPE

Paul's next instruction, "Let your reasonableness be known to everyone" (Phil. 4:5), directs our attention from the Lord, the source of our joy, to other people, who are often the source of our stress. The key term,[5] which the ESV conveys as "reasonableness," appears only five times in the New Testament.[6] "Reasonableness" is an acceptable translation, but the NIV's "gentleness" is better—or "clemency, graciousness, forbearance,"[7] or even "magnanimity."[8] The term refers to the calm and kind disposition that enables a person to offer a nonviolent, even generous, response to others' aggression. Aristotle explained "gentleness" as a willingness to forgo one's own rights according to the letter of the law.[9] So this word nicely captures the thrust of Paul's earlier exhortation about the way that Christians should treat each other: "Let each of you look not only to his own interests, but also to the interests of others" (2:4). Elsewhere Paul uses this word to teach that elders must not be quarrelsome, but *gentle* (1 Tim. 3:3). All believers should be *gentle* rather than quarreling (Titus 3:2).[10] Paul associates *gentleness* with meekness as displayed by Christ (2 Cor. 10:1).

Here Paul expands the circle of those to be treated "gently" beyond the borders of the church. We are to display such forbearing kindness to "everyone," including those who are making our lives miserable. As children of a Father who sends sunshine and rainfall on the just and the unjust (Matt. 5:45), as brothers and sisters of the beloved Son who died for us while we were his

4. John Piper, *God Is the Gospel: Meditations on God's Love as the Gift of Himself* (Wheaton, IL: Crossway, 2005), 103 (and often throughout this book).

5. Greek *to epieikes*, a neuter adjective here functioning as an abstract noun (substantive).

6. Phil. 4:5; 1 Tim. 3:3; Titus 3:2; James 3:17; 1 Peter 2:18. The cognate noun *epieikeia* also appears twice, in Acts 24:4 and 2 Corinthians 10:1.

7. *BAGD*, s.v. *epieikeia* and *epieikēs*, 292.

8. Gerald F. Hawthorne and Ralph P. Martin, *Philippians*, rev. ed., WBC 43 (Nashville: Thomas Nelson, 2004), 244.

9. Aristotle, *Nichomachean Ethics*, 5.10.8, cited in G. Walter Hansen, *The Letter to the Philippians*, PNTC (Grand Rapids: Eerdmans, 2009), 288n40. See John Reumann, *Philippians: A New Translation with Introduction and Commentary*, AYB (New Haven, CT: Yale University Press, 2008), 611–12.

10. James says that wisdom from above is peaceable and gentle (*epieikēs*) (James 3:17).

enemies (Rom. 5:10), believers should extend kindness rather than retaliation to those who harass and oppress them.

Paul cinches this summons to gentleness with a promise, or a reminder, or both: "The Lord is at hand." This brief statement can be interpreted in two ways. On the one hand, most recent interpreters understand it to refer primarily, if not exclusively, to the eschatological "nearness" of Jesus' second coming. Although no one knows the timing of Christ's second coming, the New Testament assures us that our Lord will not needlessly delay, but will come "soon" (Luke 18:7–8; Rev. 22:7, 12, 20). James urged suffering believers to wait patiently, "for the coming of the Lord is at hand" (James 5:7–8). Paul has just reminded the Philippians that we "await" the future appearance of the Lord Jesus Christ from heaven (Phil. 3:20–21). Because our coming Lord will give joy beyond our wildest imaginations, we can now be gentle in hope.[11]

On the other hand, Paul may intend us to understand "the Lord is at hand" as an assurance of Christ's nearness to us *even now* through his indwelling Holy Spirit, as the psalmists affirm: "The LORD is near to the brokenhearted and saves the crushed in spirit" (Ps. 34:18; see 145:18). Paul has mentioned the heart-transforming work that God is performing in and among believers (Phil. 1:6; 2:12). So "the Lord is at hand" may motivate our gentleness by assuring us that, even now as we undergo injustices, we are not alone.[12] Or Paul may intend "the Lord is at hand" to convey both truths: the Lord is *near now* by his Spirit, bringing aid in our sufferings; and he is *coming soon* in his glory, bringing suffering to an end.[13]

ANTIDOTE #4: PRAYING WITH THANKS

Paul's "let everyone see your gentleness" shows a better way to respond to trouble than lashing out at other people. Now his "replace worry with prayer"

11. Taking the future-eschatological view are: J. B. Lightfoot, *Saint Paul's Epistle to the Philippians* (1913; repr., Grand Rapids: Zondervan, 1953), 160; Ralph P. Martin, *Philippians*, NCBC (Grand Rapids: Eerdmans, 1980), 155; Moisés Silva, *Philippians*, 2nd ed., BECNT (Grand Rapids: Baker, 2005), 198–99; Hansen, *Philippians*, 289; and Reumann, *Philippians*, 635–36.

12. John Calvin, *The Epistles of Paul the Apostle to the Galatians, Ephesians, Philippians and Colossians*, ed. David W. Torrance and Thomas F. Torrance, trans. T. H. L. Parker (Grand Rapids: Eerdmans, 1965), 288–89, held that Paul was focusing on the Lord's present "nearness" by the Spirit and his providence.

13. Holding that Paul intends "the Lord is at hand" in both a current and a future sense are O'Brien, *Philippians*, 490; Sinclair B. Ferguson, *Let's Study Philippians* (Edinburgh: Banner of Truth, 1997), 103–4 (citing Ps. 34:18); Gordon D. Fee, *Paul's Letter to the Philippians*, NICNT (Grand Rapids: Eerdmans, 1995), 407–8; and Hawthorne and Martin, *Philippians*, 244–45.

(Phil. 4:6–7) turns our hearts back toward God, urging us to approach him not with grumbling or questioning (see 2:14) but with gratitude and expectant petition. Paul is echoing Jesus' Sermon on the Mount. Having invited us to address God as "Our Father" (Matt. 6:9), Jesus went on to show that God's children do not need to worry over life's necessities:

> Therefore do not be anxious, saying, "What shall we eat?" or "What shall we drink?" or "What shall we wear?" For the Gentiles seek after all these things, and your heavenly Father knows that you need them all. But seek first the kingdom of God and his righteousness, and all these things will be added to you. (Matt. 6:31–33)

Here Paul compiles a rich inventory of prayer-vocabulary—prayer, supplication, requests—to emphasize the freedom of access that is ours to bring *every sort of concern* to our Father. We "make our requests known" to him, obviously, not because he would be ignorant of them unless we informed him, but rather because speaking them aloud expresses our dependence and trust that he cares for us personally and delights in his children's speech.

Notice the ingredient of "thanksgiving" that is to be blended with our requests. Gratitude preserves our prayers from going sour with complaint or degenerating into a list of self-centered demands. Thanksgiving is the natural response to a generous gift, freely bestowed. In polite society, we sometimes say "thank you" just because it is expected. But *real thanksgiving* bubbles up from the heart when we are delighted by a gift beyond anything we expected, unearned and undeserved. That is the thanksgiving that must permeate our prayers as we bring our requests to the Father.

If we were to ask Paul the reasons to thank God, no doubt he could go on for hours. But if we asked him to pick *the very best gift*, he would take us to his explosion of amazement in 2 Corinthians 9:15: "Thanks be to God for his inexpressible gift!" He would explain that the gift that goes beyond words is God's Son, Jesus, as he wrote to the Romans: "He who did not spare his own Son but gave him up for us all, how will he not also with him graciously give us all things?" (Rom. 8:32). So this antidote to anxiety is to feast your heart on God's gracious gift of Christ to the point that you burst forth in thanks, as you bring your worrisome problems—rejection, resistance, recession, or relationships—to your loving heavenly Father. Rather than fretting like orphans left to fend for themselves, you can bring your griefs with your gratitude to God, confident that, whatever his answer on the particulars, his peace will guard your heart in Christ Jesus.

Antidote #5: Pondering Christ's Character

The promises in Philippians 4:7 and 9, with the wordplay connecting "the peace of God" and "the God of peace," show that Paul's directives in 4:8–9 are linked to his previous summons to joy in the Lord, gentleness in hope, and prayer with thanksgiving. Moreover, these last two prescriptions for our worrywart hearts are bound to each other by their parallel structure: first a list, then a command "ponder these things" (4:8), and then another list, followed by "practice these things" (4:9).[14]

Paul knows that the thoughts that occupy our minds and the images that capture our imaginations shape our characters and find expression in our behavior. As Israel's ancient sage observed, "Keep your heart with all vigilance, for from it flow the springs of life" (Prov. 4:23). Jesus confirmed that the heart's secret thoughts are the fountain from which our outward actions flow (Mark 7:14–23). So Paul speaks first of letting our minds dwell on qualities that reflect the perfections of our Creator (Phil. 4:8), and then he calls us to practice the pattern that we have heard in the gospel and seen in those who live Christ-focused lives (4:9).

Paul tells us to think about, or to ponder, "whatever is true, . . . honorable, . . . just, . . . pure, . . . lovely, . . . commendable, . . . any excellence, . . . anything worthy of praise." Many of these words are rare in Paul's letters and in the New Testament as a whole. Although Paul uses the adjective *true* rarely,[15] he often insists that God's truth—God's utterly trustworthy and accurate portrayal of reality—must control believers' minds and, consequently, our behavior. Because "the truth . . . in Jesus" means that we have shed the control of deceitful desires, our legacy from the original Adam, and have been clothed with "righteousness and holiness" characterized by truth (Eph. 4:20–24), our conduct must follow suit: "having put away

14. The parallel is more obvious in the Greek word order, in which "these things" immediately follows the items in each list, preceding the imperative verbs: "*these things* ponder [ESV: 'think about']," "*these things* practice."

15. In addition to Philippians 4:8, it appears in Romans 3:4 (of God), 2 Corinthians 6:8 (of apostles), and Titus 1:13 (of a Cretan prophet's assessment of his countrymen).

falsehood, let each one of you speak the truth with his neighbor" (4:25). Paul uses *honorable* or its related noun six other times[16] (and he is the only New Testament author to do so), always with reference to men and women whose spiritual maturity, dignity, and authority make them worthy of others' respect. In this context, *just* refers not to the legal standing that belongs to believers through Christ's imputed righteousness (as Paul used a related noun in Philippians 3:9; see Rom. 1:17). Rather, it describes that which conforms to God's perfect norm of equity. For example, masters must treat their slaves "*justly* and fairly" (Col. 4:1). What is "pure" is free from defilement or pollution. It includes sexual purity and fidelity (Titus 2:5; 1 Peter 3:2; and, metaphorically, 2 Cor. 11:2–3), but it extends well beyond the realm of sexuality. In Philippians 1:17 Paul used a related term to describe the unworthy motives of some who preached Christ "out of rivalry, not *sincerely*."[17] *Lovely* appears nowhere else in the New Testament. Its uses elsewhere in ancient Greek suggest that it refers to the quality that warrants and attracts admiration. Similarly, *commendable* makes its only New Testament appearance here, although Paul once uses a related noun to describe the varying responses to his ministry, "through slander and *praise*" (2 Cor. 6:8).[18] Paul uses the term *excellence* only here, where the context suggests that he has in view ethical integrity, the same sense that it has in 2 Peter 1:5, in the Greek Septuagint,[19] and often in the moral literature of the Greco-Roman world.[20] Finally, Paul commends to our reflection "anything worthy of praise"—a term that he uses eight times elsewhere to refer to praise directed toward God (Phil. 1:11; Eph. 1:6, 12, 14) or toward human beings who deserve commendation (Rom. 2:29; 13:3; 1 Cor. 4:5; 2 Cor. 8:18; see 1 Peter 1:7; 2:14). Paul urges us to fix our thoughts on themes that are not only intrinsically virtuous because God approves them but also visibly virtuous, attracting the approval of human beings who care about integrity, purity, and justice.

16. 1 Tim. 2:2; 3:4, 8, 11; Titus 2:2, 7.

17. In Philippians 1:17 (ESV), "sincerely" represents the Greek adverb *hagnōs*, a cognate of the adjective *hagnos* in 4:8. For Paul's other uses of the adjective, see 2 Corinthians 7:11; 1 Timothy 5:22. It also appears in James 3:17; 1 John 3:3.

18. In Philippians 4:8 (ESV), "commendable" represents *euphēmos*. In 2 Corinthians 6:8, Paul uses the cognate noun *euphēmia*.

19. Wisd. 4:1; 5:13; 8:7.

20. The Greek *aretē* also appears in 1 Peter 2:9; 2 Peter 1:3, to refer to God's praiseworthy attributes and acts (see LXX Isa. 42:8, 12; 43:12, 21; 63:7; Hab. 3:3).

Where these terms do appear more frequently (than in the New Testament) is in Jewish wisdom literature and in the great pagan thinkers of the ancient world: Plato, Aristotle, and the Stoics.[21] When these philosophers discussed the qualities of character that the good person should display—integrity, justice, self-control, prudence, courage, and so on—they used many of the terms that Paul uses here. The moral decay in Greco-Roman society in the first century was a sordid preview of the decline of the West in the twenty-first century: kinky sex (homosexual practice, sexual abuse of children, bestiality, and so on), self-indulgent luxury, and brutal violence masquerading as entertainment. Nevertheless, there were morally sensitive thinkers in that day, as there are today. The ancient world even had its equivalents to *The Book of Virtues*, which William J. Bennett, former U.S. Secretary of Education, assembled to try to recapture America's imaginations with a vision of the good, the true, and the beautiful.[22] Paul seems to be picking up terms from those ancient "books of virtues," perhaps to implement Jesus' instruction, "Let your light shine before others, so that they may see your good works and [*recognizing their goodness*, they will] give glory to your Father who is in heaven" (Matt. 5:16).

Paul realizes that not everything that is considered "lovely" or "commendable" by society at large would meet with God's approval, so he adjusts his grammar slightly at the end, calling us to exercise discernment: "if there is any *excellence*, if there is anything *worthy of praise*, think about these things." The term rendered "think about" is not the one that Paul has used so frequently in this epistle (*phroneō*), but instead one that expresses "taking into account" or assessing that which deserves approval.[23] As we survey the virtues celebrated in society, we must blend our appreciation with discernment. Paul is applying complementary truths that Scripture teaches elsewhere: (1) though the human family is fallen and flawed by sin, God's common grace still sustains even in unbelievers a sense of what is true, honorable, just, pure, and praiseworthy; yet (2) true virtue can be defined

21. Fee, *Philippians*, 415, mentions the *Discourses* of the Greek Stoic Epictetus and the *Moral Essays* of the Roman Stoic Seneca.

22. William J. Bennett, ed., *The Book of Virtues: A Treasury of Great Moral Stories* (New York: Simon and Schuster, 1993).

23. Greek *logizomai*, in the sense of "*evaluate, estimate*" (*BAGD*, 476). Fee, *Philippians*, 419, argues that Paul's grammatical shift in concluding the list of virtues signals believers' responsibility to exercise Christ-centered discernment with respect to the values of the surrounding society.

only by its supreme standard, the character of our infinitely holy Creator, revealed to us in his Scriptures and in his Son. Therefore, in the next verse, as Paul's exhortation moves from thought to action, from pondering to practice, he places the *common* virtues just listed into the *specific framework* of the gospel of Christ.

Antidote #6: Practicing Paul's Pattern

Whereas the points to ponder in verse 8 were character traits, the list in verse 9 focuses on the *means of communication* by which the Philippians encountered the gospel and observed its life-changing power. "What you have learned and received" sums up the message that Paul and Silas brought to Philippi. That message is good news ("gospel"), and its subject is Christ himself (Phil. 1:5, 15, 17–18). Christ's divine and human person and his redemptive mission (2:6–11) were the only theme that Paul cared to convey (1 Cor. 2:1–2; Col. 1:28).

What they had "heard and seen" in Paul was the fruit of God's grace, as the Holy Spirit had caused the gospel to take root in his life. This second pair of verbs alludes to the situation that Paul mentioned in Philippians 1:30, namely, that the believers of Philippi had *seen* Paul suffer in the past, while he was among them, but now *heard* from a distance that he still suffered. In his current imprisonment and legal crisis, Paul is prepared for whichever outcome God has planned for him, affirming one single-minded resolve: "it is my eager expectation and hope that . . . with full courage now as always Christ will be honored in my body, whether by life or by death. For to me to live is Christ, and to die is gain" (1:20–21). Paul has shown his friends how faith in Jesus works out in practice, in the midst of life's trials. Just as Christ is the Savior who captivates Paul's belief, so also Christ is the Lord who controls Paul's behavior.

What the Philippians heard and saw in Paul was the effect of Jesus Christ transforming a selfish, sinful man into the beauty of his own image, in holiness and love. When we read the virtues listed in verse 8 in the light of Paul's gospel-focused message and lifestyle in verse 9, we see that he is not just saying, "Think good thoughts, like the upright pagans." Rather, he is calling us to ponder the dimensions of *Christ's* perfection. Christ is the standard of truth, honor, justice, purity, beauty, and praiseworthiness.

So Paul sets the pattern for how *pondering* Christ's perfections progresses on to *practicing* them in daily living. Fixing and feasting your minds on Jesus must ignite the fire of your will and motivation, so you are eager to express your love for him by loving others. In that move from *ponder* to *practice*, God's Spirit quietly conforms our desires to the mind-set of Jesus, so we are no longer preoccupied with our safety or rights, no longer intimidated by threats, no longer paralyzed by anxiety. As trusting children, we learn to let our wise and mighty Father deal with factors beyond our control. Set free from that burden of protecting ourselves from harm and loss, we begin to practice the self-forgetting servanthood of Jesus, as we have seen it reflected in people such as Paul.

So Paul does close with a parental "to-do" list of sorts: rejoice in the Lord, be gentle in hope, pray with thanks, ponder and practice the beauties of Jesus. We could even add to these that, in some circumstances, there are other practical steps that we may take to address the occasions of our anxiety. If you have accrued looming debt by living beyond your means, beginning to practice responsible stewardship before the Lord—spending less, paying off overdue bills, and saving more—may be God's means to bring some relief to your stress. If the strain in your relationship with another believer has persisted because you are afraid to speak the truth in love or too proud to confess sin and seek forgiveness, consider your discomfort God's prod, prompting you to pay the price of gospel-grounded peacemaking.

Yet the ultimate antidote to anxiety is not to be found in what *we* do but in what *God has done and is doing* for us. Appropriately, therefore, we close our survey of God's antidotes to anxiety by returning to the twin promises about the peace of God and the God of peace in verses 7 and 9.

Antidote #7: Protected by the Peace of God and the God of Peace

Christ our Champion promises the protection of God's peace through the presence of the God of peace (Phil. 4:7, 9). As we stop wasting energy in futile worry and turn instead to thankful prayer, "the peace of God . . . will guard your hearts and your minds in Christ Jesus." The word *guard*[24] is a military term that often refers to a soldier's duty to ensure that prisoners do not escape (see 2 Cor. 11:32; Gal. 3:23). (Did Paul glance at the guard sitting next to him and the chain that bound them together?) Another purpose of a military guard was to protect a target of attack, as Roman forces had stepped in to keep an angry mob in the Jerusalem temple from tearing Paul apart (Acts 21:27–36). That is the scenario that Paul paints here. Our hearts and minds are under attack and need God's protection.[25] The Philippians faced intimidation that threatened their hearts with fear, perhaps by threatening their bodies with harm (Phil. 1:28). So Paul promises that peaceful calm will replace worry when we pray to God with thanks for grace already given.

This peace "surpasses all understanding" because, as one scholar put it, "believers experience it when it is unexpected, in circumstances that make it appear impossible: Paul suffering in prison, the Philippians threatened by quarrels within and by enemies without."[26] For me, this description of God's peace as "surpassing all understanding" will always bring to mind a summer afternoon—Monday, July 9, 1973—in sweltering New Jersey. Our first son had been born two days before, to our great joy. But early that Monday morning I was awakened by a phone call from our new pediatrician, notifying me that our infant had suffered a seizure, that neonatal meningitis was suspected, that antibiotics had been started, and that he was being transported by ambulance to Children's Hospital in Newark. Throughout the day I shuttled between home, the hospital in which my wife had given birth, and the neonatal nursery ward in Newark. In the heat and humidity our Volkswagen's engine developed vapor lock and refused to start. A church member lent me a car. That car's exhaust pipe became partly dislodged, stranding me on a side street, miles from my home, my wife, and our newborn son. As I waited for a friend to retrieve me and the second disabled car of the day, suddenly, surprisingly, I realized that my heart was at peace. Though it was out of character for me, I was not torn up with fear for my little boy, nor cursing the crummy cars that had let me down, nor even resenting the sticky heat. My heavenly Father, knowing that I needed

24. Greek *phroureō*.

25. Compare 1 Peter 1:4–5: "you, who by God's power are being guarded [*phroureō*] through faith for a salvation ready to be revealed in the last time."

26. Silva, *Philippians*, 196.

him right then, had wrapped my heart in the protective armor of his peace, his Spirit's kind and calming presence . . . and who could explain it?[27]

Do not for one moment think that the explanation lay in my towering confidence in the goodness and sovereignty of God! I believed then and believe now that our God is both infinitely good and infinitely sovereign. But more often than I care to admit before, after, and even close by the surprising sweetness of that summer afternoon, what I *know* to be true of my Father has not set my heart free from turmoil. Jesus himself was troubled as he anticipated the cross (John 12:27). Paul included "anxiety for all the churches" not in a list of his sins or failures of faith, but rather in an inventory of his sufferings as a servant of Christ (2 Cor. 11:23–28). Do not confuse feeling anxiety with lacking all faith in Christ and his care! The military mission of God's peace, as it occupies its guard post to protect believers' hearts and minds, is not to numb us to life's pains or to blind us to its threats. It is to draw our troubled hearts to the truth that will strengthen us to stand: we are "in Christ Jesus" (Phil. 4:7), and not even the worst that life or death throws at us can sever us from his love (Rom. 8:34–39).

Now, when Paul speaks of "the peace of God," he has in view not only a mellow state of mind for individuals (as I experienced so wonderfully that afternoon). He also refers to two other dimensions of peace, one deeper and one wider. The *deeper* peace is God's reconciling mercy that ended the hostility between us rebels and himself. About this deeper peace Paul writes to the Roman Christians, "Therefore, since we have been justified by faith, we have peace with God through our Lord Jesus Christ" (Rom. 5:1). Because God has made peace with us through his Son, no danger of the present or threat of the future can separate us from his love (8:35–39). If you think you can achieve lasting peace of mind by mantras or meditation, without receiving peace with your Maker through humble trust in Christ and his cross, you are self-deceived. The peace that lasts through time and into eternity is found only in the peacemaking mission of Jesus the Son of God.

God's peace also extends *wider* than my personal peace of mind or yours. Christ's peacemaking mission to reconcile us to his Father has created a community of peacemakers and peacekeepers. When our hearts and minds

27. Over the next month, in response to the countless prayers of many people in many places God used an experimental antibiotic to eliminate the infection from our son's spinal fluid. Today he is a husband, a father, and a Bible translator.

are guarded by God's peace, our motives are ruled by his reconciling, unifying love as we relate to others. This is the oneness to which Paul just called Euodia and Syntyche. Paul instructed believers to bear with "one another and, if one has a complaint against another, [to] forgiv[e] each other; as the Lord has forgiven you, so you also must forgive. And above all these put on love, which binds everything together in perfect harmony. And *let the peace of Christ rule* in your hearts" (Col. 3:13–15). Patience, forgiveness, love, harmony are all about interpersonal relationships among believers. In this family context, "Let the peace of Christ rule in your hearts" means that Christ's peace, rather than personal self-interest, must set the agenda for our attitudes and interactions. The Philippians, too, need the power of God's peace to protect their hearts from the self-centered focus that has fed their worries and prolonged their friction, as do we.

Notice the "package" in whom God's protective peace is delivered: it comes to us "in Christ Jesus." Paul has kept Jesus in view all along. He is the Lord in whom we rejoice, the Lord who is near, our greatest reason for thanks, the apex of virtue, the theme of Paul's preaching and practice. Here Paul speaks his name. God's peace comes to us "in Christ Jesus," as we rest in his saving work, completed for us, and as we trust in his living person, now praying for us at God's right hand and present among us by his Spirit. Paul directs our attention toward Christ himself as the ultimate Peacemaker (see Eph. 2:14–19).

In Philippians 4:9, Paul inverts the wording of verse 7, speaking now not of "the peace of God" protecting us but of "the God of peace" present with us. The God of peace is with us now through the indwelling of his Holy Spirit. God's peace is not a prescription electronically transmitted from a doctor's office to a pharmacy, to be picked up and self-administered by the invalid. No, this divine Physician of our souls makes house calls! The peace of God guards our hearts because the God of peace comes near us by his Holy Spirit. The Immanuel promise, "God with us," which we celebrate at Christmas, did not apply only to the thirty-three years that Jesus walked this earth. It is still in force. Before his death, Jesus promised, "I will not leave you as orphans; I will come to you" (John 14:18). After his resurrection, he assured us, "I am with you always, to the end of the age" (Matt. 28:20).

Admittedly, the troubles in the world are easy to see, whereas God's presence by his Spirit is invisible. But as Jesus told Nicodemus, though we cannot see the wind or discover its origin, we hear its roar and see trees swaying by its power (John 3:8)! So the unseen Spirit of Christ shows his presence in our lives in many surprising ways, making us calm when we expect to be panicky, and moving us to serve others when we once lavished all care only on ourselves.

The Only Cure

Do you need antidotes to anxiety? There are lots of remedies on the market, I suppose. You could consult your physician, or check the self-help section at your bookstore. But only one cure was designed by the Manufacturer who knows how you are put together from the inside out, the One who knows why your heart is unsettled by the uncontrollable factors of life. To find the peace that you long for, to silence the worries that keep you awake at night, what you need is nothing less than God himself as your Friend and Father, your ever-present Protector.

You need to find your joy in the Lord, whether there are figs on the tree and grapes on the vine or not. Fix your hope on Jesus' coming, and you will find the strength to react to hostility with gentleness rather than retaliation. Set your heart's "anxiety alarm" so that when you start wallowing in worry, you know that it is time to rehearse God's good gifts—especially Jesus, God's great gift—and then bring the hassles that make you fret to your Father. Instead of exhausting your mental energy on the futile "what if?" treadmill, focus your thoughts on the true, honorable, just, pure, lovely, commendable, excellent, praiseworthy Son of God, who loved you and gave himself for you.

17

Cracking the Contentment Code

Philippians 4:10–13

I know how to be brought low, and I know how to abound. In any and every circumstance, I have learned the secret of facing plenty and hunger, abundance and need. I can do all things through him who strengthens me. (Phil. 4:12–13)

Financially tough times, such as economic recession, put people under intense pressures. Those pressures can reveal hairline cracks in our characters that would have been invisible in more comfortable conditions. As the visible props on which we often lean crumble, people are forced to face the truth that sources from which we seek satisfaction and security are surprisingly fragile.

Tight budgets in crunch times pose two temptations for Christians and non-Christians alike: *anxiety* and *discontent*. Anxiety frets over the question "Will I have enough to survive, to meet my family's basic needs?" To such fear Anxiety typically answers, "Maybe not," or even "Probably not." So Anxiety tends to close its fist around what little it has, afraid that hands that give too freely will end up empty. Discontent asks, "Do I have enough to make me happy, to satisfy my desires?" To that question Discontent *always* answers, "No, not enough, not yet." Therefore, Discontent, like Anxiety, keeps a tight grip on its treasures, and Discontent resents the fact that it doesn't have even more.

The followers of Jesus in first-century Philippi were "on the edge" financially. A half-dozen years before Paul wrote to the Philippians, he pointed out to the wealthier church in Corinth the "extreme poverty" and the "severe test of affliction" endured by believers in Philippi and other Macedonian cities (2 Cor. 8:2). For many people, terms such as "extreme poverty" and "severe test of affliction" hit close to home. Perhaps for you any downturn in the economy is not just a matter of trends and statistics reported in newspaper stories or analyzed in online blogs. Perhaps you have been downsized or you had to let employees go. Perhaps you have lost your home or been threatened with foreclosure. Perhaps your retirement reserves have been decimated. You are feeling the pressures of economic scarcity, just as the Philippians experienced twenty centuries ago.

What Paul found so refreshing about the Philippian Christians was that their ongoing economic crisis did not keep them from giving generously to help needy believers far away, in Judea. In fact, they had begged for the privilege of giving (2 Cor. 8:1–5)! Now, six or seven years later, they still refused to let financial pressures compress their hearts into a hard, heavy nucleus of self-protective miserliness. They had dispatched their trusted messenger Epaphroditus to deliver an ample donation toward Paul's daily living expenses in Roman custody (Acts 28:30). Among Paul's purposes for writing this warm letter of friendship was to express his gratitude for the Philippians' gift and for the affection that had prompted it. This is the final theme to which he turns before closing the epistle (Phil. 4:10–20).

Paul's "thank-you note," coming as it does in the final lines of the letter, has struck many students of Scripture as rather odd. Some have labeled this section an expression of "thankless thanks,"[1] since the apostle seems to express gratitude out of one side of his mouth while emphasizing, out of the other, that he was quite content without the gift. Some find the place-

1. John Reumann, *Philippians: A New Translation with Introduction and Commentary*, AYB (New Haven, CT: Yale University Press, 2008), 685n31, attributes this expression to a German scholar, C. Holsten (1876). See also M. R. Vincent, *A Critical and Exegetical Commentary on the Epistles to the Philippians and to Philemon*, International Critical Commentary (Edinburgh: T & T Clark, 1897), 146, who strongly refuted Holsten's interpretation.

ment, so late in the letter, a signal that Paul's expression of gratitude is more obligatory than sincere. Some speculate that Paul's tepid "thanks, but no thanks" response reflects discomfort with a donation that could undermine his authority by elevating the Philippian congregation as his patrons and demoting Paul to the status of their dependent. Some even suggest that these appreciative sentences originally belonged to a completely different document (though there is no evidence that any copy of Philippians ever existed without this passage at this place, or that this passage ever circulated apart from the epistle as a whole).[2]

In our next study, we will explore Paul's rationale for placing his thanks for the Philippians' gift at the end of his letter. For our current consideration of Philippians 4:10–13, we need to grasp why Paul starts by expressing joy for the Philippians' renewed thoughtfulness—demonstrated in their gift—and then seems, so abruptly, to turn ungrateful and ungracious by insisting that he was quite content without it. Happily, the explanation is not far to seek. It is not to be found in alleged interpersonal tensions between Paul and the Philippians, nor in a supposed preoccupation by Paul with keeping his superior status in the church's social "pecking order."

Quite simply, Paul shifts from expressing his joy over the Philippians' donation (Phil. 4:10) to affirming his contentment (4:11–13) for two reasons. First, he removes any regret that his friends may feel over the significant time delay between their initial resolve to help their beloved apostle ("you were . . . concerned") and the actual arrival of their donation ("you had no opportunity"). They do not need to wonder whether Paul has been moping in self-pity, doubting their love and resenting their delay. Second, and more importantly, Paul—ever the pastor and spiritual parent—is again using his own situation and his response to it, as he has done throughout the epistle, as a "case study" in the Christ-centered response to adversity. This time the adversity in view is financial exigency and all that accompanies it—being "in need," being "brought low," and experiencing "hunger." Paul will show his beloved spiritual children how to weather the careening

2. Gordon D. Fee, *Paul's Letter to the Philippians*, NICNT (Grand Rapids: Eerdmans, 1995), 424–25, cogently and vividly refutes the groundless proposal that 4:10–20 was originally part of a different letter: "The linguistic ties between this passage and much that has preceded are so significant that any dismembering of this letter destroys the very 'magic' that makes it work so well as a letter of friendship. . . . One must dance with unlimited sidesteps to remove this passage from the rest of the letter."

roller-coaster trajectory of life—lack and lowliness to abundance and plenty and back again—with the inner equilibrium that comes from the God "who strengthens me."

Clues in the context of our passage suggest that the Philippian Christians need Paul's example on the issue of their mind-set toward money. Though they have displayed openhanded generosity from the depths of their poverty, both for believers in Judea and for Paul, they also feel the pressure of their tight finances. If they were not tempted to *worry*, Paul would not have wasted papyrus to prescribe prayer as an antidote to their anxiety: "Do not be anxious about anything, but in everything by prayer and supplication with thanksgiving let your requests be made known to God. And the peace of God, which surpasses all understanding, will guard your hearts and your minds in Christ Jesus" (Phil. 4:6–7). Moreover, Paul's insertion of "thanksgiving" alongside their requests implies that they need an antidote not only to *Anxiety* but also to *Discontent* concerning God's provision. We might assume that such generous givers could not be troubled by a lack of contentment. But Paul does not waste words telling people to avoid traps[3] that would never tempt them. If his Philippian friends need Paul's call to thanksgiving, it must be because they are tempted to thanklessness, a telltale symptom of discontent. This finds confirmation in Paul's earlier instruction, "Do all things without grumbling or questioning" (2:14). The pressures they face are creating fissures in their relationships with each other ("grumbling") and in their relationship with God ("questioning" his providence).

Before we investigate Paul's "secret" of staying content in any and every circumstance, we should recognize that it is not only those facing tight budgets who are tempted to be discontented with what they have. Even when we live well above the poverty line, surrounded by far more than the bare necessities, it is easy to feel that we don't have enough stuff, or *good* enough stuff, or *new* enough stuff, or *fast* enough stuff. Affluent people, who never wonder where the next meal is coming from, can be very discontented with what they have. Advertisers know how to tease our hungry hearts, how to make us feel that our lives will be unfulfilled unless we have the latest thing in information technology, a more luxurious car, a more spacious, comfortable

3. In 1 Timothy 6:9 Paul uses the vivid imagery of a snare to capture unwary wildlife to describe the seductive danger of the love of money.

home, or whatever else may be on our wish lists. We are told that a society's economic health actually depends on cultivating material discontent, and economists express alarm when consumers prefer saving over spending. If everybody felt as Paul feels in his chains, content with so little material wealth, the growth of global economies might stall. So it is not only those on the edge of destitution who need to learn the secret "contentment code" that Paul cracks open here. This secret is for everyone who wrestles with the question "How much is enough?" It can even cure the toughest case of discontent: the avarice of the affluent!

Joy in Your Concern, Not Your Cash

Paul is deeply and genuinely grateful for the funds that Epaphroditus brought from Philippi. He rejoices not only because the gift reveals the Philippians' affection for Paul but also because it displays their devotion to Christ. Before their gift arrived, times had been tough for Paul in terms of cash flow, as he acknowledges frankly in mentioning "my trouble" in verse 14. It had been quite some time since the Philippian church—one of the few that consistently partnered with Paul financially—had managed to send a donation for his rent and other expenses. So Paul is eager to convey his gratitude, but this thank-you note has to be worded delicately, to avoid leaving wrong impressions in the Philippians' minds.

Paul opens the subject of their financial contribution by stressing that he "rejoiced . . . greatly" when their gift arrived (Phil. 4:10). This is the last appearance of the much-used verb *rejoice* in this epistle, and Paul boosts its intensity with the adverb *greatly*, as well as the crucial phrase "in the Lord." He has just called his friends to "rejoice in the Lord," adding that—because their joy is grounded in the Lord and not their circumstances—they must rejoice "always" (4:4; see 3:1). Now he again sets the pace, affirming that he, too, has "rejoiced in the Lord." Because the Lord—not their gift—is the deep source of his joy, he has rejoiced "greatly." Moreover, he has rejoiced and can rejoice in any and all circumstances, not only in his current plenty but also in his prior need.

Lest his friends mistake his meaning in describing his great joy that their gift arrived "at length," Paul immediately adds two clarifying footnotes.

First, Paul hastens to say that he knows that his friends had him in mind[4] all along, even though he has described their donation as a sign that their concern for Paul has "revived," like buds and blossoms reappearing on bare trees after a long winter of dormancy.[5] Paul has known all along that their care for him had not really gone into hibernation. To his friends at Philippi, he was never "out of sight, out of mind." It was just that they had not found a way to get the funds to Paul sooner. So he doesn't want them to imagine him grumbling in his imprisonment, "Why have the Philippians, so generous in the past, now forgotten me?"

Paul's second footnote is in verse 11, where he stresses that, though he did in fact rejoice over their gift, it was *not because he was in need*. He amplifies this point in verse 17, where he insists that what he is seeking is not their gift (for his personal benefit) but rather the fruit that will accrue and be "credited" to their account by God himself. Their gift will, in the end, do them more good than it is doing Paul. This is what strikes some readers as a grudging, ungrateful way to say "thank you": "Thanks so much for your gift, but I didn't really need it."

Perhaps this reminds you of the way one of your relatives—a great-aunt or a grandmother, perhaps—reacts when you have found the perfect gift for her. First she says, "Oh, you really shouldn't have!" Then she may go on about why it was too expensive, or why she will never use it. You begin to wonder why you bothered in the first place. So some New Testament scholars, trying to find friction in Paul's mellow relationship with the Philippians, think they hear an edge to Paul's voice, a subtle signal of rebuke in Paul's words, "Not that I am speaking of being in need." That suspicious way of reading these words would mean that when Paul says, "I rejoiced . . . greatly," he is faking it or being ironic. Yet hypocrisy doesn't fit Paul's character, nor the genuine contentment that he goes on to express. And sarcasm doesn't fit Paul's friendship with the folks in

4. The Greek verb rendered "concern for me" twice in Philippians 4:10 is again *phroneō*, which Paul uses so frequently in this epistle. Sometimes *phroneō* emphasizes the cognitive, but here its sense fits more in the emotive/affectionate "neighborhood" of the verb's wide semantic range, as in 1:7: "It is right for me to *feel* [*phroneō*] this way about you all, because I hold you in my heart."

5. Greek *anathallō*. See John Calvin, *The Epistles of Paul the Apostle to the Galatians, Ephesians, Philippians and Colossians*, ed. David W. Torrance and Thomas F. Torrance, trans. T. H. L. Parker (Grand Rapids: Eerdmans, 1965), 291–92; G. Walter Hansen, *The Letter to the Philippians*, PNTC (Grand Rapids: Eerdmans, 2009), 306; *BAGD*, 54, offers the gloss "*cause to grow* or *bloom again*."

Philippi. Rather, Paul wants his Philippian friends to know that his joy over their gift is not simply because their friendship has proved profitable to him. The gift itself he can take or leave; what he treasures is the affection behind the gift.

Paul also emphasizes that he was content before their gift arrived in order to emphasize his deep satisfaction in God's providence, whether it entails abundance or abasement. Paul does not want his friends to imagine that in his "trouble" he has been sulking over *their* neglect, or resenting *God's* plan and provision. To echo his exhortation in Philippians 2:14, Paul has not been "grumbling" about his fellow believers, nor has he been "questioning" God's care and good purposes.

Loving parent that he is, Paul sees the Philippians' generosity as presenting to him another "teaching moment." He can give them a gift of wisdom that will fortify them for the poverty and hardship they are facing. Paul takes the occasion of the Philippians' gift to share with them the secret that has kept him content and joyful both in poverty and in affluence. So again he opens a window in his heart, inviting us to look inside and watch how he applies the gospel to the ups and downs of fluctuating finances.

To learn the lesson that Paul has to teach us, we need to understand first what true contentment means, and then the process by which such contentment is acquired.

Content with Where I Am and What I Have, but Not with Who I Am—Yet!

So far our discussion has focused on money, whether lots or little, because the occasion of Paul's words is his gratitude for a financial gift. But in verse 12 Paul starts his list of the extremes of experience in which he has found contentment with a contrast of opposites that are not limited to cash: "I know how to be brought low, and I know how to abound." Being "brought low" is not merely living below the poverty line. It has more to do with status than with income.[6] This was clear when Paul used the same word in chapter 2 to describe Christ's self-humbling when, instead of exploiting his equality with the Father, he became a

6. Greek *tapeinoō*. Hansen, *Philippians*, 311, explains: "'To be humbled' means to lose prestige or status, to be humiliated, to be subject to strict discipline."

human being, took a servant's status, "made himself low,"[7] and submitted to death on a cross (Phil. 2:6–8).

Paul has a purpose in beginning where he does as he paints the spectrum of experience in which he finds contentment. Because Christ has so humbled himself to redeem Paul's life and transform Paul's heart, the apostle is content not only to forgo funds but also to forfeit honor and to occupy a low place where others fail to notice or respect him. He practices the counsel that he has given his Philippian friends, humbly counting others more significant than himself.

Sometimes being content in a situation in which nobody appreciates you is as hard as being content on a tight budget. Don't you know people who have more than they need materially, but still work and overwork because what drives them is not the size of their salary but the acclaim of the industry? Or people who want an office in the church rather than quiet opportunities to serve, so that others will recognize their abilities? Is that driven honor-seeker you? Or can you sincerely sing this very countercultural hymn of Christian humility:

> I would not have the restless will
> That hurries to and fro,
> Seeking for some great thing to do,
> Or secret thing to know;
> I would be treated as a child,
> And guided where I go.
>
> I ask thee for the daily strength,
> To none that ask denied,
> A mind to blend with outward life,
> While keeping at thy side,
> Content to fill a little space,
> If thou be glorified.[8]

If you find in yourself a "restless will that hurries to and fro, seeking for some great thing to do or secret thing to know," Paul would show you a higher

7. Greek *tapeinoō*, which the ESV renders "humbled himself" in Philippians 2:8.

8. Anna L. Waring, "Father, I Know That All My Life" (1850), in *Trinity Hymnal* (Suwanee, GA: Great Commission Publications, 1990), no. 559.

path, a path of lowliness in which you will be "content to fill a little space," as long as God is glorified—that "now as always Christ will be honored in my body, whether by life or by death" (Phil. 1:20).

With the next pairs in verse 12—"facing plenty and hunger, abundance and need"—Paul turns the spotlight directly on material resources. He is content with what he has. He reduces things to the bare essentials, to the question whether there is food on the table. The ESV's "facing plenty" represents a Greek verb that conveys eating one's fill, consuming food to the point of complete satisfaction.[9] In effect, Paul says, "I can be stuffed or starved; either way, I am content." When I hear Paul say that he can respond to near-starvation without resenting God's providence or envying the well fed, I am embarrassed by how little it takes to unsettle my contentment: a car that needs repairs, a computer that boots up slowly, a tough steak. How about you?

We might be surprised to hear Paul say that his contentment extends to times when he has plenty to eat and is on top of the world. "Well, of course," we think, "why should he not be content when he has all he needs?" But a moment's reflection brings to mind people who have far more than they need but who still cannot bring themselves to say, "Enough!" Money and what it can buy can be as addictive as narcotics: the more you have, the more you feel you need in order to get the same "high."[10] For many of us, the challenge is not to be content when we have nothing. After all, we have never had *nothing*. The challenge is to be content when we have *more* than we need but *less* than we want. John Calvin wisely observed:

> He who knows how to use abundance soberly and temperately with thanksgiving, prepared to part with everything whenever it may please the Lord, giving also a share to his brother according to his ability, and is also not puffed up, that man has learned to excel and to abound. This is an excellent and rare virtue, and much greater than the endurance of poverty.[11]

9. *BAGD*, 883, suggests as glosses of *chortazō*: "*feed, fill, satisfy*; pass.: *eat one's fill, be satisfied*." For other uses of *chortazō*, see Luke 9:17; 15:16; Rev. 19:21.

10. Hansen, *Philippians*, 312: "the need to learn contentment in the midst of wealth may not be immediately apparent. Biblical wisdom, however, teaches that 'the surfeit of the rich will not let them sleep' (Eccles. 5:12) . . . . Wealthy people tend to 'set their hopes on the uncertainty of riches' rather than on God (1 Tim. 6:10, 17)."

11. Calvin, *Galatians, Ephesians, Philippians, and Colossians*, 292.

The secret of Christian contentment not only enables us to receive small gifts from our Father's hand with gratitude. It also empowers us to receive great wealth from our Father's hand without having our hearts stolen away from the Giver by his gifts. Especially in times of material abundance, you must coach yourself and cultivate in yourself the wise perspective on reality that God's Word provides. When tempted to rest your sense of security in your investments, hear Jesus say to you, "One's life does not consist in the abundance of his possessions" (Luke 12:15). Heed God's wise counsel: "Do not toil to acquire wealth . . . . When your eyes light on it, it is gone, for suddenly it sprouts wings, flying like an eagle toward heaven" (Prov. 23:4–5). You must "keep your life free from love of money, and be content with what you have," because God has promised what funds can never offer: "'I will never leave you nor forsake you.' So we can confidently say, 'The Lord is my helper; I will not fear; what can man do to me?'" (Heb. 13:5–6). This ever-living, never-leaving Lord is "Jesus Christ . . . the same yesterday and today and forever" (13:7). The secret of contentment that Paul has learned and wants to teach is to be "at home" with our place, whether low or high, and "at home" with our goods, whether few or many.

To get a complete and balanced understanding of what Paul means by *contentment*, we must recall another perspective that Paul presented earlier in the epistle. Without that perspective, we will confuse Christian contentment with a mellow, laid-back, "whatever" approach to life. As we have seen, Paul is *anything but* complacent or unmotivated! Paul can take or leave honor or dishonor, full meals or famines. His joy and contentment are not dependent on the living conditions that his loving Father has ordained for him. But Paul cannot, will not, be complacent in his pursuit of the great prize on which his heart is set. He is not content—yet—with *who he is*. He has not yet reached the goal and received the full prize for which he runs with all his might: "that I may know [Christ] and the power of his resurrection, and may share his sufferings, becoming like him in his death, that by any means possible I may attain the resurrection from the dead" (Phil. 3:10–11). Paul knows his Savior and rests in Jesus' righteousness, but these tastes of redeeming grace have only whetted Paul's appetite for the full feast awaiting him at the Lord's return. Paul presses on for "the prize of the upward call of God in Christ Jesus" and invites all who would be mature to join his lifelong quest (3:12–15).

God's Word infects us with a holy discontentment. There are things in life worth longing for and striving for with all our might. Food and clothes and technology are not those things. The treasure that is not yet fully in our grasp, that should still stir our discontentment, is the fullness of God's grace in Jesus. If you trust him, your sins are forgiven and his righteous record is yours. His Spirit has begun a good work in you (Phil. 1:6), renewing you from the inside out. But the best is yet to come! Be content with where you are and what you have, but discontent with who you are—so far!

NOT SELF-SUFFICIENCY, BUT CHRIST-SUFFICIENCY

Another dimension of what Paul means by *contentment* is expressed in verse 13: "I can do all things through him who strengthens me." I remember that when I was a youngster, I saw this verse on wall plaques and daydreamed of great exploits, if only I could build up a muscular faith: "Wow! Jesus will make me Superman, faster than a speeding bullet, more powerful than a locomotive, able to leap tall buildings in a single bound!" Now that I have studied this declaration in its context, I realize that it is not God's blank check, signed and waiting for me to fill in the amount of strength I want, to achieve impossible deeds. The "all things" that Paul can handle in verse 13 are the range of situations that he has just described in verse 12. Where we read "in any and every circumstance," the Greek says "in each thing and in all things"[12]—in the whole range of situations, from being stuffed to being starved, from riding high to crawling low.

Although Paul's "I can do all things" declaration is not God's carte blanche for my youthful delusions of grandeur, the apostle's words do cast a very distinctive light on the meaning of contentment for those who trust Christ. The fact is that Christ's apostle has borrowed the word *content*[13] from the ancient Stoic philosophers, and then twisted it inside out. The word's origin and its contemporary usage gave it the meaning "self-sufficient." It conveyed the ideal of self-contained independence

12. In verse 12 the ESV's "in any and every circumstance" is *en panti kai en pasin*. In verse 13 "all things" again represents *panta*.

13. Greek *autarkēs*, which appears only here in the New Testament. Paul uses the related noun in 1 Timothy 6:6–8: "There is great gain in godliness with contentment [*autarkeia*] . . . . If we have food and clothing, with these we will be content [*arkesthēsometha* "be satisfied" = consider to be enough]." See also Hebrews 13:5.

that Stoicism advocated. The Stoics claimed that the wise person realizes that every experience, whether pleasurable or painful, is part of an interconnected matrix permeated by Reason. Thus it is pointless to resent illness or injustice. The key to contentment, said the Stoics, was to become emotionally self-sufficient by insulating oneself from the variables of pain and pleasure. One scholar sums up the Stoic conception of contentment this way: "By the exercise of reason over emotions, the Stoic learns to be content. For the Stoic, emotional detachment is essential in order to be content."[14] Whereas the Stoics believed that intellectual aloofness could provide protection from emotional distress, Paul refuses to insulate his heart from sorrow by keeping people and their hurts at arm's length. He rejoices with people and weeps over them.

Moreover, here he twists the Stoics' favorite term, *self-sufficiency*, inside out. His capacity to handle life's ebbs and flows is not self-generated. It comes from outside Paul, from "him who strengthens me." Paul's contentment is found not in self-sufficiency, but in Christ's sufficiency. Paul has learned the secret of real contentment, which was portrayed beautifully by the prophet Jeremiah:

> Blessed is the man who trusts in the LORD,
> whose trust is the LORD.
> He is like a tree planted by water,
> that sends out its roots by the stream,
> and does not fear when heat comes,
> for its leaves remain green,
> and is not anxious in the year of drought,
> for it does not cease to bear fruit. (Jer. 17:7–8)

How can a tree keep its green leaves in the summer heat and bear fruit in years without rainfall? Not because the tree itself contains an internal spring of water, but because it is planted by a flowing stream. If you are trusting in Jesus Christ, you are this stream-irrigated tree, just as Paul was.

The older Greek manuscripts and more recent English versions based on them, such as the ESV, do not name "Christ" as the One "who gives me strength." Later manuscripts and the translations based on them, such as the

14. Hansen, *Philippians*, 310. See also Fee, *Philippians*, 431–32.

KJV, include the name.[15] The authenticity of the shorter reading, "him who strengthens me," is supported by the age of the copies that contain it, which makes them closer to the original manuscript. Also, it is more plausible to envision why copyists might insert the name *Christ*, making explicit what Paul left implicit, than it is to suppose that they would delete the name if they had found it in earlier manuscripts. In any case, the spiritual instinct of the copyists who inserted (or retained) the name *Christ* was exactly right: Christ himself is Paul's source of strength. Elsewhere the apostle speaks of "him who has given me strength, Christ Jesus our Lord" (1 Tim. 1:12; see Col. 1:28–29). Paul is far from self-sufficient, but the all-sufficient Christ is Paul's source of strength. Paul has just promised that "the Lord is at hand" and "the God of peace will be with you" (Phil. 4:5, 9). Because Paul knows the nearness of Christ even in his captivity, Paul can receive whatever God's providence brings his way, whether painful or pleasant, with a deep joy "in the Lord." How can we learn the secret and skill of Christian contentment that sustained Paul?

CONTENTMENT IS A LEARNED SKILL AND A SHARED SECRET

In verses 11 and 12 Paul uses four verbs to communicate how he has acquired the contentment that enables him to rejoice in the Lord in plenty or in want. He writes:

> "I have learned . . .
> I know . . .
> I know . . .
> I have learned the secret."

The double occurrence of "I know"[16] shows the result of the learning process indicated in the first and fourth verbs—"I have learned" and "I have learned the secret." Paul can say, "*I know* how to be brought low" and again "*I know* how to abound" because he has gone through a learning process and been

15. "Christ" is absent from such fourth-, fifth-, and sixth-century Greek manuscripts as Sinaiticus (original copyist), Alexandrinus, Vaticanus, and Claramontanus (original copyist), as well as later sources. It was apparently inserted by later correctors of Sinaiticus and Claramontanus and subsequently appears in many manuscripts belonging to the Byzantine textual tradition.

16. Greek *oida*.

initiated into a secret that gives him a Christ-centered perspective on his fluctuating situation.

The fact that Paul has "learned"[17] contentment shows that his calm response to life's ups and downs is a skill honed through practice. The author to the Hebrews uses the same term, writing that Christ himself, "although he was a son, *learned* obedience through what he suffered" (Heb. 5:8). The eternal Son of God entered the world ready to fulfill the Father's will (10:5–10), but his holy resolve was tested and proved through his obedient suffering. In this sense he "learned" in practice what obedience entailed, and what it cost. Christ-centered contentment is not preinstalled on our hearts, like a software program preloaded into a new computer. Nor is Christian contentment injected in a single dose, as though it were a vaccine that could make us immune to a complaining spirit. *It takes practice.* Contentment grows over time, as we face adverse situations—in finances, health, relationships, or other areas—and seek Christ's strength to release our grip on his gifts, while we strengthen our grasp on his grace.

Yet cultivating Christian contentment is not merely a matter of following an exercise regimen to reprogram our attitudes. Contentment is a secret that has been shared with Paul *by Another.* Our version's "I have learned the secret" represents a single Greek word, which could also be translated "I have been initiated."[18] This is the only place in the whole New Testament that this word appears. In Paul's day it was associated with the bizarre initiation rituals of the pagan "mystery religions." (In fact, the verb is related to the Greek noun *mystērion*, from which we get *mystery* in English.)

Mystery religions, such as those devoted to Dionysus and Mithra, hid their best secrets behind closed doors, keeping them away from the curious eyes of uncommitted observers. Only the "insiders," who had undergone initiation, got in on the deepest mysteries. On the other hand, Paul and the other heralds of Christ were broadcasting God's good news, God's mystery, right out in the open, for anyone to hear:

> By the open statement of the truth we would commend ourselves to everyone's conscience in the sight of God. And even if our gospel is veiled, it is

17. Greek *emathon*, second aorist active of *manthanō*.

18. Greek *memuēmai*, perfect passive of *mueō*, for which *BAGD*, 529, proposes the gloss "*initiate (into the mysteries)*." Thus Paul's use of the passive voice should be understood as "I have been initiated."

> veiled only to those who are perishing. In their case the god of this world has blinded the minds of the unbelievers, to keep them from seeing the light of the gospel of the glory of Christ, who is the image of God. For what we proclaim is not ourselves, but Jesus Christ as Lord, with ourselves as your servants for Jesus' sake. For God, who said, "Let light shine out of darkness," has shone in our hearts to give the light of the knowledge of the glory of God in the face of Jesus Christ. (2 Cor. 4:2–6)

The mystery of the gospel is an "open secret" concerning public events: Jesus, the Son of God, became man, lived a perfectly obedient life, then died a criminal's death under God's wrath (not for his own sins but for others' offenses), rose from the dead, ascended to heaven, rules now, and will return in glory. There it is, God's most wonderful secret, right out there on the open market . . . no passwords, no secret handshakes, no going down into a pit to be showered with the warm blood of a freshly slaughtered bull as in Mithraism.

Nevertheless, here Paul picks up a piece of mystery-religion jargon, cleans it up, and uses it to get our attention. He implies that *there is a secret* to contentment, a code to be cracked that will enable you to weather the best of times and the worst of times. Contentment in Christ is a kind of "insider knowledge." Yet the great difference between Jesus and Mithra is that in Christianity the boundary between "outsiders" and "insiders" can be crossed simply by *believing the very public gospel* that Paul preached—by entrusting your life to the crucified and risen God-man, Jesus the Messiah. Christ himself is the secret to contentment—not a mystical Christ hidden behind secret rituals or visionary experiences, but the historical Jesus who lived and died and rose again, who is now proclaimed openly among the nations. The better we get to know Christ, the more we discover that he is the One who satisfies our hearts.

Food and shelter are necessities for existence on this earth. Extra food and comfortable shelter, as well as cars and computers and the other extras that many of us enjoy—these are not necessary, but they are nice. Yet none of this can quench your heart's thirst because, at the core of who you are, you were made for friendship with the living God. When you're tempted to think that there is something else, anything else, that you "just have to have" to make life worth living, that is the time to remind yourself of the secret.

By faith in the gospel of God's Son, you have been initiated as an "insider." You are in on the secret. You have Christ at the center of your life, and in the end he is all that you need!

Contentment Entails Exerting Strength

Does all of this sound like too facile and cheap a solution to the real-world shortages and crises that keep you awake at night? After the horror of World War II, Americans—whose cities and countryside, unlike those of Europe and Asia, bore few scars from the devastation—were in the mood for an optimistic outlook and economic resurgence. The composer-lyricist team of Richard Rodgers and Oscar Hammerstein II caught the wave in a series of Broadway musicals that prescribed "happy talk" to make our dreams come true (*South Pacific*, 1949), "whistling a happy tune" in the face of our fears (*The King and I*, 1951), mustering "confidence in confidence alone" to confront daunting challenges, and remembering our "favorite things" when the dog bites and the bee stings (*The Sound of Music*, 1959). At the same time, Norman Vincent Peale's *The Power of Positive Thinking* appeared (1952). The cheeriness of Hammerstein's hymns to hope-for-hope's-sake and Peale's mind-over-matter advice may sound naive today, but the self-counsel that they offered to that postwar generation still resonates with many people. Why else would hurting and desperate people around the world grasp at preachers' promises that unwavering faith and generous donations will bring prosperity, healing, and overall well-being, not in some distant future but here and now?

Is Paul simply offering an ancient form of "happy talk"? No, Paul is a sober realist, and his closing word on contentment, before he resumes his thanksgiving for the Philippians' gift, shows that the contentment he commends requires that we flex the mental and spiritual muscle that Christ has given us by his indwelling Spirit: "I *can do* all things through him who *strengthens* me" (Phil. 4:13).

Christ is the source of Paul's strength and ours, but we must not ignore Paul's "I can do"—or, as Paul's Greek says, "I have power" or "prevail over" every circumstance.[19] Paul uses a term that has "strength" built into it in

19. Greek *ischyō*. *BAGD*, 383, offers glosses such as "*be able*," but also "*have power, be mighty*," and "*prevail*." Paul uses the verb only once elsewhere, in Galatians 5:6, where it means "*be valid*."

order to remind us that Christian contentment is not a sedative. Christian contentment is something that we fight for. We must exert effort to wage war against the temptation to complain, to envy others, to fixate on what is uncomfortable and inconvenient and downright wrong in our circumstances. We strive to focus instead on the faithfulness and mercy and strength of our God. Paul flexes his mental muscle to remind himself often that in Christ he already has the supreme treasure, and that he is racing toward a goal that will mean an even greater experience of his Savior's grace and glory. And again, Paul is not striving in his own strength or racing in his own energy. The key to his patience in the present and his hope for the future is the presence of the Christ who gives him strength.

CONCLUSION

How can you follow Paul's lead, winning the war over both Anxiety and Discontent, even in financially tight times? Focus your mind on the truth that, if you are trusting in Jesus, the living God is with you and at work in you, through the unseen but very real and very powerful presence of the Holy Spirit (Phil. 1:19).

The more you direct your heart toward Christ's presence and power, the less you will waste your mental and emotional energy on the stuff that doesn't last. You will be able to keep a light grip on what you do have, and you won't fret over what you don't have. God will keep the amazing promise that Paul issued just before our text, "The God of peace will be with you" (Phil. 4:9). And you will want to invest the resources that he does entrust to you in ways that enable others to see in you glimpses of the generosity of Jesus himself and the contentment that he imparts to those who trust him wherever he leads, whether through plenty or through poverty.

18

The Puzzling Partnership of Getting by Giving

Philippians 4:14—23

I have received full payment, and more. I am well supplied, having received from Epaphroditus the gifts you sent, a fragrant offering, a sacrifice acceptable and pleasing to God. And my God will supply every need of yours according to his riches in glory in Christ Jesus. To our God and Father be glory forever and ever. Amen. (Phil. 4:18–20)

Congregations that have a robust budget for world missions have a big heart for the gospel. After all, Jesus told us, "Where your treasure is, there your heart will be also" (Matt. 6:21), so it is not surprising that Christians who *give generously* to spread the gospel among all nations *care passionately* about the global expansion of Christ's kingdom. What do the percentages and proportions of expenditures in your congregation's annual budget reveal about the location of your hearts? Does your church have a heart that embraces the world? Do you, as a follower of Jesus and recipient of his grace, have such a heart? The first-century church at Philippi certainly did! What was it that opened their hearts wide, so that they gladly opened their wallets wide as well?

As Paul recalls his first encounter with the folks at Philippi over a decade ago, he easily identifies the factor that ignited their zeal for missions, which in turn spurred their generous giving for missions. He describes those early days as "the beginning of the gospel" (Phil. 4:15), resuming his recollection at the beginning of the epistle, when he expressed his thanks to God "because of your partnership in the gospel from the first day until now" (1:5). God's good news has broken into their lives, and they have never been the same. Through the message of Christ's self-humbling, sacrificial death, and resurrection to endless life and lordship, these Gentiles emerged from their pagan pasts and entered the citizenship of heaven itself. They became Paul's partners in the gospel, first, because they and he shared together in the gift of grace that God bestows through faith in the good news of Christ. Second, they were eager to partner with Paul financially to see this good news reach other people who needed to hear the "word of life" (2:16). The Philippians gave so gladly for missions because they had received grace so freely through the gospel.

Despite the Philippians' "extreme poverty" (as Paul characterized it in 2 Cor. 8:2), their financial support for Paul's gospel mission was outstanding for its consistency and its generosity. Paul recalls that when he left Macedonia, heading south to Achaian cities such as Athens and Corinth, the Philippian Christians promptly sent a gift as his partners "in giving and receiving" (Phil. 4:15).

Whenever Paul moved into a new city or region, he worked hard to distance himself from those who purveyed philosophical or religious insights for a price. As early as the fifth century before Christ, Greek Sophists professionalized the wisdom industry by charging sizable fees to tutor young aristocrats. The practice of marketing wisdom for payment met suspicion in some Athenian circles and was critiqued by Socrates.[1] Yet still in Paul's day the Greek-speaking intelligentsia apparently assumed that the value of a teacher's content could be gauged by the price he accepted

1. W. K. C. Guthrie, *A History of Greek Philosophy*, vol. 3, *The Fifth-Century Enlightenment* (Cambridge: Cambridge University Press, 1969), 35–40.

from students, whether charged as fees or accepted as "gifts."[2] Therefore, the apostle had to defend his practice of refusing to charge a rate commensurate with the priceless worth of his message: "Or did I commit a sin in humbling myself . . . because I preached God's gospel to you free of charge?" (2 Cor. 11:7).[3] Not only his message but also his method differed sharply from his competitors in the Greek-speaking marketplace of ideas. Paul was not just another entrepreneur in "the business" for the money to be made: "we are not, like so many, peddlers of God's word" (2:17). Because Paul's message announced God's free gift, his method fit that message: "What then is my reward? That in my preaching I may present the gospel free of charge" (1 Cor. 9:18). When Paul reached a new town, therefore, his policy was to support himself and his team by making tents, or to cover his expenses from contributions from congregations that were already established. We see both of his support strategies at work in Acts 18, which records the planting of the church in Corinth. Upon Paul's arrival in that prosperous and profligate Achaian metropolis, Paul first joined the tent-making business of Aquila and Priscilla, spending his Sabbaths in the synagogue teaching about Jesus the Messiah (Acts 18:1–4). But when "Silas and Timothy arrived from Macedonia, Paul was occupied with the word" (18:5). The arrival of these colleagues enabled Paul to proclaim the gospel not only on the Sabbath but on other days of the week as well. Paul and the Corinthians knew what had freed Paul up from long days at the workbench: "When I was with you and was in need, I did not burden anyone, for the brothers who came from Macedonia supplied my need" (2 Cor. 11:9). Silas and Timothy, along with others, had delivered the gift from Philippi that Paul mentions in our text.

Actually, the Philippians had expressed their generosity even earlier than the donation that Silas and Timothy had brought to Corinth. Paul says in our text, "Even in Thessalonica you sent me help for my needs once and again" (Phil. 4:16). Thessalonica was the next major Macedonian city west of

2. Peter Marshall, *Enmity in Corinth: Social Conventions in Paul's Relations with the Corinthians*, Wissenschaftliche Untersuchen zum Neuen Testament 2, Reihe 23 (Tübingen: J. C. B. Mohr / Paul Siebeck, 1987), 226–30.

3. Ralph P. Martin, *2 Corinthians*, WBC 40 (Nashville: Thomas Nelson, 1986), 344, proposes that the critique of Paul may be attributable to the Sophists' legacy of academic professionalism and the Greek contempt for manual labor (since Paul often worked with his hands to provide his team's needs, Acts 20:34).

Philippi, and in Thessalonica Paul's ministry lasted most of a month ("three Sabbath days," with weekdays on either side of those Sabbaths, presumably) (Acts 17:1–9). The ESV's wording implies that the Philippians sent Paul at least two contributions during those few weeks that he stayed in Thessalonica. The distance between Philippi and Thessalonica on the Via Egnatia was about 160 kilometers (100 miles). Although it is possible that couriers from Philippi made that trip at least twice in the three or four weeks in question, Paul's Greek could also be interpreted this way: "*even* in Thessalonica, *and* [then] once and twice you sent help for my needs." Paul's point would then be first to emphasize how quickly the Philippians began sending contributions ("even in Thessalonica") and how consistently they continued to do so ("once and twice"—we might say "time and again") as he preached elsewhere.[4] Acts 16 shows that as soon as the Lord opened Lydia's heart to believe the gospel, she immediately opened her home to extend hospitality to Paul's missionary team. When Paul and company left Philippi, the whole church showed the same grateful generosity in response to God's grace, sending at least one gift, and perhaps more, with a courier who caught up to the missionaries even before they left Macedonia.

This is a shared history of which Paul and the Philippians were aware; he says, "You Philippians yourselves know" (Phil. 4:15). He takes space to rehearse it, though, for a reason. There has been a hiatus in their donations for Paul's mission, and Paul wants them to be reassured that he has not at all construed their "financial silence" as a sign of their indifference, either toward himself or toward the gospel. Paul has not forgotten their "track record" from their very infancy in Christ, so they can rest assured that he never doubted their commitment and their readiness to express it in costly ways.

At last now they have found a concrete way to convey their love for Jesus and his missionary. Epaphroditus, their "messenger and minister to [Paul's] need" (Phil. 2:25), has arrived with their donation to help Paul meet his financial obligations. Paul, in turn, is writing not only to update them about his situation and offer advice about theirs, but also to express his appreciation

4. Peter T. O'Brien, *The Epistle to the Philippians*, NIGTC (Grand Rapids: Eerdmans, 1991), 535, calls attention to the presence of the Greek conjunction "and" (*kai*, untranslated in ESV) preceding "once" and persuasively argues that Paul's meaning is: "For *both* when I was in Thessalonica *and* more than once in other places you sent me gifts to meet my needs" (emphasis added, and both emphasized words represent the Greek conjunction *kai*).

for their gift and, even more so, for the heart behind their gift. This letter is, among other things, *a missionary's thank-you note.*

Yet Paul's acknowledgment of the gift at the very end of the letter (Phil. 4:10–20), just before the closing greetings (4:21–23), has struck many readers as a strange expression of gratitude. Since the 1870s, in fact, some critical scholars have referred to this passage as "thankless thanks," suggesting that Paul was ungracious and ungrateful in the timing and the wording of his acknowledgment. As we saw in the previous chapter, some think they hear in Paul's declaration that he is always content, whether in want or in abundance, an implied rebuke or reassertion of his independence. In fact, Paul asserts his constant contentment in Christ to make it clear that he was neither grumbling against them nor questioning God's providence before their gift arrived. Rather, he finds strength for every situation in Christ, and he is showing his Philippian friends that Jesus' sustaining grace is sufficient for their troubles, too.

Some scholars suspect that Paul deferred this topic to the end because the Philippians' donation put him in an awkward or unpleasant situation. But those suspicions ignore the fact that Paul has been referring to the Philippians' loving support throughout the epistle. In the letter's opening lines he alluded to their contribution when he expressed joy for "your *partnership* in the gospel from the first day until now" (Phil. 1:5). In the New Testament the Greek term *koinōnia*, which the ESV rightly renders "partnership," often has financial associations (Acts 2:42–44; Rom. 15:26; 2 Cor. 8:4; 9:13). This word group conveys that connotation when Paul returns to it in our present text, where he tells his friends that they have done well to "share" his trouble, as they have often in times past "entered into partnership" with him in giving and receiving.[5] Again in the middle of the letter Paul has alluded to the gift in his report on Epaphroditus's health crisis. The Philippian church had sent the gift to "minister to my need" (Phil. 2:25) and "to complete what was lacking in your service to me" (2:30). Of course, Epaphroditus did more than deliver a check. He supported Paul's work in other ways as well. But the financial relief that Epaphroditus delivered was part of the package that eased Paul's circumstances.

5. The ESV "share my trouble" represents the Greek verb *sygkoinōneō* (see *sygkoinōnos* in Phil. 1:7). "Enter into partnership" reflects the Greek verb *koinōneō*.

Now in the conclusion, as a kind of grand finale, Paul spells it out explicitly: his friends gave a gift so lavish that it not only covered the bills, but made Paul's accounts overflow (Phil. 4:18). Paul has saved discussion of their gift for last not because he does not appreciate it but because he does. As New Testament scholar Gordon Fee observes:

> He concludes the letter on the same note with which it began (1:3–7)—their mutual partnership . . . in the gospel—thus placing this matter in the emphatic, climactic position at the end. When read aloud in the gathered community, these will be the final words that are left ringing in their ears: that their gift to him has been a sweet-smelling sacrifice, pleasing to God.[6]

Paul's deep gratitude over their gift is the treasure that he wants his Philippian friends to take away from hearing his letter read in their midst.

But he wants them to hear far more than his personal appreciation. In fact, he helps them and us, twenty centuries later, to see that giving for missions is so much more than having pity on poor missionaries who depend on others' largesse. It is participating in a partnership. So here Paul compiles terms that belong to the ancient business world. In a phrase that appears both in the realm of friendship and in references to records kept by accountants, he speaks of "the matter of giving and receiving" (Phil. 4:15 NIV).[7] Likewise, his "I have received full payment" (4:18) is found on ancient receipts for rent and taxes, meaning "paid in full."[8]

By the end of verse 18, however, Paul transposes these *commercial* terms into the imagery of *worship*. In a larger perspective, his friends' donation to cover his rent and groceries does not just benefit Paul. It is actually "a fragrant offering, a sacrifice acceptable and pleasing to God" (4:18). Their missions offerings are not just about Paul's hunger and his friends' pity.

6. Gordon D. Fee, *Paul's Letter to the Philippians*, NICNT (Grand Rapids: Eerdmans, 1995), 423.

7. The ESV renders *logon doseōs kai lēmpseōs* simply as "giving and receiving," but the NIV's translation of *logos* as "matter" helpfully reflects for English readers the presence of this Greek noun, which includes in its semantic range the concept of "*computation, reckoning*" or even "*account*" (*BAGD*, 478, s.v. *logos* 2.a). Gerald F. Hawthorne and Ralph P. Martin, *Philippians*, rev. ed., WBC 43 (Nashville: Thomas Nelson, 2004), 257, translate it "an accounting of expenditures and receipts." See also G. Walter Hansen, *The Letter to the Philippians*, PNTC (Grand Rapids: Eerdmans, 2009), 318.

8. *BAGD*, 84, identifies *apechō* with *panta* as a commercial technical term meaning "*receive* a sum *in full* and give a receipt for it." See James Hope Moulton and George Milligan, *The Vocabulary of the Greek Testament: Illustrated from the Papyri and Other Non-Literary Sources* (1930; repr., Grand Rapids: Eerdmans, 1976), 57.

Their gift brings glory and pleasure to God himself! This perspective transforms giving toward missions from duty or charity into privilege and joy. You get to give your hard-earned dollars for missions not only because your missionaries need the funds, but also because the living God has brought you into partnership with himself, and through your giving you can bring glory "to our God and Father . . . forever and ever" (Phil. 4:20)!

The Three-Way Partnership

Paul and his friends, and you and your missionaries, belong to a three-way partnership[9] that unites grateful Christians who give, gospel heralds who go, and the living God who graciously guides them both. God the Senior Partner has the final say on policy, strategy, and operations in this enterprise. That fits the business model we are familiar with in the marketplace. What makes this gospel partnership unique is that the Senior Partner amply provides the resources for his "junior partners," and then he compensates us from the fullness of his grace and "his riches in glory in Christ Jesus" (Phil. 4:19).

I glimpsed the privileges of partnership while strolling the United Airlines concourse at Chicago's O'Hare airport some years ago. Overhead were huge posters, each showing a smiling face. These men and women all had the friendliest smiles. After all, the motto of the airline on which I was flying was "The Friendly Skies®"! Then I noticed that each smiling United staff member had a title: *Pilot/Owner*, or *Mechanic/Owner*, or *Flight Attendant/Owner*. I recalled the news that as a result of recent labor negotiations between the airline and the unions, United was now owned by its employees. It was an *owner* (a mechanic) who found the defect that grounded the plane for my connecting flight home. He deserved our thanks: he had kept the skies

9. "What is unique to Paul's relationship with the Philippians is that their 'partnership' with him was not so much 'one on one,' as it were, but *a three-way bond*—between him, them, and Christ (and the gospel). This third factor results in a considerable 'skewing' of the convention [of Hellenistic contractual friendship]. Left intact is reciprocity and mutuality; 'skewed' is the form these take in Paul. The discussions in Greco-Roman literature of 'giving and receiving' in relation to friendship indicate that they often ended up in (sometimes destructive) one-up-manship. . . . But in Christ, Paul's relationship with the Philippians has been 'leveled out' in its own divine way." Fee, *Philippians*, 444. "Once again the threefold cord of the relationship of God, Paul, and the Philippian church is tightly woven. . . . Gifts to one another are sacrifices to God. When friends give to one another, they give pleasure to God." Hansen, *Philippians*, 324.

friendly, and all of us safe. In the air an *owner* served us soft drinks and snacks. The point of the posters was that everyone at United has "ownership," a vested interest, in making the airline excellent. No one at United can say, "Don't blame me; I just work here." They are all partners!

That's the perspective on partnership that Paul shows his generous, missions-minded friends in Philippi: by God's grace, they are privileged to partner with Paul as participants in God's global gospel enterprise, to bring the peoples of the world to the foot of his throne in worship.

So let's explore this text from the perspective of the three partners—the missionary, the church, and the Lord himself—surveying what each partner invests and the dividends that each receives from their collaboration. The close relationship of Paul and the Philippians, so long ago and far away, will give us a greater grasp of the honor and obligation that Christ's grace bestows on every one of us today. God sent his beloved Son not only to redeem us from his wrath but also to bring us into the "family business" as partners who share his grace both as receivers and as investors who pass it along to others.

The Missionary's Investment and Dividend

Paul's role in the three-way partnership represents all heralds of the gospel, whether they are missionaries planting churches far from home, pastors serving established congregations, military chaplains, evangelists on college campuses, or those who spread the Word in other ways. What is the investment—what is the price to pay—for those whom Jesus calls to bring his good news to others? What is the return on their investment, the dividend, that keeps them motivated in their mission?

The missionary's *investment* is implied in Paul's mention in Philippians 4:15 of "the beginning of the gospel." This expression would bring to his first readers' memories the events recorded in Acts 16, when he and Silas, with Luke and Timothy, first brought the good news of Christ to Philippi. If Lydia and her friends and the jailer and his family are in the congregation as this letter is read, they will recall how Paul's team "invested" God's Word into their minds and hearts, whether at the women's prayer gathering by the river or in answering the terrified jailer's question, "What must I do to be saved?" with the life-transforming direction, "Believe in the Lord Jesus, and you will be saved, you and your household" (Acts 16:31). Missionaries invest God's Word into others' lives.

While investing the good news of Christ, Paul also paid the price of suffering. He and Silas were accused of disloyalty to Rome, arrested, beaten, and imprisoned, as the Philippian believers saw (Phil. 1:30). To enlist the Philippians as partners in God's amazing grace, Paul paid a heavy price. He lost his comfort, his reputation, and his freedom.

These two forms of investment—the joy of telling good news and the pain of suffering—are inextricably bound together, as much as we might wish that missionaries and evangelists and pastors could keep the joy and avoid the pain. Paul knew well that those who herald the news of life through the suffering Savior are called to share the Savior's sufferings. He wrote to the Colossians, "I rejoice in my sufferings for your sake, and in my flesh I am filling up what is lacking in Christ's afflictions for the sake of his body, that is, the church" (Col. 1:24). He told the Corinthians that Christ's life spreads to others as Christ's death is experienced by those who proclaim that life (2 Cor. 4:7–12). Paul has set the pace for the Philippians by expressing his desire to know two things that always belong together: the power of Christ's resurrection and the partnership of his sufferings (Phil. 3:10).

The suffering aspect of the missionary's investment might include not only prison, violence, or death, but also the daily inconveniences and interruptions of being "on call" to people in need. Ajith Fernando, director of Youth for Christ in Sri Lanka, testified in an essay in *Christianity Today* that he is humbled by the faith and courage of Sri Lankan pastors whom he is privileged to teach. These pastors are "assaulted and accused falsely; stones are thrown onto their roofs; their children are given a hard time in school; and they see few genuine conversions." When Fernando receives invitations to travel and speak in the West, he is "able to 'use my gifts' and spend most of my time doing things I like. But when I resume being a leader in Sri Lanka's less-efficient culture, frustration hits me. . . . As a leader [here], I am the bond-servant (*doulos*) of the people I lead. . . . My schedule is shaped more by their needs than by mine."[10] To have your personal agenda trumped by others' urgent needs or thwarted by

10. Ajith Fernando, "To Serve Is to Suffer," *Christianity Today* (August 2010). Accessed at http://www.christianitytoday.com/globalconversation/august2010/.

traffic jams or government bureaucracy or electrical outages—these, too, are investments of suffering for the sake of the gospel. God's sustaining grace enables his servants to endure even mundane, daily frustrations with patience and with joy.

Many missionaries have left family, all things familiar (words, foods, customs, and so on), and countless conveniences that most of us take for granted. Cross-cultural living abounds with distractions, delays, and frustrations. As the missionaries' partners in prayer, ask the Lord to keep their hearts focused on their calling as Jesus' partners to invest the gospel of grace into the lives of others. Paul stitched his share of tents, but he always kept in view his goal: to make more partners in God's grace through the gospel. Pray that they, like Paul, will gladly pay the price of suffering for Christ, whatever form that suffering takes. It might not be flogging, chains, or martyrdom as it was for Paul and is for our brothers and sisters elsewhere in the world today. Instead, your missionaries may suffer inconvenience and frustration, loneliness, suspicion, ridicule, or rejection. Whatever form their suffering takes, pray that your missionaries will meet it with joy, knowing that partnering in Jesus' sufferings is also sharing in his resurrection power.

What *dividend* has Paul received for his investment of gospel truth and suffering into the lives of his friends at Philippi? Obviously, one "return" is the Philippians' monetary support! The Philippians' faithfulness in meeting his needs keeps Paul free to focus on spreading the good news. It is also one means that enabled him to give, free of charge, the news of God's free gift of forgiveness and new life. As congregations are established, Paul expects that Christians will express their gratitude for grace and the ministry of their teachers: "One who is taught the word must share [*koinōneō*] all good things with the one who teaches" (Gal. 6:6). Just as Israel's priests could eat from the temple sacrifices, so "the Lord commanded that those who proclaim the gospel should get their living by the gospel" (1 Cor. 9:9–14). The believers at Philippi have gone above and beyond from the start. They have not only supported their own overseers but also contributed beyond their congregation to free Paul up to preach Christ's free grace free of charge in other towns. Now again in Rome, Paul's need has been filled by the Philippians' donation. In fact, their gift has done more than cover a deficit: Paul marks this receipt not merely "paid in full,"[11] but overflowing.[12] He is so "well supplied"[13] that his reserves spill over the top of his cup (Phil. 4:18).

Although the financial contribution that Epaphroditus brought was a real help, the funds are not the dividend that Paul treasures most from his investment. In verse 17 he states why he is so excited over their gift: "Not that I seek the gift, but I seek the fruit that increases to your credit." What thrills Paul about the Philippians' gift is not what it does for him, but what it does for his friends in Philippi! Compared to their thriving in the grace of Christ, the money itself does not really matter one way or the other, except as a means of ministry that frees Paul to offer God's free grace to others free of charge. God's Spirit has so radically realigned Paul's desires and delights, freeing him from self-centeredness, that he counts his brothers and sisters themselves as his most treasured dividend. Paul has said this in various ways throughout this letter. He has said, "I yearn for you all with the affection of Christ Jesus" (Phil. 1:8). Although a speedy death by Caesar's sword would be better for Paul personally, ushering him immediately into Christ's presence, for the sake of his friends he chooses to remain alive and in service "for your progress and joy in the faith" (1:23–25). They are his joy and crown, whom he loves and longs for (4:1). Paul feels the same affection for believers elsewhere. He tells the Thessalonians, for example: "For what is our hope or joy or crown of boasting before our Lord Jesus at his coming? Is it not you? For you are our glory and joy" (1 Thess. 2:19–20). He even told the Corinthian Christians, who often gave him grief, "I seek not what is yours but you" (2 Cor. 12:14). He was not after their money; he was after their hearts. The apostle John also found his greatest delight in the spiritual well-being of those whom he had introduced to God's love. He wrote to Gaius: "I have no greater joy than to hear that my children are walking in the truth" (3 John 4). This is the dividend that Christ's missionaries treasure most highly: seeing the Word take root and bear fruit in transformed lives.

Since Paul sought such dividends, what are the implications for the missionaries with whom we partner, and for us as well? Pray that your missionar-

11. Greek *apechō panta.*

12. Greek *perisseuō*, of which *BAGD*, 651, offers the glosses "*have an abundance, abound, be rich.*" Philippians 4:18 represents this verb in the very brief "and more." In verse 12 the same verb is rendered "to abound."

13. Greek *plēroō*, for which *BAGD*, 670–71, offers such glosses as "*make full, fill.*" Paul will use this verb again in the promise of Philippians 4:19 (ESV: "will supply").

ies will seek their joy not merely in generous monetary contributions, but in the people whom they serve on the field and in their partners here at home. Living cross-culturally can be expensive, especially during global recession. Ask God to keep your missionaries aligned to Jesus' heart—not desperate for money, but finding joy as they see Jesus transform people by grace. As you ask God to help your missionaries to treasure Christ's people more than their contributions, ask yourself, "Where is my heart's true treasure? Is Christ himself the source of my security, so that I am free to spend myself and my substance to help others encounter his love and thrive in his love?"

THE CHURCH'S INVESTMENT AND DIVIDEND

The Philippians' *investment* in the gospel partnership was obviously financial. Paul sums up their history in terms such as "partnership with me in giving and receiving," and he documents the gifts they sent him both in Thessalonica and in Corinth (Phil. 4:15–16). Comfortable Western Christians feel discomfort with this truth, but Jesus insists on being Master of your MasterCard (or VISA or other credit card). There are so many religious marketers who prey on gullible givers that it is understandable for our defenses to go up when preachers start talking about money—unless they say up front, as Paul does here, "I don't need yours." But the remedy to preachers who talk too much about money or who do so with the wrong motives is not to avoid discussing money at all.

Money is a deeply spiritual matter. The Lord Jesus himself unveiled the spiritual issues involved in money management: "Do not lay up for yourselves treasures on earth, where moth and rust destroy and where thieves break in and steal, but lay up for yourselves treasures in heaven . . . . For where your treasure is, there your heart will be also" (Matt. 6:19–21). Where your funds are invested shows where your heart really lies, so the question is: "Where is your heart?"

What Paul finds so refreshing about his friends at Philippi is that, though they might have money worries and need to apply the antidote of thankful prayer (Phil. 4:6), still their hearts are open to others, and so are their wallets. They are poor in funds but rich in joy and therefore eager to relieve others' needs, whether hungry fellow believers in Judea or a missionary on a tight budget in Rome. Their gift is their way of "shar[ing] my trouble" (v. 14), of becoming Paul's partners in pain. The donation delivered by Epaphroditus after his arduous, life-threatening journey is not a hastily jotted check or an electronic funds transfer zipped off painlessly to fulfill a duty and relieve a conscience. No, their offering actually hurts because they are going without, so that Paul can go without a little less than he has been.

When we turn to consider the church's *dividend*, we notice something bizarre about God's business plan: his junior partners receive dividends *before* they make any investment! The Philippians have given their hearts and their funds to Paul because they have *first of all* received from God a priceless gift. That's the point that Paul made in opening his letter, when he introduced the partnership theme: "you are all partakers ['partners'[14]] with me of *grace*" (Phil. 1:7). This is definitely a bizarre business. The Senior Partner lavishes so generous a dividend on the junior partners, Paul and the Philippians and our missionaries and us, *before* we have invested anything! That is why Paul calls it *grace*. God loves first and gives first, and his love lavished on us through Christ evokes our reaction of love and generosity because we want others to taste this grace. In Philippians 2:1 Paul reminds us who initiated this partnership, mentioning Christ's encouragement, God's love, and the Holy Spirit's "partnership."[15] God is heavily invested in us through the life and death of Christ and the powerful presence of his Spirit! He paid the greatest price, his beloved Son, when we were his enemies. He paid so much not just to acquire us as property but to adopt us as sons and heirs and partners.

Our text also speaks of other *dividends* that we receive as Paul's partners and partners of Paul's Lord, Jesus. The apostle says that the Philippians' donation will yield "fruit that increases to your credit" (Phil. 4:17). In first-century commerce, it appears that the term *fruit* was borrowed from agriculture to serve as a general economic metaphor for monetary profit "harvested" from any form of labor or investment.[16] The Philippians can rest assured that their contribution has not really diminished their resources. Rather, they have put those resources on deposit with the Creator of all things. Their investment is secure, and it will bring a return beyond their imaginations.

14. Paul uses a compounded noun form, *sygkoinōnous*, anticipating his use of the compounded verb, *sygkoinōneō*, in Philippians 4:14.

15. Greek *koinōnia*. See also 2 Cor. 13:14.

16. O'Brien, *Philippians*, 538 (cautiously); and Hawthorne and Martin, *Philippians*, 271, who call attention to the reappearance of *logos* (as in 4:15) in the economic sense of "account."

Paul will use up the funds to buy groceries and pay the rent, of course. But in the Lord's treasury the Philippians' gift is an investment that continues to accrue interest. Proverbs promised: "Whoever is generous to the poor lends to the LORD, and he will repay him for his deed" (Prov. 19:17). Jesus reaffirmed this promise when he announced that those who extend hospitality to those too poor to return dinner invitations "will be repaid at the resurrection of the just" (Luke 14:13–14). God is the "cosigner" on the debts that his needy children cannot repay. He will more than compensate their benefactors in the end. Jesus the Son of Man counts kindness extended to his needy siblings as kindness shown to himself, and he will distribute his dividends accordingly (Matt. 25:31–40).[17]

Yet Paul is not merely extending the hope of a *long-term* "payoff" for our current investment in the causes of world missions and mercy. The promise[18] in verse 19 applies not only to a distant day, when Jesus will wipe every tear from our eyes, but also to the present when we are still in need: "And my God will supply every need of yours according to his riches in glory in Christ Jesus."

This verse is sometimes misinterpreted by "health and wealth" preachers, who promise us material abundance if we exercise faith by making donations to "the ministry." The cruel, though unstated, implication of their inflated promises is that those who remain poor have only themselves to blame: their unbelief and stinginess have locked up heaven's treasuries. If we hear Paul's promise in those diminished terms, our expectations are much too small!

17. See 1 Timothy 6:17–19: those blessed with material wealth "in this present age" must not "set their hopes on the uncertainty of riches, but on God, who richly provides us with everything to enjoy." As they share their wealth with those in need, they will be "storing up treasure for themselves as a good foundation for the future, so that they may take hold of that which is truly life."

18. In most Greek manuscripts, Philippians 4:19 is a promise: "my God *will supply* [Greek future indicative, *plerōsei*] every need of yours." In a few ancient manuscripts, however, the verb is in the optative mood (*plerōsai*), making the statement into a prayer expressing Paul's wish for his friends: "*may* my God *supply* every need." Some commentators favor the optative (wish-prayer), believing that Paul would not promise that his readers' every need would be fully supplied in this age, since he has just acknowledged the extremes of poverty and plenty that believers must currently endure with contentment (4:11–13). Others believe that Paul is announcing a firm promise, but one that will be fulfilled only in the consummation at Christ's return (interpreting "in glory" as strictly eschatological). The early and widespread manuscript evidence supporting the future indicative, "shall supply," shows that the ESV is correct to render this verse as a promise, not merely a prayer. Its ultimate fulfillment is in the new heavens and earth (Rev. 21:4), but believers already experience a foretaste (firstfruits [Rom. 8:23] and pledge [2 Cor. 1:22; 5:5; Eph. 1:14]) of that coming glory in the present through the indwelling Holy Spirit and God's faithful providence.

Paul promises his friends so much more than mere monetary reimbursement or affluence. Wealthy people may still live constricted lives, fearful and fretful that the financial reserves on which their lives depend may sprout wings and fly away, as Proverbs 23:5 warns. No amount of money can fill up their "every need," for what they need is a heart set free from the illusion that a safe haven can be constructed out of the fragile and fickle materials listed on a balance sheet.

Paul's promise to the Philippians, who were poor in funds but rich in faith (James 2:5), resembles his reassurance to the wealthier Christians in Corinth:

> He who supplies seed to the sower and bread for food will supply and multiply your seed for sowing and increase the harvest of your righteousness. You will be enriched in every way to be generous in every way, which through us will produce thanksgiving to God. For . . . the saints . . . will glorify God . . . , while they long for you and pray for you, because of the surpassing grace of God upon you. (2 Cor. 9:10–14)

At first hearing, this talk of multiplied seed and increasing harvest might sound like the "prosperity now" message that fills the media airwaves. But listen more closely. The "seed" that God would multiply, to set the Corinthian Christians free to give joyfully, is not mere money. It is "the surpassing grace of God" that pries open their hearts and their fists, so that they gladly cast forth their funds to help others. In the same way, Paul promises the poor but generous Philippians that his generous God will meet their every need. God will do so as he is meeting Paul's needs, enabling him to rest content in feast or in famine (Phil. 4:10–13). No one can deprive us of the treasure that we really need: Jesus, "who strengthens me."

How do the Philippians' investment in Paul's gospel mission and the dividends they were reaping speak to us today? First, they challenge us to be honest to God and to ourselves. How does your investment in the global advance of the kingdom of Christ through the gospel compare with your investment in, say, the stock market, or information technology, or fast food? Jesus said that the flow of your "treasure" tracks the direction of your heart. Draw a pie chart showing the distribution of your dollars. Consider what it reveals about your ultimate allegiance. If the chart exposes a misaligned heart or undisciplined spending habits, now is the time to repent, cast yourself on God's mercy, and make concrete, even costly "course corrections" in where and how you direct the resources that Christ entrusts to you as his steward.

Second, the eternal perspective in which Paul places the Philippians' partnership for the gospel urges us to be strategic. Having "audited" our investments according to the "kingdom accounting criteria," we need to consider which investments promise the most reliable and valuable rate of return. Jesus told a parable about a wealthy farm owner who imagined that bumper crops and bigger barns spelled security for years to come, only to have his strategic business plans cut short by death, God's sudden summons to the ultimate audit (Luke 12:13–21). Even if our health holds up and our financial reserves are sufficient to sustain a comfortable retirement, why settle for paltry dividends when Jesus promises infinitely greater reward in the form of lives eternally transformed by his gospel of grace?

Finally, the Philippian church's "investment strategy" challenges us to be trusting. Are you waiting until you have more "margin" in the bank before you give more for missions? How much "cushion" in reserve is enough to free you to open your heart and your checkbook wider?

I do not mean that paying your bills and providing prudently for your family both now and in the future are signs of selfishness or unbelief. We should do all those things as faithful stewards of God's money. But we also need to cultivate a holy "greed" for the *spiritual* prosperity of Christ's reign over the earth, so that we seize opportunities to put our money where our hearts are—at the ends of the earth.

God's Investment and Dividend

Finally, we turn to the Senior Partner in this three-way alliance, the Lord whose wealth of mercy and glory funds the whole enterprise. What does God in Christ invest in this partnership, and what return does he receive from it?

The infinite reserve from which God makes his investment is stated by Paul in his promise that God will meet the Philippians' every need "according to his riches in glory in Christ Jesus" (Phil. 4:19). The Creator's wealth is so far beyond our imagination that it makes any of today's billionaires look like Africa's poorest orphan. But Paul's point is that this infinitely rich Ruler of all invests in people like us "in Christ Jesus." The beloved Son is the Father's dearest treasure, the Father's most costly investment in human lives. Because we are united to this Son by grace, through faith, the Father opens his treasury of untold wealth to paupers like us.

In 2 Corinthians 8 and 9 Paul tries to persuade the comfortable Corinthians to open their narrow hearts and their fat billfolds to help other believers in need. As he builds the case for trust and generosity, he plays his trump card toward the beginning and again at the end of his discussion. Early on, he insists that he is not browbeating them into doing their duty but inviting them to respond in love to God's greatest gift: "For you know the grace of our Lord Jesus Christ, that though he was rich, yet for your sake he became poor, so that you by his poverty might become rich" (2 Cor. 8:9). At the end he closes his argument with the celebratory outburst, "Thanks be to God for his inexpressible gift!" (2 Cor. 9:15).

Likewise, Paul closes this letter to the poor but generous Philippians with the reminder of that same grace, extended to us by that same Lord: "The grace of the Lord Jesus Christ be with your spirit" (Phil. 4:23). Can you hear how this benediction flows from the whole discussion of giving and receiving? Yes, of course, Paul *always* closes his letter with a benediction that mentions the grace of Christ. But that does not make this a meaningless formula. The grace of the Lord Jesus Christ is just the note that he wants to leave ringing in his friends' ears and hearts! God's riches in glory are opened to us in Christ Jesus. They flow into our lives, free of charge, because the beloved Son paid the price to erase our debt to God's justice and to usher us into the family, to share his inheritance with us.

What's in it for God? What dividend does God receive for his costly investment, to bring debtors like us into partnership with himself? Paul's concise answer is the doxology that he inserts to conclude his discussion of the partnership: "To our God and Father be glory forever and ever. Amen" (Phil. 4:20). The majestic Creator of all things would have received glory forever from radiant cherubim and seraphim, even without paying the terrible price of sending his Son to the cross. God's flawless heavenly servants would have adored him endlessly for his limitless power and wisdom, his holiness and justice, and so much more. But there is one dimension of the great Creator's worthiness to be worshiped that, it seems, could not be displayed in any other way than for the Son to become the Servant, to walk down the path of obedience to death on a cross, and then to ascend to life and lordship above all earthly powers. The divine attribute that could be displayed only through the gospel is God's amazing grace. The apostle Peter tells us that angels long to lean over and look into our salvation (1 Peter 1:12). They stand as outside observers, marveling at a dimension of their Master that they themselves have never experienced: *grace*, mercy to those who deserve wrath, warm welcome to those who deserve rejection and destruction. For this costly investment of grace, to rescue and recruit us as his junior partners, "our God and Father" will receive increasing glory forever and ever.

God's great return on his costly "investment" is that he will be glorified not only *for* redeeming rebels but also *by* this redeemed and transformed people! Therefore, Paul switches from commercial terminology to sacrificial-worship images in verse 18. The Philippians' gift is "a fragrant offering, a sacrifice acceptable and pleasing to God." From the day that Noah emerged from the ark and presented a burnt offering in thanks to God for salvation from the flood, the Bible has spoken of "fragrant offerings," that is, those with a "pleasing aroma" (Gen. 8:21; see Ex. 29:18, 25, 41; Lev. 1:9, 13, 19; etc.). This description evokes the delicious aroma of sizzling beef on the grill as a sensory analogy to the Lord's delight in the sincere worship of the people whom he has rescued. What truly delights the God of the universe, of course, is not the aroma of charbroiled beef but the devotion of thankful hearts (Ps. 51:16–17). The Philippians' generous donation to Paul is just such an expression of heartfelt love—not only for the apostle, but also for the Lord himself.[19] God's "return" on his investment is the pleasure he takes from seeing humans, whom he first created to be and bear his image, reflecting his own generosity toward each other, so that "the earth will be filled with the knowledge of the glory of the LORD as the waters cover the sea" (Hab. 2:14; see Isa. 11:9).

God is pleased with his "dividend" as *all the earth's peoples* glorify him *together*. Paul's choice of language brings into view these aspects of the people who are the Lord's most treasured dividend: that we are *global*, and that we are *one*. Did you notice the subtle shift of pronouns, as Paul moves from the Philippians' dividend in verse 10 to God's dividend in verse 20? His promise to believers implies the personal relationship that Paul himself

19. For sacrificial imagery applied to Christian obedience generally, see Romans 12:1: "I appeal to you therefore, brothers, by the mercies of God, to present your bodies as a living sacrifice, holy and acceptable to God, which is your spiritual worship." Sacrificial imagery is applied to generosity toward others in Hebrews 13:16: "Do not neglect to do good and to share what you have, for such sacrifices are pleasing to God."

enjoys with his Savior: "*My* God will supply every need of yours" (Phil. 4:19). But this intimate relationship is one that he shares with all believers, so it is "*our* God and Father" who will receive glory forever (4:20). God is not merely "*my* God" or "*your* God"—he is "*our* God and Father." The gravitational force of God's grace pulls us together into one harmonious family, despite our differences of background.

Paul reinforces our unity in his closing greetings. He asks his Philippian friends to extend his warm greeting to one another, omitting no one: "Greet every saint in Christ Jesus" (Phil. 4:21). Paul ends where he began. In opening he addressed his friends as "all the saints in Christ Jesus who are at Philippi" (1:1). Here he greets "every saint in Christ Jesus" and sends them greetings from "all the saints" in Rome. By throwing his arms wide to embrace *each and every believer* who bears Christ's name, Paul returns to the tender issue of unity that he addressed in chapter 2 and earlier in chapter 4. On Paul's behalf Euodia must embrace Syntyche, and vice versa. Their respective partisans, if they have them, must do the same. Anyone who has neglected others' interests through competition and conceit must care with humility for those whom he or she has ignored.

The oneness of heart and mind that characterizes this new family of God spans the globe and crosses boundaries of class and ethnicity. Therefore, Paul extends greetings from the church at Rome, not only from "the brothers" who serve closely with him but also from "all the saints" (Phil. 4:21–22). One demonstration of the gospel's power is that the company of those Roman saints includes "those of Caesar's household." Within the imperial palace itself, perhaps serving Caesar's meals or tending his gardens, are those who eagerly await the return from heaven of the one true Savior and Lord, Jesus Christ (3:21–22). The global family of Jesus, the greater Lord, is being gathered through the gospel, even among the emperor's retinue.

The Privilege of Partnership

So how would a church like Philippi respond to a "thank-you note" like this? Would they say to themselves, "Well, when Paul finally got around to mentioning our gift, I was afraid he was going to ask for more. I am so relieved that he doesn't need any more, so we can keep what little we have for ourselves"? I doubt it! If the Philippians reacted like that, it would mean that they had not gotten Paul's point. If your own reaction, when you hear missionaries report their need for financial support, is one of reluctance, you have not grasped the privilege of your partnership with the living God and the ambassadors of his expanding kingdom. We are partners with Jesus in his gospel enterprise, and therefore partners with missionaries like Paul. We are not mere employees—we are owners!

Doesn't your partnership in grace make you even more eager to partner with those who are carrying the Word out to the growing edges of the kingdom of grace? You are offering sacrifices that smell so sweet to God himself! You are investing his resources (they are not yours, after all!) in ways that will bring him endless eternal glory and pleasure, and that will expand your own capacity to enjoy God forever. So take to heart Paul's perspective on the privilege of partnering with God. Think prayerfully about how this puzzling partnership of getting by giving might overhaul your priorities and reshape your whole budget. And start with the costly investment that God made to bring you into his family business as his partner, his child, beloved forever: the gift of Jesus the Son. Who knows where it might go from there?

Select Bibliography of Commentaries Cited or Consulted

Calvin, John. *The Epistles of Paul the Apostle to the Galatians, Ephesians, Philippians and Colossians.* Edited by David W. Torrance and Thomas F. Torrance. Translated by T. H. L. Parker. Grand Rapids: Eerdmans, 1965.

Edwards, Mark J., ed. *Galatians, Ephesians, Philippians.* Ancient Christian Commentary on Scripture: New Testament 8. Downers Grove, IL: InterVarsity Press, 1999.

Fee, Gordon D. *Paul's Letter to the Philippians.* New International Commentary on the New Testament. Grand Rapids: Eerdmans, 1995.

Ferguson, Sinclair B. *Let's Study Philippians.* Edinburgh: Banner of Truth, 1997.

Hansen, G. Walter. *The Letter to the Philippians.* Pillar New Testament Commentary. Grand Rapids: Eerdmans, 2009.

Hawthorne, Gerald F., and Ralph P. Martin. *Philippians.* Rev. ed. Word Biblical Commentary 43. Nashville: Thomas Nelson, 2004.

Lightfoot, J. B. *Saint Paul's Epistle to the Philippians.* 1913. Reprint, Grand Rapids: Zondervan, 1953.

Martin, Ralph P. *Philippians.* New Century Bible Commentary. Grand Rapids: Eerdmans, 1980.

O'Brien, Peter T. *The Epistle to the Philippians.* New International Greek Testament Commentary. Grand Rapids: Eerdmans, 1991.

Reumann, John. *Philippians: A New Translation with Introduction and Commentary.* Anchor Yale Bible. New Haven, CT: Yale University Press, 2008.

Silva, Moisés. *Philippians.* 2nd ed. Baker Exegetical Commentary on the New Testament. Grand Rapids: Baker, 2005.

Index of Scripture

Philippians

EXTRA-BIBLICAL CITATIONS

Index of Subjects and Names

Available in the Reformed Expository Commentary series

1 Samuel, by Richard D. Phillips
1 Kings, by Philip Graham Ryken
Esther & Ruth, by Iain M. Duguid
Daniel, by Iain M. Duguid
Jonah & Micah, by Richard D. Phillips
Zechariah, by Richard D. Phillips
The Incarnation in the Gospels, by Daniel M. Doriani, Philip Graham Ryken, and Richard D. Phillips
Matthew, by Daniel M. Doriani
Luke, by Philip Graham Ryken
Acts, by Derek W. H. Thomas
Galatians, by Philip Graham Ryken
Ephesians, by Bryan Chapell
Philippians, by Dennis E. Johnson
1 Timothy, by Philip Graham Ryken
Hebrews, by Richard D. Phillips
James, by Daniel M. Doriani

Forthcoming

Ecclesiastes, by Douglas Sean O'Donnell
John, by Richard D. Phillips
1 Peter, by Daniel M. Doriani